Natural Interaction with Robots, Knowbots and Smartphones

Joseph Mariani • Sophie Rosset
Martine Garnier-Rizet • Laurence Devillers
Editors

Natural Interaction with Robots, Knowbots and Smartphones

Putting Spoken Dialog Systems into Practice

Springer

Editors
Joseph Mariani
IMMI-CNRS & LIMSI-CNRS
Orsay
France

Sophie Rosset
LIMSI-CNRS
Orsay
France

Martine Garnier-Rizet
IMMI-CNRS & ANR
Orsay
France

Laurence Devillers
LIMSI-CNRS & University
Paris-Sorbonne IV
Orsay
France

ISBN 978-1-4614-8279-6 ISBN 978-1-4614-8280-2 (eBook)
DOI 10.1007/978-1-4614-8280-2
Springer New York Heidelberg Dordrecht London

Library of Congress Control Number: 2013947791

Printed on acid-free paper

Springer is part of Springer Science+Business Media (www.springer.com)

Preface

The International Workshop on Spoken Dialog Systems (IWSDS) series provides an international forum for the presentation of research and applications and for lively discussions among researchers as well as industrialists, with a special interest to the practical implementation of spoken dialog systems in everyday applications.

Following the success of IWSDS'09 (Irsee, Germany), IWSDS'10 (Gotemba Kogen Resort, Japan), and IWSDS'11 (Granada, Spain), the Fourth IWSDS'12 took place at the castle of Ermenonville, near Paris (France), on November 28–30, 2012.

This book consists of the revised versions of a selection of the papers that were presented at the IWSDS'12 conference.

Spoken dialog has been a matter of research investigations for many years. The first spoken language processing systems aimed at providing such an interaction between humans and machines. It slowly appeared that the problem was much more difficult than it was initially thought, as it involves many different components: speech recognition and understanding, prosody analysis, indirect speech acts, dialog handling, maintenance of the communication with verbal or nonverbal events such as backchannels, speech generation and synthesis, multimodal fusion and fission. Social interaction among humans is characterized by a continuous and dynamic exchange of information carrying signals. Producing and understanding these signals allow humans to communicate simultaneously on multiple levels. The ability to understand this information, and for that matter adapt generation to the goal of the communication and the characteristics of particular interlocutors, constitutes a significant aspect of natural interaction. It shows that it is actually very complex to develop simple, natural interaction means.

Even if the research investigations kept on being conducted, it induced a shift of interest to easier tasks, such as voice command, voice dictation, or speech transcription. However, scientific achievements in language processing now result in the development of successful applications such as IBM Watson, the Evi, Apple Siri, Google Voice Action, Microsoft Bing Voice Search, Nuance Dragon Go!, or Vlingo for access to knowledge and interaction with smartphones, while the coming of domestic robots advocates for the development of powerful communication means with their human users and fellow robots.

We put this year workshop under the theme “Towards a Natural Interaction with Robots, Knowbots and Smartphones,” which covers:

- Dialog for robot interaction (including ethics)
- Dialog for open-domain knowledge access
- Dialog for interacting with smartphones
- Mediated dialog (including multilingual dialog involving speech translation)
- Dialog quality evaluation

We enjoyed the invited Keynote Talks of Jérôme Bellegarda (Apple, USA), Alex Waibel (Karlsruhe Institute of Technology (Germany) and Carnegie Mellon University (USA)), Axel Buendia (SpirOps) and Laurence Devillers (LIMSI-CNRS and University Paris-Sorbonne, France) and Marilyn Walker (UCSC, USA) on those topics. We also had an invited talk on the conclusions of the SemDial workshop on the semantics and pragmatics of dialog, which took place in Paris in September 2012, by its organizer, Jonathan Ginzburg (University Paris Diderot). We warmly thank all of them.

We also encouraged the presentation and discussion of common issues of theories, applications, evaluation, limitations, general tools, and techniques. We particularly welcomed papers that were illustrated by a demonstration.

This book first includes several parts on the implementation of spoken dialog systems for various areas of application and especially those related to the main topics of the conference: smartphones, robots, and knowbots. It then has a part on spoken dialog systems components and a final one on spoken dialog management.

The first part deals with spoken dialog systems in everyday applications. First, Jérôme Bellegarda from Apple Inc. presents the Siri experience, which has had a tremendous impact in the actual use of spoken interaction on personal assistants. He introduces the two major semantic interpretation frameworks, statistical and rule-based, discusses the choices made in Siri, and speculates on how the current implementation might evolve in the near future. Hansjörg Hofmann and colleagues from Daimler AG depict the development of speech-based in-car human-machine interaction for information exchange. The permanent use of smartphones impacts the automotive environment, necessitating an intuitive interface in order to reduce driver distraction. They investigate two different dialog strategies, command-based or conversational speech dialog, and different graphical user interfaces, one including an avatar. Those prototypes are evaluated regarding usability and driving performance. Alan Black and Maxine Eskenazi address the problem of developing spoken dialog systems with controlled users, who may not act as real users, in a study related to a task of providing bus information hosted at Carnegie Mellon University. They report on several lessons learned from the experience and provide recommendations on various approaches, including crowdsourcing. Daniel Sonntag and Christian Schulz from DFKI describe the use of a multimodal multi-device infrastructure for collaborative decision-making in the medical area: the Radspeech industrial prototype. In their study, two radiologists use two different mobile speech devices (Apple iPhone and iPad) and collaborate via a connected large screen installation, jointly using pointing and spoken interaction.

The second part presents five examples of spoken dialog prototypes and products in different domains such as crosslingual communication, city exploration and services, or ambient intelligence environments.

First, Feiyu Xu and colleagues from Yocoy and DFKI LT Lab (Germany) describe Yochina, a mobile multimedia and multimodal crosslingual dialog system. The mobile application combines language technologies such as speech synthesis, template-based translation, and dialog to offer language and travel guide without depending on an Internet connection. A novel strategy of linking provided knowledge with covered communication situations is explained. Yochina is available for two language pairs: English to Chinese and German to Chinese. Johan Boye and colleagues from KTH and Liquid Media (Sweden) address the challenging problem of giving navigation instructions to pedestrians through a spoken dialog approach rather than a map-based approach. It means interpreting and generating utterances within a rapidly changing spatial context even though the pedestrian's position, speed, and direction are uncertain due to possible GPS errors. They present the results of a user experiment conducted in Stockholm. The paper by Nieves Ábalos and colleagues from the Department of LSI, University of Granada, and from Systems Laboratory, University Rey Juan Carlos (Spain), deals with a multimodal dialog system to enable user control of home appliances in an Ambient Intelligence environment (lights, TV, etc.). It describes the interaction of Mayordomo, a multimodal dialog system which uses either spontaneous speech or a traditional GUI, with Octopus, a system which enables AmI applications through a file-based service access. Sunao Hara and colleagues from the Graduate School of Information Science at the Nara Institute of Science and Technology (Japan) depict a toolkit for multi-agent server-client spoken dialog systems: *tankred on rails (ToR)*. iTakemaru is the client software for mobile phones. It provides a speech-guidance service handling one main agent and multiple subagents. It allows the client to obtain more information thanks to the communication between the main agent and the subagents based on a server-to-server communication. The last paper of this part describes a voice portal based on the VoiceXML standard to provide the citizens with municipal information (city council, city services, etc.). The authors, David Griol and colleagues from the Computer Science Department, Carlos III University of Madrid, and the Department of Languages and Computer Systems, University of Granada (Spain), give the results of both a subjective evaluation, through quality assessments, and an objective evaluation (successful dialogs, average number of turns per dialog, confirmation rate, etc.).

The third part (Multi-domain, Crosslingual Spoken Dialog Systems) deals with model adaptation when facing changes of languages or domains.

Teruhisa Misu and colleagues, from the National Institute of Information and Communication Technology, address a very actual issue of cross-domain/cross-language portability of dialog systems. They present an approach for extending a language model designed for one task in a given language to another task by using resources in other languages or tasks using statistical machine translation systems. They propose a selection mechanism to automatically extract relevant parts in those resources, based on a spoken language understanding module corresponding to the source language and task. Pierre Lison, from the University of Oslo, addresses the problem of online learning of dialog policy. The proposed approach relies on probabilistic rules (in order to simplify the inference) and on a Monte Carlo sampling method to determine the best action to perform. Injae Lee and colleagues, from the Pohang University of Science and Technology (Korea) and the Institute for Infocomm Research (Singapore), address the problem of the domain selection for a multiple-domain dialog system. The proposed approach includes a domain preselection, which provides, for each user utterance, a list of possible domains associated with scores. Then a content-based filtering method is performed on the domain candidate list to select the final domain. The experimental results show an improvement in terms of accuracy and processing time compared to more standard approaches.

The fourth part deals with dialog for robot interaction, including ethics.

First, Alex Buendia from the French SpirOps SME and Laurence Devillers from LIMSI-CNRS and University Paris-Sorbonne address the challenges for going from informative cooperative dialogs to long-term social relationship with a robot. They aim at exploring the ability of a robot to create and maintain a long-term social relationship through more advanced dialog techniques. They expose the social, psychological, and neural theories used to accomplish such complex social interactions. From these theories, they build a consistent, computationally efficient model to create a robot that can understand the concept of lying and have compassion: a robotic social companion. Taichi Nakashima, Kazunori Komatani, and Satoshi Sato from the Graduate School of Engineering at Nagoya University in Japan propose the integration of multiple sound source localization results for speaker identification in a multiparty dialog system. They present a method of identifying who is speaking more accurately by integrating the multiple sound source localization results obtained from two robots. The experimental evaluation revealed that using two robots improved speaker identification compared with using only one robot.

Ina Wechsung, Patrick Ehrenbrink, Robert Schleicher, and Sebastian Möller from the Quality and Usability Lab of the Berlin Telekom Innovation Laboratories at the Technical University of Berlin investigate the social facilitation effect in human-robot interaction. The current study indicates that a higher degree of human likeness results in a social inhibition effect. In this experiment, the reported differences were caused by the appearance of the robot, whereas its synthetic voice was kept constant. After the social inhibition as well as the uncanny valley effect could be confirmed for this setup, it would be interesting to study whether the same effect can also be observed for voices with different degrees of anthropomorphism. Emer Gilmartin and Nick Campbell from the Speech Communications Lab, Trinity College Dublin, present how to build a chatty robot. Their work describes the design and implementation of a robot platform for the extraction of data and acquisition of knowledge related to spoken interaction, by capturing natural language and multimodal/multisensorial interactions using voice-activated and movement-sensitive sensors in conjunction with a speech synthesizer.

Takaaki Sugiyama, Kazunori Komatani, and Satoshi Sato from the Graduate School of Engineering at Nagoya University tackle the novel problem of predicting when a user is likely to begin speaking to a humanoid robot. Clément Chastagnol, Céline Clavel, Matthieu Courgeon, and Laurence Devillers from LIMSI-CNRS show how to design an emotion detection system for a socially intelligent human-robot interaction. This work is part of the French ANR ARMEN project that aims at designing and building a prototype for a robotic companion (RC) for the elderly and disabled people. In their paper, Kristiina Jokinen and Graham Wilcock from the University of Helsinki present ongoing work on multimodal interaction with the Nao robot, including speech, gaze, and gesturing. It also describes the interaction with the Nao robot from the point of view of constructive dialog modeling and demonstrates how the framework can be applied to the WikiTalk open-domain interaction. Finally, Ridong Jiang, Yeow Kee Tan, Dilip Kumar Limbu, Tran Anh Dung, and Haizhou Li from the Institute for Infocomm Research in Singapore describe a component pluggable dialog framework, which is domain-independent, cross-platform, and multilingual, and its application to the interface with social robots, showing a shorter development cycle while improving the system robustness, reliability, and maintainability.

The last two parts of this book are about the development of specific aspects of dialog systems. The fifth part (Spoken Dialog Systems components) is about specific components while the sixth specifically concerns the dialog management module.

In the fifth part, Martin Heckmann, from the Honda Research Institute Europe, investigates the use of acoustic and visual cues to detect prominent (e.g., corrected) words in an utterance. The experiment shows that when using only the fundamental frequency as an acoustic feature, the improvement of the classification is interesting when combining to this acoustic feature the visual features but that when all possible acoustic features are used, the combination with visual features allows for a less important gain. Bart Ons and colleagues, from ESAT-PSI (KU Leuven), address the problem of robustness of a direct mapping between an acoustic signal and a command in the context of a learning system. The proposed approach is based on a supervised nonnegative matrix factorization. The results show that this learning approach is robust to label noise. Rafael Torres and colleagues, from the Nara Institute of Science and Technology and from the Institute of Statistical Mathematics in Tokyo, present a work on topic classification of spoken user utterances received by a guidance system. They specifically study a semi-supervised approach, using a transductive support vector machine and the impact of the inclusion of unlabeled examples during the training process of the classifier. Experimental results show that this approach can be useful for taking advantage of unlabeled samples, which are simpler to obtain than labeled ones.

Yoo Rhee Oh and colleagues, from the Spoken Language Processing Team, Electronics and Telecommunications Research Institute (ETRI, Korea), address the problem of the decoding of nonnative speech. Most automatic speech recognition systems have to face one important problem: speakers can be nonnative and then the performance of the system decreases. The proposed decoding strategy consists in decoding speech with both native and nonnative speakers models and selecting, based on the likelihood scores, which model to use for each frame to decode. The experimental results show a reduction of the word error rate. Marcela Charfuelan and Geert-Jan Kruijff, from DFKI GmbH, are interested in analyzing speech under stress. They address the problem of acoustical analysis of stress in a USAR database and examine a range of acoustical cues which are annotated by two annotators into the categories of neutral, medium, or high stress. Analysis results show that traditional prosody and acoustic features are robust enough to discriminate among the different types of stress and neutral data.

In the sixth part, Marilyn Walker and colleagues, from the University of California at Santa Cruz, address the problem of adapting the answers of dialog agents to a particular user, either within the context of a single interaction or over time. A general spoken language generation framework is presented along with dynamic generation for task-oriented dialog systems and most importantly expressive generation. Stefan Ultes and colleagues, from the Institute of Communications Technology (University of Ulm), address the problem of an interaction quality estimator in spoken dialog systems. They describe how conditioned hidden Markov models (CHMM) can be used to estimate the interaction quality of a spoken dialog system, developed for the "Let's Go Bus Information System." Unfortunately using CHMM does not allow for improvements in the results compared to standard approaches such as HMM or SVM. Fabrizio Morbini and colleagues, from the Institute for Creative Technologies (University of Southern California), present a dialog manager based on the information-state update approach that performs forward inference and exploits local dialog structures. This approach is related to plan-based approaches of dialog management with the addition of rewards attributed to specific states. Two examples of implementation are described. Zoraida Callejas and colleagues, from the University of Granada, Carlos III University of Madrid, and the Quality and Usability Lab (Deutsche Telekom Laboratories), are interested in using user profiles to implement intelligent dialog systems. They proposed an approach to cluster user profiles using interaction parameters and overall quality prediction. They provide experimental results related to young and senior user groups and to users with high vs. low technical skills. The general conclusion is that a better grouping of users should distinguish between three groups and not four: young users with high technical affinity, senior users with low technical affinity, and a third group considering the remaining users.

Etsuo Mizukami and Hideki Kashioka, from the National Institute of Information and Communications Technology (NICT), introduce an extension to the dialog mechanism of grounding, called the extended grounding networks. They implemented this extended grounding network using the concept of contribution topics, in the context of touristic information systems. The contribution topics are units of achievement corresponding to discourse segments. Senthilkumar Chandramohan and colleagues, from Supelec, CNRS-Georgia Tech and University of Avignon/LIA-CERI, present a work developed in the context of stochastic-based dialog management. They describe a coadaptation framework and a method to learn optimal dialog policies by taking into account the adaptation of users to systems over time. Experimental results show that this coadaptation framework is a robust approach for facilitating dialog evolution. Lasguido and colleagues, from the Nara Institute of Science and Technology and the Faculty of Computer Science (Universitas Indonesia), are interested in non-goal-oriented dialog systems. In this framework, they present a method, based on the example-based dialog management approach, for developing a dialog manager by generalizing from examples from drama television (the Friends TV show) in order to achieve more natural dialog interaction. The main problem in such an approach is to select the useful examples. They propose a tri-turn unit for dialog extraction and semantic similarity analysis techniques to ensure that the content extracted from drama script files forms an appropriate dialog example.

Klaus-Peter Engelbrecht, from the Quality and Usability Lab, Telekom Innovation Laboratories (TU Berlin), presents a causal user model for user simulation as it is used for spoken dialog systems development. The approach is based on connectionist models of human behavior. The objective of this work is to generate user simulators which are more meaningful and portable across tasks. The presented approach relies on parameters of the model that are related to the characteristics of the users and the task, and the model is useful to explain why a specific behavior is observed. Finally, Sanat Sarda and colleagues, from Nanyang Technological University, are interested in providing real-time feedback about an ongoing conversation to speakers. The system extracts various kinds of information such as speaking time, speaker turns, and duration. This information is then displayed in real time. This is somehow a monitoring system on ongoing conversations. The extracted information is then displayed in different ways to the speakers using icons, animation, etc. Haruka Majima and colleagues, from the Graduate School of Information Science at Nara Institute of Science and Technology, the Graduate School of Natural Science and Technology at Okayama University, and the Department of Statistical Modeling at the Institute of Statistical Mathematics (Japan), present a method for detecting invalid inputs for a spoken dialog system. Invalid inputs include background voices, which are not directly uttered to the system, and nonsense utterances. The main idea is to feed the decision method with different features like signal-noise ratio, utterance duration, and bag of words (BOW) when available. They compare two different methods, one based on SVM and the other on maximum entropy. The SVM-based methods reached an F-measure of 0.870 while the ME-based one obtained a $F = 0.837$. This has to be compared to the baseline method (GMM-based) which reached $F = 0.817$.

Finally, we wish to thank the IWSDS Steering Committee, the members of the IWSDS 2012 Organizing Committee and Scientific Committee, the participating and supporting organizations, and our sponsors: ELSNET (the European Language and Speech Network), ELRA (the European Language Resources Association), and the QUAERO project.

Orsay, France

Joseph Mariani
Sophie Rosset
Martine Garnier-Rizet
Laurence Devillers

Contents

Part I
Spoken Dialog Systems in Everyday Applications

Chapter 1
Spoken Language Understanding for Natural Interaction: The Siri Experience

Jerome R. Bellegarda

Abstract Recent advances in software integration and efforts toward more personalization and context awareness have brought closer the long-standing vision of the ubiquitous intelligent personal assistant. This has become particularly salient in the context of smartphones and electronic tablets, where natural language interaction has the potential to considerably enhance mobile experience. Far beyond merely offering more options in terms of user interface, this trend may well usher in a genuine paradigm shift in man-machine communication. This contribution reviews the two major semantic interpretation frameworks underpinning natural language interaction, along with their respective advantages and drawbacks. It then discusses the choices made in Siri, Apple's personal assistant on the iOS platform, and speculates on how the current implementation might evolve in the near future to best mitigate any downside.

1.1 Introduction

In recent years, smartphones and other mobile devices, such as electronic tablets and more generally a wide variety of handheld media appliances, have brought about an unprecedented level of ubiquity in computing and communications. At the same time, voice-driven human-computer interaction has benefited from steady improvements in the underlying speech technologies (largely from a greater quantity of labeled speech data leading to better models), as well as the relative decrease in the cost of computing power necessary to implement comparatively more sophisticated solutions. This has sparked interest in a more pervasive spoken language interface, in its most inclusive definition encompassing speech recognition, speech synthesis, natural language understanding, and dialog management.

J.R. Bellegarda (✉)
Apple Inc., One Infinite Loop, Cupertino, CA 95014, USA
e-mail: jerome@apple.com

J. Mariani et al. (eds.), *Natural Interaction with Robots, Knowbots and Smartphones: Putting Spoken Dialog Systems into Practice*, DOI 10.1007/978-1-4614-8280-2_1,

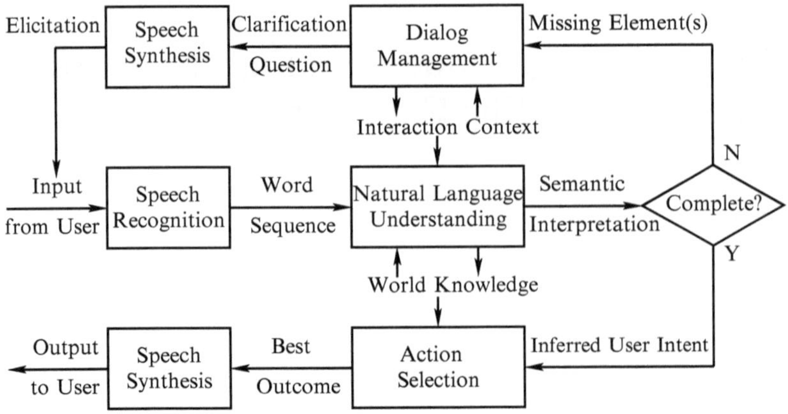

Fig. 1.1 Overview of "intelligent personal assistant" interaction model

To wit, multiple voice-driven initiatives have now reached commercial deployment, with products like Apple's Siri [1], Google's Voice Actions [8], Microsoft's Bing Voice Search [13], Nuance's Dragon Go! [15], and Vlingo [21]. The well-publicized release of Siri in Apple's iPhone 4S, in particular, may have heralded an irreversible shift toward the "intelligent personal assistant" paradigm: just say what you want, and the system will automatically figure out what the best course of action is. For example, to create a new entry on his/her calendar, the user may start the interaction with an input like:

Schedule a meeting with John Monday at 2pm (1.1)

The system then has to recognize that the user's intent is to create a new entry and deal with any ambiguities about the attributes of the entry, such as who will be invited (*John Smith* rather than *John Monday*) and when the meeting will take place (*this coming Monday* rather than *last Monday*).

An overview of the underlying interaction model is given in Fig. 1.1. The speech utterance is first transcribed into a word sequence on which to perform natural language understanding, leading to a semantic interpretation of the input. In case any element is missing, dialog management relies on interaction context to elicit the relevant information from the user. Once the semantic interpretation is complete, task knowledge guides the selection of the best action for the situation at hand. Finally, the selected outcome is conveyed to the user. Success in this realm is measured in subjective terms: *how well* does the system fulfill the needs of the user relative to his/her intent and expectations? Depending on the task, "well" may variously translate into "efficiently" (with minimal interruption),"thoroughly" (so the task is truly complete), and/or "pleasantly" (as might have occurred with a human assistant).

Of course, many of the core building blocks shown in Fig. 1.1 have already been deployed in one form or another, for example, in customer service applications with automatic call handling. Wildfire, a personal telephone assistant, has similarly been available since the mid-1990s [22]. Yet in most consumers' perception, at best the resulting interaction has not been significantly more satisfying than pressing touch-tone keys. So how to explain the growing acceptance of Siri and similar systems? While the interaction model of Fig. 1.1 has not suddenly become flawless, it has clearly matured enough to offer greater perceived flexibility. Perhaps a key element of this perception is that the new systems strive to provide a direct answer whenever possible, rather than possibly heterogeneous information that may contain the answer, as in the classical search paradigm.

Arguably, the most important ingredient of this new perspective is the accurate inference of user intent and correct resolution of any ambiguity in associated attributes. While speech input and output modules clearly influence the outcome by introducing uncertainty into the observed word sequence, the correct delineation of the task and thus its successful completion heavily hinges on the appropriate semantic interpretation of this sequence. This contribution accordingly focuses on the two major frameworks that have been proposed to perform this interpretation and reflects on how they each contribute to the personal assistant model.

The material is organized as follows. The next section describes the statistical framework characteristic of data-driven systems, while Sect. 1.3 does the same for the rule-based framework underpinning expert systems and similar ontology-based efforts. In Sect. 1.4, we focus on Siri as an example and discuss in particular how the choices adopted proved critical to a successful deployment. Finally, the article concludes with some prognostications regarding the next natural stage in the evolution of the user interface.

1.2 Statistical Framework

1.2.1 Background

Fundamentally, the statistical approach to semantic interpretation is aligned with the data-driven school of thought, which posits that empirical observation is the best way to capture regularities in a process (like natural language) for which no complete *a priori* model exists. This strand of work originated in speech recognition, where in the 1980s probabilistic models such as hidden Markov models were showing promise for reconstructing words from a noisy speech signal [16]. Applying similar probabilistic methods to natural language understanding involved the integration of data-driven evidence gathered on suitable training data in order to infer the user's intent.

The theoretical underpinnings for this kind of reasoning were first developed in the context of a partially observable Markov decision process (POMDP) [17]. The key features of the POMDP approach are (1) the maintenance of a system of beliefs, continually updated using Bayesian inference, and (2) the use of a policy whose performance can be quantified by a system of associated rewards and optimized using reinforcement learning via Bellman's optimality principle [10]. Note that Bayesian belief tracking and reward-based reinforcement learning are mechanisms that humans themselves appear to use for planning under uncertainty [6]. For example, experimental data shows that humans can implicitly assimilate Bayesian statistics and use Bayesian inference to solve sensorimotor problems [11].

This in turn motivated the application of the POMDP framework to spoken dialog systems, to similarly learn statistical distributions by observation and use Bayes' rule to infer posteriors from these distributions [24]. However, this proved challenging in practice for several reasons. First, the internal state is a complex combination of the user's goal, the user's input, and the dialog history, with significant uncertainty in the user's utterances (due to speech recognition errors) propagating uncertainty into the other entities as well. In addition, the system action space must cover every possible system response, so policies must map from complex and uncertain dialog states into a large space of possible actions.

1.2.2 Current State of the Art

Making the POMDP framework tractable for real-world tasks typically involves a number of approximations. First, state values can be ranked and pruned to eliminate those not worth maintaining. Second, joint distributions can be factored by invoking some independence assumptions that can be variously justified from domain knowledge. Third, the original state space can be mapped into a more compact summary space small enough to conduct effective policy optimization therein. Fourth, in a similar way, a compact action set can be defined in summary space and then mapped back into the original master space [23].

As an example, Fig. 1.2 shows a possible POMDP implementation for the meeting scheduling task associated with (1.1). It illustrates one time step of a (partial) dynamic Bayesian network, in which the (hidden) system state and (observed) event are represented by open and shaded circles, respectively, while the (observed) command executed by the system is denoted by a shaded rectangle. The state is decomposed into slots representing features such as *person* (indexed by p), *date* (indexed by d), *location*, and *topic* (not shown). Each slot comprises information related to user goal, user input, and dialog history so far. In this simple example, the only dependence modeled between slots is related to the person information. This configuration, known as a "Bayesian update of dialog state" (BUDS) system [20], retains the ability to properly represent system dynamics and to use fully parametric models, at the cost of ignoring much of the conditional dependency inherent in real-world domains.

Because the state of the system (encapsulating the intent of the user) is a hidden variable, its value can only be inferred from knowledge of the transition probabilities between two successive time instants and the observation probabilities associated with the observed event. This leads to a belief update equation of the form:

$$b_{t+1} = K \cdot O(o_{t+1}) \cdot T(c_t) \cdot b_t , \tag{1.2}$$

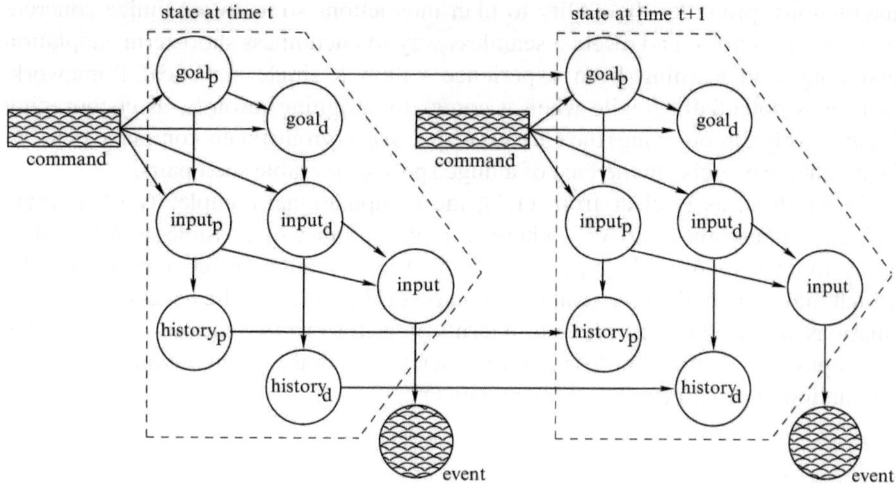

Fig. 1.2 (Partial) dynamic Bayesian network for meeting scheduling task

where the N-dimensional vector $b = [b(s_1) \dots b(s_N)]^T$ is the belief distribution over N possible system states s_i, $O(o)$ is a diagonal matrix of observation probabilities $P(o|s_i)$, and $T(c)$ is the $N \times N$ transition matrix for command c. Given some assumed initial value b_0, (1.2) allows the belief state to be updated as each user input is observed. Since the actual state is unknown, the action taken at each turn must be based on the belief state rather than the underlying hidden state.

This mapping from belief state to action is determined by a policy $\pi : b \longrightarrow c$. The quality of any particular policy is quantified by assigning rewards $r(s,c)$ to each possible state-command pair. The choice of specific rewards is a design decision typically dependent on the application. Different rewards will result in different policies and most likely divergent user experiences. However, once the rewards have been fixed, policy optimization is equivalent to maximizing the expected total reward over the course of the user interaction. Since the process is assumed to be Markovian, the total reward expected in traversing from any belief state b to the end of the interaction following policy π is independent of all preceding states. Using Bellman's optimality principle, it is possible to compute the optimal value of this value function iteratively. As mentioned earlier, this iterative optimization is an example of reinforcement learning [18].

1.2.3 *Trade-Offs*

From a theoretical perspective, the POMDP approach has many attractive properties: by integrating Bayesian belief monitoring and reward-based reinforcement learning, it provides a robust interpretation of imprecise and ambiguous human interactions, promotes the ability to plan interactions so as to maximize concrete objective functions, and offers a seamless way to encompass short-term adaptation and long-term learning from experience within a single statistical framework. Still, it is potentially fragile when it comes to assigning rewards, as encouraging (respectively discouraging) the correct (respectively wrong) state-command pair can be a delicate exercise in the face of a huge space of possible such pairs.

In addition, as is clear from (1.2), the computational complexity of a single inference operation is $\mathcal{O}(N^2)$, where N is the number of possible system states. Thus, for even moderately large values of N exact computation becomes intractable, which makes it challenging to apply to real-world problems. The necessary approximations all have drawbacks, be it in terms of search errors, spurious independence assumptions, quantization loss from master to summary space, or imperfect user simulation to generate reinforcement data [7].

1.3 Rule-Based Framework

1.3.1 Background

In contrast with the systems just mentioned, the rule-based framework does not attempt to leverage data in a statistical way. At its core, it draws its inspiration from early expert systems such as MYCIN [4]. These systems, relying on an inference engine operating on a knowledge base of production rules, were firmly rooted in the artificial intelligence (AI) tradition [12]. Their original purpose was to create specialized agents aimed at assisting humans in specific domains (cf., e.g., [14]). Agent frameworks were later developed to create personal intelligent assistants for information retrieval. In this context, the open agent architecture (OAA) introduced the powerful concept of delegated computing [5]. This was later extended to multi-agent scenarios where distributed intelligent systems can model independent reactive behavior (cf., e.g., [19]).

In the early to mid-2000s, DARPA's PAL (perceptive assistant that learns) program attempted to channel the above efforts into a learning-based intelligent assistant comprising natural language user interaction components layered on top of core AI technologies such as reasoning, constraint solving, truth maintenance, reactive planning, and machine learning [3]. The outcome, dubbed CALO for the Cognitive Assistant that Learns and Organizes, met the requirements for which it was designed, but because of its heterogeneity and complexity, it proved difficult for nonexperts to leverage its architecture and capabilities across multiple domains. This sparked interest in a more streamlined design where user interaction, language processing, and core reasoning are more deeply integrated within a single unified framework [9].

An example of such framework is the "Active" platform, which eschews some of the sophisticated AI core processing in favor of a lighter-weight, developer-friendly version easier to implement and deploy [9]. An application based on this

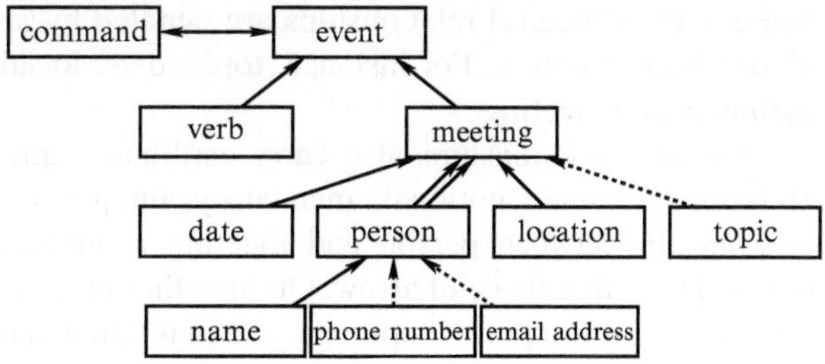

Fig. 1.3 Active ontology for meeting scheduling task

framework consists of a set of loosely coupled services interfacing with specialized task representations crafted by a human expert. Using loosely coupled services eases integration of sensors (cf. speech recognition, but also vision systems, mobile or remote user interfaces, etc.), effectors (cf. speech synthesis, but also touch user interfaces, robotics, etc.), and processing services (such as remote data sources and other processing components).

1.3.2 Current State of the Art

In the "Active" framework, every task is associated with a specific "active ontology." Whereas a conventional ontology is a static data structure, defined as a formal representation for domain knowledge, with distinct classes, attributes, and relations among classes, an active ontology is a dynamic processing formalism where distinct processing elements are arranged according to ontology notions. An active ontology thus consists of a relational network of concepts, where concepts serve to define both data structures in the domain (e.g., a meeting has a date and time, a location, a topic, and a list of attendees) and associated rule sets that perform actions within and among concepts (e.g., the date concept derives a canonical date object of the form: `date(DAY, MONTH, YEAR, HOURS, MINUTES)` from a word sequence such as *Monday at 2pm*).

Rule sets are collections of rules where each rule consists of a condition and an action. As user input is processed, data and events are inserted into a fact store responsible for managing the life cycle of facts. Optional information can be specified to define when the fact should actually be asserted and when it should be removed. As soon as the contents of the fact store changes, an execution cycle is triggered and conditions evaluated. When a rule condition is validated, the associated action is executed. The active ontology can therefore be viewed as an execution environment.

To fix ideas, Fig. 1.3 shows the active ontology for the meeting scheduling task associated with (1.1). The active ontology consists of a treelike structure defining the structure of a valid command for this task. The command operates on a complete event concept representing the action of scheduling a meeting. The meeting concept itself has a set of attributes comprising one or more persons, a topic, a location and a date. Structural relationships are denoted by arrows, which relate to a "has a" ontological notion. For instance, topic, date, location, and person concepts are members of a meeting.

Structural relationships also carry cardinality information and record whether children nodes are optional, mandatory, unique, or multiple. For instance, the relationship between person and meeting is multiple and mandatory, which is denoted by a double solid arrow. On the other hand, the relationship between topic and meeting is unique and optional, which is denoted by a single dashed arrow. This structure is used to provide the user with contextual information. In the example of (1.1), as the location node is linked as mandatory, the user will be asked to provide a location. Through this mechanism, the active ontology not only generates a structured command but also builds dynamic information to interactively assist the user.

As alluded to earlier, concepts incorporate various instantiations of canonical objects. For example, *Monday at 2pm* and *tomorrow morning* are two instances of date objects in the date concept. These objects relate to a "is a" ontological notion. To the extent that rule sets can be specified to sense and rate incoming words about their possible relevance to various concepts, this makes the domain model portable across languages. In addition, it has the desirable side effect of making the approach insensitive to the order of component phrases.

1.3.3 Trade-Offs

Pervasive in the above discussion is the implicit assumption that language can be satisfactorily modeled as a finite state process. Strictly speaking, this can only be justified in limited circumstances, since, in general, the level of complexity of human languages goes far beyond that of context-free languages. Thus, rule-based systems may be intrinsically less expressive than data-driven systems.

In addition, an obvious bottleneck in their development is the specification of active ontologies relevant to the domain at hand. For the system to be successful, each ontology must be 100 % complete: if an attribute is overlooked or a relationship between classes is missing, some (possibly rare) user input will not be handled correctly. In practice, this requires the task domain to be sufficiently well-specified that a human expert from the relevant field is able to distill it into the rule base. This so-called knowledge engineering is typically hard to "get right" with tasks that are highly variable or subject to a lot of noise.

On the plus side, once the ontology correctly captures the whole domain structure, deployment across multiple languages is relatively straightforward. Since a near-exhaustive list of relevant word patterns is already included inside each concept and word order is otherwise largely ignored, only individual surface forms have to be translated. This makes this approach paradoxically similar in spirit to (data-driven) bag-of-words techniques such as latent semantic mapping [2].

1.4 The Siri Experience

1.4.1 Scope and Behavior

Siri was originally formed as a startup company to leverage the results of the CALO project within a much tighter effort with a commercial focus. Its architecture adopted the "Active" platform described earlier as the intermediate layer between mobile I/O and web services. Initial efforts centered on creating tools to make it easier to develop the necessary domain modules and associated active ontologies. The aim was to alleviate many of the issues linked to the ontology bottleneck. With such tools in place, along with the necessary software engineering work required to support them, it became less tedious for human experts to create active ontologies for a given task domain. Upon acquisition by Apple, the Siri team further concentrated on integration and scalability issues.

One important aspect of the Siri experience is the system's response to user requests that fall outside of the well-defined domains it knows about. A typical example is a query such as "Siri, do you love me?", for which a factual response is besides the point. When faced with such out-of-domain queries, most systems typically exhibit a rather "clinical" behavior, with responses like: "*Sorry, I don't understand what you mean.*" While technically adequate, such behavior lacks humanness. In contrast, Siri tries to provide somewhat more entertaining and/or whimsical responses. Not only does this policy inject a bit of sassiness into the system, it also makes the out-of-domain fall-back more palatable to the user.

To illustrate, Table 1.1 gives some examples of such requests under three different scenarios, along with some of the answers provided. Note that giving the same input three times in a row results in three different answers, as the same response would likely be annoying and otherwise destroy the illusion of anthropomorphism. Imbuing the assistant with such socially adept behavior substantially contributed to giving Siri its unique personality. We believe that this strategy proved critical to a successful deployment.

Table 1.1 Examples of out-of-domain requests

User:	Should I go to bed, Siri?
Siri:	*I think you should sleep on it*
User:	I love you!
Siri:	*That's nice. Can we get back to work now?*
User:	I love you!!
Siri:	*Oh, I bet you say that to all your Apple products*
User:	I love you!!!
Siri:	*All you need is love. And your iPhone*
User:	You are useless...
Siri:	*I don't really like these arbitrary categories*

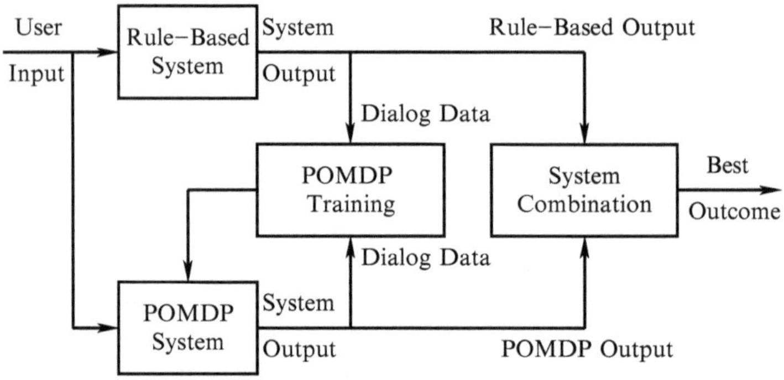

Fig. 1.4 Toward the convergence of rule-based and statistical frameworks

1.4.2 *Likely Evolution*

Like all rule-based systems, Siri suffers from some of the pitfalls mentioned in Sect. 3.3. It is legitimate to ask whether mitigating this downside might involve integrating the associated top-down outlook with the bottom-up outlook adopted by the statistical framework. This possibility unfolds naturally from the inherent complementarity in the respective advantages and drawbacks of the two approaches. Whereas ontology specification requires upfront labor-intensive human expertise, data-driven systems can be run in completely automated fashion. On the other hand, rule-based systems can be deployed right away, while the statistical framework calls for a large amount of suitable training data to be collected beforehand. On the flip side, the former is much more amenable to leveraging know-how across languages, thus enabling rapid deployment in multiple languages, while in the latter every language essentially involves the same amount of effort.

Complementarity between the frameworks, moreover, goes beyond a mere data-vs-knowledge distinction. Whereas rule-based systems are generally sensitive to noise, in principle the POMDP approach can cope with various sources of uncertainty. Yet its elegant optimization foundation assumes specification of suitable rewards, which are probably best informed by empirical observation, and thus rules derived therefrom. In addition, POMDP systems typically involve deleterious approximations to reduce the computational complexity inherent to the sophisticated mathematical machinery involved. In contrast, the AI framework may be intrinsically less expressive but tends to exhibit a more predictable behavior.

Such complementarity bodes well for an eventual convergence between the two approaches, perhaps by way of the virtuous cycle illustrated in Fig. 1.4. First, the deployment of a rule-based system such as Siri provides some real-world dialog data that can be used advantageously for POMDP training, without the difficulties inherent to data collection via user simulation. This in turn enables the deployment of a statistical system like BUDS, which further provides real-world data to refine POMDP models. Such large-scale data collection potentially removes one of the big limiting factors in properly handling uncertainty. It thus becomes possible to combine the rule-based and statistical outputs to come up with the best outcome, based on respective confidence measures for both systems (which may vary over time). By enabling more robust reasoning and adaptation, this strategy should considerably strengthen the cognitive aspects of natural language understanding.

1.5 Conclusion

In this contribution, we have examined the emerging deployment of the "intelligent personal assistant" style of interaction. Under this model it is critical to accurately infer user intent, which in turn hinges on the appropriate semantic interpretation of the words uttered. We have reviewed the two major frameworks within which to perform this interpretation, along with their most salient advantages and drawbacks. Ontology-based systems, such as Siri, are better suited for initial deployment in well-defined domains across multiple languages, but must be carefully tuned for optimal performance. Data-driven systems based on POMDP have the potential to be more robust, as long as they are trained on enough quality data.

The inherent complementarity between these two frameworks sets the stage for the two to converge toward a more cognitive mainstream user interface, which will take intelligent delegation to the next level across many more usage scenarios. Under that hypothesis, the personal assistant model ushers in the next natural stage in the evolution of the user interface: as depicted in Fig. 1.5, the desktop, browser, and search metaphors of past decades thus lead to a new solve metaphor focused on context and tasks. The underlying assumption is that the user will increasingly get used to expressing a general need and letting the system fulfill it in a stochastically consistent manner. This development will likely be a key stepping stone toward an ever more tangible vision of ubiquitous intelligence.

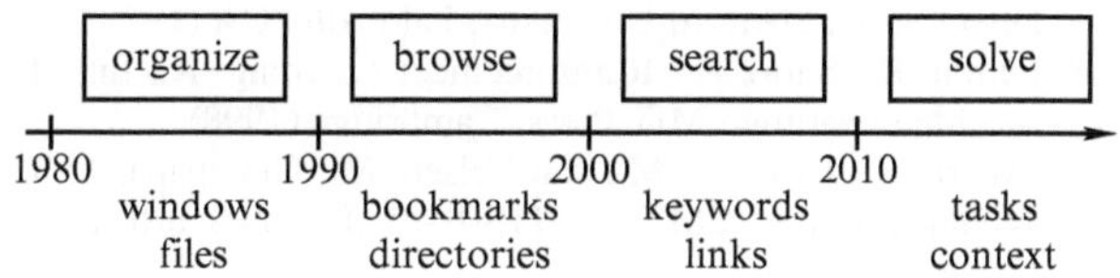

Fig. 1.5 Natural stages in the evolution of the user interface

References

1. Apple Inc. http://www.apple.com/iphone/features/siri.html. Accessed Oct 2011
2. Bellegarda, J.R.: Latent semantic mapping. In: Deng, L., Wang, K., Chou, W. (eds.) Signal Processing Magazine, Special Issue on Speech Technology and Systems in Human-Machine Communication, vol. 22(5), pp. 70–80, Sep 2005
3. Berry, P., Myers, K., Uribe, T., Yorke-Smith, N.: Constraint solving experience with the CALO project. In: Proceedings of Workshop on Constraint Solving Under Change and Uncertainty, pp. 4–8 (2005)
4. Buchanan, B.G., Shortliffe, E.H.: Rule–Based Expert Systems: The MYCIN Experiments of the Stanford Heuristic Programming Project. Addison–Wesley, Reading (1984)
5. Cheyer, A., Martin, D.: The open agent architecture. J. Auton. Agents Multi-Agent Syst. **4**(1), 143–148 (2001)
6. Fu, W.-T., Anderson, J.: From recurrent choice to skill learning: a reinforcement-learning model. J. Exp. Psychol. Gen. **135**(2), 184–206 (2006)
7. Gasic, M., Keizer, S., Mairesse, F., Schatzmann, J., Thomson, B., Young, S.: Training and evaluation of the HIS POMDP dialogue system in noise. In: Proceedings of 9th SIGdial Workshop Discourse Dialog, Columbus, OH (2008)
8. Google Mobile. http://www.google.com/mobile/voice-actions (2008)
9. Guzzoni, D., Baur, C., Cheyer, A.: Active: a unified platform for building intelligent web interaction assistants. In: Proceedings of 2006 IEEE/WIC/ACM International Conferences on Web Intelligence and Intelligent Agent Technology, IEEE Computer Society, 2006
10. Kaelbling, J.L., Littman, M., Cassandra, A.: Planning and acting in partially observable stochastic domains. Artif. Intell. **101**, 99–134 (1998)
11. Kording, J.K., Wolpert, D.: Bayesian integration in sensorimotor learning. Nature **427**, 224–227 (2004)
12. Laird, J.E., Newell, A., Rosenbloom, P.S.: SOAR: an architecture for general intelligence. Artif. Intell. **33**(1), 1–64 (1987)
13. Microsoft Tellme. http://www.microsoft.com/en-us/Tellme/consumers/default.aspx (2008)
14. Morris, J., Ree, P., Maes, P.: SARDINE: dynamic seller strategies in an auction marketplace. In: Proceedings of ACM Conference on Electronic Commerce, pp. 128–134 (2000)
15. Nuance Dragon Go! http://www.nuance.com/products/dragon-go-in-action/index.htm (2011)
16. Rabiner, L.R., Juang, B.H., Lee, C.-H.: An overview of automatic speech recognition, Chapter 1. In: Lee, C.-H., Soong, F.K., Paliwal, K.K. (eds.) Automatic Speech and Speaker Recognition: Advanced Topics, pp. 1–30. Kluwer Academic Publishers, Boston (1996)
17. Sondik, E.: The optimal control of partially observable markov decision processes. Ph.D. Dissertation, Stanford University, Palo Alto, CA (1971)
18. Sutton, R., Barto, A.: Reinforcement Learning: An Introduction. Adaptive Computation and Machine Learning. MIT Press, Cambridge (1998)
19. Sycara, K., Paolucci, M., van Velsen, M., Giampapa, J.: The RETSINA MAS Infrastructure. Technical Report CMU- RI-TR-01-05, Robotics Institute Technical Report, Carnegie Mellon University, 2001
20. Thomson, B., Schatzmann, J., Young, S.: Bayesian update of dialogue state for robust dialogue systems. In: Proceedings of International Conference on Acoustics Speech Signal Processing, Las Vegas, NV (2008)
21. Vlingo Mobile Voice User Interface. http://www.vlingo.com/ (2008)
22. Wildfire Virtual Assistant Service, Virtuosity Corp. http://www.wildfirevirtualassistant.com (1995)
23. Williams, J., Young, S.: Scaling POMDPs for spoken dialog management. IEEE Trans. Audio, Speech Lang. Process. **15**(7), 2116–2129 (2007)
24. Williams, J., Poupart, P., Young, S.: Factored partially observable Markov decision processes for dialogue management. In: Proceedings of 4th Workshop Knowledge Reasoning in Practical Dialogue Systems, Edinburgh, UK (2005)

Chapter 2
Development of Speech-Based In-Car HMI Concepts for Information Exchange Internet Apps

Hansjörg Hofmann, Anna Silberstein, Ute Ehrlich, André Berton, Christian Müller, and Angela Mahr

Abstract The permanent use of smartphones impacts the automotive environment. People tend to use their smartphone's Internet capabilities manually while driving, which endangers the driver's safety. Therefore, an intuitive in-car speech interface to the Internet is crucial in order to reduce driver distraction. Before developing an in-car speech dialog system to a new domain, you have to examine which speech-based human-machine interface concept is the most intuitive. This work in progress report describes the design of various human-machine interface concepts which include speech as main input and output modality. These concepts are based on two different dialog strategies: a command-based and a conversational speech dialog. Different graphical user interfaces, one including an avatar, have been designed in order to best support the speech dialog strategies and to raise the level of naturalness in the interaction. For each human-machine interface concept a prototype which allows for an online hotel booking has been developed. These prototypes will be evaluated in driving simulator experiments on usability and driving performance.

2.1 Motivation

The number of mobile Internet accesses has increased enormously within the last years. The permanent use of smartphones and their Internet capabilities also impacts the automotive environment. In order to be "always connected," people tend to use their smartphone's Internet access manually while driving. However, the manual

H. Hofmann (✉) • A. Silberstein • U. Ehrlich • A. Berton
Daimler AG, Ulm, Germany
e-mail: hansjoerg.hofmann@daimler.com; anna.silberstein@daimler.com; ute.ehrlich@daimler.com; andre.berton@daimler.com

C. Müller • A. Mahr
DFKI, Saarbrücken, Germany
e-mail: christian.mueller@dfki.de; angela.mahr@dfki.de

J. Mariani et al. (eds.), *Natural Interaction with Robots, Knowbots and Smartphones: Putting Spoken Dialog Systems into Practice*, DOI 10.1007/978-1-4614-8280-2_2,

use of smartphone distracts the driver from driving and endangers the driver's safety [4]. Therefore, the development of an intuitive and non-distractive in-car speech interface to the Web is essential in order to increase the driver safety [12].

Before developing a new speech dialog system (SDS) in a new domain, developers have to examine how users would interact with such a system. A previous Internet user study revealed that the human-machine speech interaction styles vary depending on the type of Internet activity [5]. The subjects were presented graphically depicted Internet tasks which had to be solved orally. The tasks were categorized according to Kellar's Web information task classification [7]:

- Information Seeking: e.g., fact finding
- Information Exchange:
  - Transactions: e.g., hotel booking
  - Communications: e.g., sending a Facebook message

The analysis of the speech data revealed that a natural communication style occurred most frequently in information seeking tasks. As for information exchange tasks subjects used natural communication and command-based speaking style equally. This result is valid for transaction and communication tasks. Because of the equal frequency of occurrence we have to examine which speech dialog strategy is the most suitable for performing information exchange tasks before starting to develop a SDS.

This paper reports from work in progress in which different in-car SDS are compared. The SDS are based on different speech dialog strategies, a command-based and a conversational dialog, which will be evaluated on usability and distraction. The systems have been developed for German users and allow for performing an Internet activity for information exchange (using the example of a hotel booking service) by speech. As common in in-car SDS the speech interaction is supported by a graphical user interface (GUI). Different GUIs are designed in order to support the respective dialog strategy and to raise the level of naturalness. This research is conducted within the scope of the EU FP7 funding project GetHomeSafe.[1]

The following paper is structured as follows: Sect. 2.2 gives an overview on previous studies on this research topic. In Sect. 2.3 the functionality of the hotel booking service is explained. Section 2.4 presents the different human-machine interaction (HMI) concepts which are developed within this research work. Here, the different speech dialog strategies and the different GUI concepts are explained. Section 2.5 describes the planned experiments to be conducted in the near future in order to evaluate the different HMI concepts and, finally, conclusions are drawn.

[1] http://www.gethomesafe-fp7.eu.

2.2 Related Work

First studies on the evaluation of dialog strategies have been conducted by Devillers et al. [2] who compare two SDS allowing the user for retrieving touristic information. One dialog strategy guides the user via system suggestions, the other does not. The evaluated dialog strategies comprise the fundamental ideas our command-based and conversational dialog strategy consist of (which are explained in detail in Sect. 2.4). By applying qualitative and quantitative criteria, they conclude that user guidance is suitable for novices and appreciated by all kind of users. However, there was no GUI involved and the speech interaction was performed as primary task. Considering the driving use case other results may be achieved since the primary task is driving.

In the TALK project [10] a command-based speech dialog has been compared to a conversational dialog in the automotive environment. Here, the primary task was driving and as secondary task the driver had to control the in-car mp3-player by speech. The same GUI was used for both dialog strategies. In the field test the subjects had to use the different SDS while driving. Although the conversational dialog was more efficient, the command-based dialog was more appreciated by the subjects. According to Mutschler et al. the high error rate of the conversational strategy was the reason for the higher acceptance of the command-based dialog. The driving performance has been measured with the help of different driving data (e.g., lane keeping). There were no significant differences revealed in the driving performance when using the different SDS.

The speech recognizer quality has improved enormously within the last 5 years. Therefore, the weak speech recognition performance of Mutschler et al.'s conversational dialog may be nowadays less significant. Furthermore, the use of the same GUI for different dialog strategies could have additionally influenced the result. The GUI should be adapted to the particular dialog strategy in order to best benefit from the advantages of the respective strategy and to allow for a comparison of optimal systems. When evaluating the driving performance the averting of the driver's gaze towards the GUI has not been taken in consideration. A glance on the head unit screen could be dangerous if a cyclist would cross the street. The visual distraction can cause accidents which could not be detected with the current performance measurements. Depending on the dialog strategy the visual distraction differs which has to be examined and needs to be compared.

2.3 Functionality of the Hotel Booking Service

The chosen use case for the design of the HMI concepts is booking a hotel by speech while driving. For this purpose, the online hotel booking service HRS[2] has been

[2]http://www.hrs.com.

linked to the existing speech dialog framework. The interface and the functionality of the HRS service are briefly described.

The Web service has been linked via the provided SOAP interface into the existing framework. When sending SOAP XML requests via the interface, the service responds with the requested information encapsulated in a SOAP XML message.

The hotel service HRS allows for various hotel search functions. After having input several required parameters (e.g., location, arrival date), the service delivers a list of hotels which match the search criteria. Additionally, there is the opportunity to enter optional parameters (e.g., price range) to refine the search. The user is able to sort the result list in a certain order or filter according to desired hotel facilities (e.g., swimming pool, parking). The service offers a detailed description of each hotel. After having selected a certain hotel, it can finally be booked.

The mentioned functions have been taken into consideration for the different HMI concepts. Each concept has been designed to allow for parameter input, result list presentation, filtering, and sorting. When using the SDS prototypes, the retrieved hotel data correspond to the currently available hotel information, the booking is only simulated. HRS offers many more functions; however, these functions have not been considered when designing the HMI concepts since they would not have been of additional use to compare the different concepts and, therefore, they were not implemented.

2.4 HMI Concepts

In this section the various HMI concepts are described. First, the different dialog strategies including sample dialogs are presented. Afterwards the GUI concepts, which have been designed in order to support the speech dialog, are described with the aid of screenshots.

2.4.1 Dialog Strategy Design

Two different dialog strategies, a command-based and a conversational dialog strategy, have been designed, and prototypes have been implemented for the later evaluation.

The following technical SDS features were integrated in both prototypes: in order to speak to the system the driver has to press a push-to-activate (PTA) button. Furthermore, the driver is able to interrupt the system while prompting the user ("barge in"). State-of-the-art in-car SDS use "teleprompters" to inform the driver visually about possible commands. However, the use of "teleprompters" raises too much visual attention on the head unit screen. Therefore, the user is only informed audibly about possible commands.

The developed speech dialog prototypes have been specified for German language. However, the sample dialogs given in this section are written in English for better understanding. The characteristic of each strategy and how they differ are described in the following. When designing the different dialog strategies, we particularly focused our attention on the dialog initiative, the possibility to enter multiple input parameters, and the acoustic feedback.

2.4.1.1 Command-Based Dialog Strategy

The dialog behavior of the command-based dialog strategy corresponds to the voice-control which can be found in current state-of-the-art in-car SDS. By calling explicit speech commands, the speech dialog is initiated and the requested information is delivered or the demanded task is executed. There are several synonyms available for each command. By using implicit feedback in the voice prompts, the driver is informed about what the system has understood. After the first command the user is guided by the system and executes the steps which are suggested and displayed by the system. The GUI supports the speech dialog by showing the "speakable" commands as widgets on the screen (see Sect. 2.4.2). A sample dialog is illustrated in the following:

Driver: *Book a hotel.*
System: *Where would you like to book a hotel?*
Driver: *In Berlin.*
System: *When do you want to arrive in Berlin?*
Driver: *Tomorrow.*
System: *How long would you like to stay in Berlin?*
Driver: *Until the day after tomorrow.*

When the parameters have been input, HRS is called to retrieve the list of hotels. The user can then continue the interaction by calling certain commands.

2.4.1.2 Conversational Dialog Strategy

In the conversational dialog strategy, the dialog initiative switches during the speech interaction. The driver is able to speak whole sentences where multiple parameters can be set within one single utterance. Thereby, the dialog can run more naturally, be flexible and efficient. The driver is informed about what the system has understood by using implicit feedback. If the driver has set multiple parameters in his utterance, the system does not implicitly repeat all parameters as the system response would be too long. Therefore, the system repeats only the contextually most important parameter. The GUI does not present the "speakable" commands on the screen. In order to indicate the possible functions, icons are displayed (see Sect. 2.4.2). A sample dialog is presented in the following:

Driver: *I would like to book a hotel in Berlin.*
System: *When do you arrive in Berlin?*
Driver: *I arrive tomorrow and leave the day after tomorrow.*

As illustrated in the example, the driver can already indicate some input parameters when addressing the system for the first time. The system checks which input parameter are missing in order to send a request to HRS. The system prompts the user and collects the missing information. Although the system asks for only one parameter, the user is able to give more or other information than requested.

When the parameters have been input, HRS is called to retrieve the list of hotels. The user can now continue the interaction by speaking freely and without having to call certain commands.

2.4.1.3 Comparison of Dialog Strategies

The TRINDI ticklist from Bohlin et al. [1], which characterizes the dialog behavior of a SDS with the help of 12 Yes-No questions, gives a good overview of the implemented dialog features. Both of the SDS prototypes have been developed and differentiated corresponding to this list. The filled out TRINDI ticklist for both dialog strategies is illustrated in Table 2.1.

In this research work the most important dialog features which allow for a differentiation of both dialog strategies have been realized so far. Concerning the dialog design of the conversational dialog, we set a high value on the flexibility to input parameters by speech (e.g., Q2, Q3, Q12). Dialog features which are no beneficial characteristic of one of the dialog strategies and which do not reveal differences in the evaluation are left out to lower the development effort (e.g., Q5, Q6, Q8). Impact of the environment on the speech interaction is not in focus of this research (Q8). The dialog flow of a hotel booking dialog is linear and does not allow for context-relevant branches whereby Q11 becomes superfluous.

2.4.2 GUI Design

The different GUIs have been designed in order to support the speech dialog strategies the most and to raise the level of naturalness in the interaction. The different GUIs have been customized corresponding to the dialog strategies only as much as necessary since an objective comparison is targeted. When designing the screens we followed the international standardized AAM guidelines [3] which determine the minimum font sizes, the maximum numbers of widgets, etc., in order to minimize distraction. In the following the general differences of the different GUI concepts are described with the aid of screenshots.

Table 2.1 Characterization of speech dialog strategies on the basis of the TRINDI ticklist

	Command-based dialog	Conversational dialog
Q1:	*Is utterance interpretation sensitive to context?*	
	✔	✔
Q2:	*Can the system deal with answers to questions that give more information than was requested?*	
	✗	✔
Q3:	*Can the system deal with answers to questions that give different information than was actually requested?*	
	✗	✔
Q4:	*Can the system deal with answers to questions that give less information than was requested?*	
	✔	✔
Q5:	*Can the system deal with ambiguous designators?*	
	✗	✗
Q6:	*Can the system deal with negatively specified information?*	
	✗	✗
Q7:	*Can the system deal with no answer to a question at all?*	
	✔	✔
Q8:	*Can the system deal with noisy input?*	
	Not in scope of the research work	
Q9:	*Can the system deal with "help" sub-dialogs initiated by the user?*	
	✗	✗
Q10:	*Can the system deal with "non-help" sub-dialogs initiated by the user?*	
	✗	✔
Q11:	*Does the system only ask appropriate follow-up questions?*	
	No relevant dialog step in existent hotel booking dialog	
Q12:	*Can the system deal with inconsistent information?*	
	✗	✗

2.4.2.1 Command-Based Dialog GUI

In the command-based dialog strategy, the driver uses commands to speak to the system. In order to give the driver an understanding of the "speakable" commands, the speech dialog is supported by the GUI. For that reason the currently possible speech commands are displayed on the screen at all times, which may lead to a high visual distraction. Hence, in automotive terms the command-based speech dialog strategy is also called "speak-what-you-see" strategy.

Figure 2.1 illustrates the main screen of the hotel booking application at the beginning of the hotel booking dialog. Here, the first input parameter "destination" ("Ziel" in German) has been set by the user after being requested by the system. Afterwards the user is guided step-by-step by the system. When the driver has given the requested information, a new widget appears on the screen and the system asks the driver for the corresponding input.

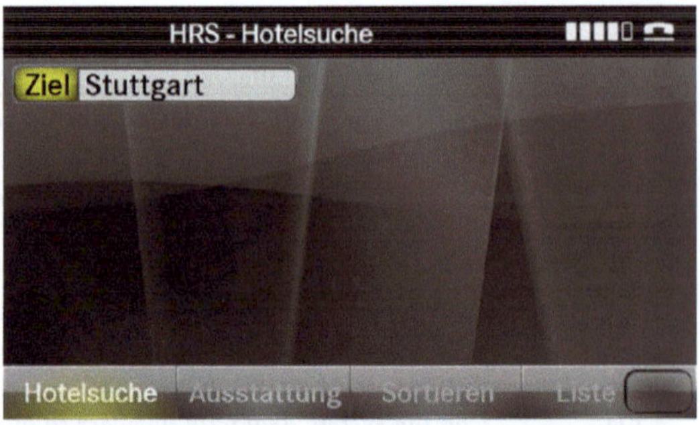

Fig. 2.1 Main screen of the command-based dialog while parameter input

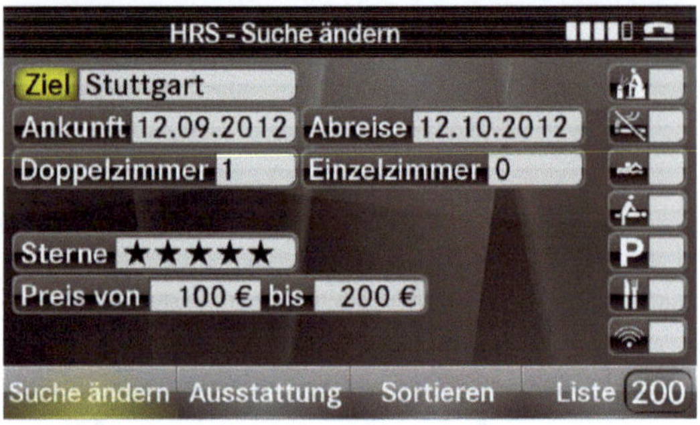

Fig. 2.2 Main screen of the command-based dialog after parameter input

When all the parameters are set and the hotel service has returned the list of hotels, the list of filters is displayed and the possible commands for changing the input parameters ("Suche ändern"), setting the hotel facilities ("Ausstattung"), sorting the result list ("Sortieren"), and presenting the result list ("Liste") become visible in the sub-function line (see Fig. 2.2). The active GUI state after receiving the list of hotels is the "Suche ändern" screen where the search parameters, which are presented in the main area of the main screen (e.g., "Ziel" or "Ankunft"), can be changed. However, the driver has several possibilities to proceed with the speech dialog by calling the other commands displayed in the sub-function line. By calling the command "Ausstattung" (or synonyms of the command) the filter sub-dialog is triggered and the hotel facility screen is displayed (see Fig. 2.3). For the presentation and the sorting of the result list there are further similar screens.

2.4.2.2 Conversational Dialog GUI

In the conversational dialog strategy, the driver can speak freely and does not have to call certain commands. There is no need to give the driver a visual feedback of the currently "speakable" commands whereby the visual distraction may be lowered. For that reason, the content on the head unit screen does not have to indicate the possible options to proceed with the speech dialog. The sub-function line which was used to indicate the available commands is replaced by only few symbols which resemble the current GUI state.

Fig. 2.3 Hotel facilities screen of the command-based dialog

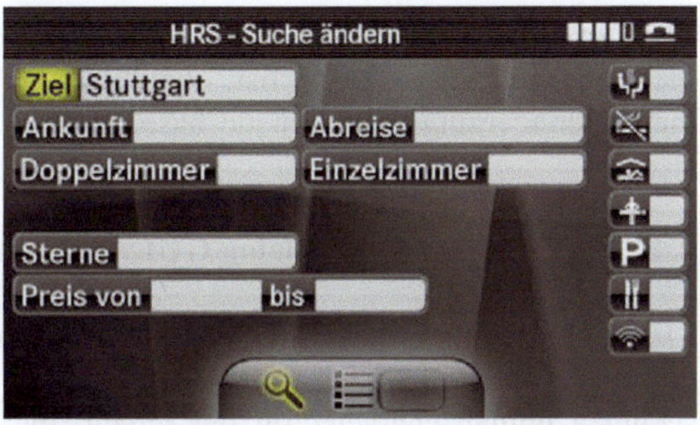

Fig. 2.4 Main screen of the conversational dialog at the beginning of the interaction

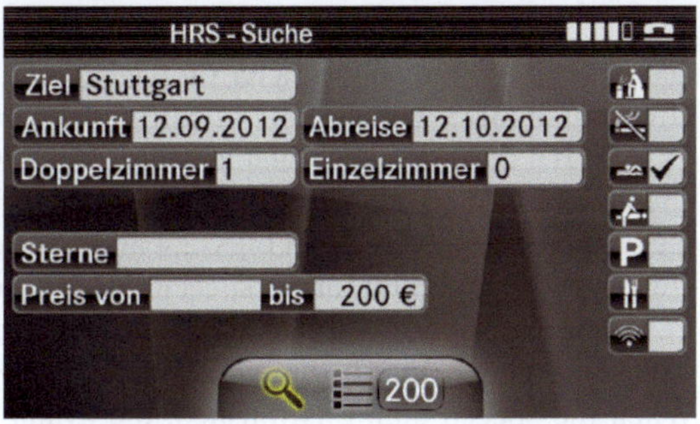

Fig. 2.5 Main screen of the conversational dialog after parameter input

Figure 2.4 shows the main screen at the beginning of the speech interaction. The user is able to input several parameters at once. He is even allowed to already set the hotel facility filters.

After having input all required parameters (and optional parameters or filters) the system calls the HRS service and retrieves a list of hotels (see Fig. 2.5). In this GUI state the driver is able to change the search parameters, change the hotel facility filters, or sort the list by speech. There are no additional screens for presenting the available filters or for the list sorting options. The alterations evoked by speech become only visible on the main screen by changing the information displayed.

Fig. 2.6 Main screen of the conversational dialog with avatar after parameter input

The symbols on the bottom of the screen resemble the GUI states for parameter input/changes and the result list. The design of the result list screen is the same as the one concerning the command-based strategy.

2.4.2.3 Conversational Dialog GUI with Avatar

The goal of using an avatar is to raise the naturalness of the HMI. By expressing gestures and mimics, the avatar contributes to a more human interaction. When seeing a human character on the screen, the driver might tend to speak more naturally, as if he would talk to a human being. This might have a positive effect on speech dialog quality and user acceptance. However, the user might be more distracted by a human character on the screen. So far, those positive and negative effects of an SDS with avatar while driving have not been examined.

The GUI concept with avatar is based on the conversational dialog GUI. A virtual character designed and developed by Charamel[3] is integrated. The avatar overlays the background illustrated in Figs. 2.4 and 2.5 but does not cover the widgets which are currently important for the speech dialog (see Fig. 2.6).

When the driver is driving without interacting with the SDS there is no avatar visible on the screen. The human agent appears when the speech dialog is initiated. When the speech dialog is finished, the avatar disappears again. In this way, the visual distraction is lowered and the driver knows when he is allowed to speak to the system. The avatar makes certain gestures to give the SDS some human character. For example, when the system asks for inputting the destination, the avatar points at the destination widget on the screen. When the user browses the hotel result list, the avatar makes a swipe gesture to support the scrolling in the list.

[3]www.charamel.de.

2.5 Evaluation

The speech-based HMI concepts that were introduced above will be evaluated with the help of formative user studies in order to test usability and driver distraction. Based on the results of the experiments, the best HMI concept will be employed in the GetHomeSafe system and will be further improved.

As a first step, a small number of subjects will test the different speech dialog strategies while performing the standard lane change task (LCT) [8]. With the help of this rather explorative test, we will prove if the actual user expectancies are met and potential system shortcomings, such as grammar deficiencies, can be corrected. As a next step, we plan to evaluate the mentioned HMI concepts by conducting a more substantial user study in the driving simulator at DFKI's "future lab" (see Fig. 2.7). We will employ the OpenDS open source driving simulation which is being developed and improved within the scope of the EU research project GetHomeSafe. In this study, the command-based dialog strategy used as reference system is tested only with GUI, whereas the conversational dialog will be presented without GUI, with GUI, and with GUI including the avatar.

As a primary driving task in the second study, we will use the ConTRe (continuous tracking and reaction) [9] task which complements the de facto standard LCT including higher sensitivity and a more flexible driving task duration without restart interruptions. Another requirement for our evaluation is a more fine-grained assessment of driver distraction, in terms of temporal resolution of performance metrics. In LCT, drivers are only once in a while directed to change the lanes (even with announcement) by conducting a rather unnatural abrupt maneuver, combined with simple lane keeping on a straight road in between. But real driving mostly demands a rather continuous adjustment of steering angle and speed without announcements, when the next demand will occur exactly and to which extend a reaction will be necessary. In order to receive more detailed results about the two diverse dialog strategies, we use a task that rather resembles continuous driving, like a car following task. Furthermore, we prefer an absolute ground truth of perfect

Fig. 2.7 DFKI driving simulator setup

Fig. 2.8 Screenshot of the ConTRe task as the first modular extension of the OpenDS simulation component

behavior for the performance metric, whereas the LCT is based on an ideal line as a generated, normative model. Another intended advantage of the ConTRe task over the LCT and also many other standard tasks is the possibility to explicitly address mental demand via an event detection task. Effects of cognitive load should be revealed above all by the achieved reaction times. Therefore, an additional discrete task was implemented as longitudinal control (gas and brake). This task should be accomplished in addition to the continuous adjustment of steering wheel angles for lateral control.

The driver's primary task in the simulator is comprised of actions required for normal driving: turning the steering wheel, as well as operating the brake and acceleration pedals. System feedback, however, differs from normal driving. In the ConTRe task, the car moves on its own with a constant speed through a predefined route on a unidirectional straight road consisting of two lanes. Turning the steering wheel moves the car laterally but no further than the edge of the carriageway. Additionally, steering manipulates a moving blue bar, which is rendered in front of the car (see Fig. 2.8). On the road ahead, the driver perceives this blue bar and another yellow bar, both moving continuously at a constant longitudinal distance in front of the car. The yellow one is called the reference bar, as it moves autonomously within the roadsides according to an algorithm. The driver controls the lateral position of the blue bar by turning the steering wheel, trying to keep it overlapping with the reference bar as well as possible. A distance metric between the reference bar and the controllable bar is recorded continuously. Effectively, on an abstract level this corresponds to a task where the user has to follow a curvy road or the exact lateral position of a lead vehicle, although correct task performance is indicated more obviously and therefore leads to less user-dependent variability.

In addition to the steering task, common gas and brake reactions are also required once in a while. However, operating the acceleration or brake pedal does not have any effect on vehicle speed. There is a traffic light placed on top of the reference cylinder containing two lights: the lower one can be lighted green, whereas the top light shines red when it is switched on. Only one of these lights appears once in a while. The red light requires an immediate brake reaction with the brake pedal, whereas green indicates that an immediate acceleration with the gas pedal should be performed. As soon as the driver reacts correctly, the light turns off (see Fig. 2.8). Reaction time as well as accuracy can be assessed.

Besides measuring driver distraction via performance metrics, we will assess subjective mental workload with the help of the DALI questionnaire [11] after each system condition. Eye tracking will be used for gaze-based distraction evaluation, including average and maximum glance duration for the different GUI variants. A qualitative assessment of the dialog strategies will be performed using the PARADISE framework [13], which appraises overall dialog quality by means of several interaction criteria (e.g., success rate, number of interaction steps). The SASSI questionnaire [6] will be used to survey subjective usability evaluation of the speech dialog variants. Previous knowledge of participants on SDS will be assessed in the very beginning as part of a biographic questionnaire.

Overall, we expect better usability evaluation for the conversational dialog conditions compared with the command-based condition. For the conversational dialog conditions, we do not expect large differences regarding usability when comparing the conditions with GUI with the conditions without GUI. However, for this comparison we expect the GUI to cause more driver distraction in terms of glances onto the GUI screen and in terms of decreased driving performance. For these metrics we expect the command-based GUI variant to perform worse than the conversational GUI. Furthermore, we expect to find longer task completion times for the command-based dialogs. If increased task duration occurs on the same level of performance decrease, the condition with shorter task duration should be chosen. When using the avatar we expect positive effects on usability. However, we expect the GUI with avatar to cause more driver distraction than the normal conversational GUI. The presented experimental investigation will help us to decide about the most preferable dialog strategy and about what kind of GUI should be employed.

2.6 Conclusions

This paper reports from work in progress in which different in-car speech-based HMI concepts are compared. For each concept a prototype which allows for an online hotel booking has been developed.

The described HMI concepts are based on different dialog strategies which include speech as main input and output modality. The speech dialog is supported by a GUI which is adapted to the respective speech dialog strategy. The first HMI concept is based on a command-based dialog strategy where the driver is able to start the speech dialog by single commands and is led step-by-step by the system afterwards. The available commands are displayed on the head unit screen. The second dialog strategy, the conversational dialog, allows the driver to speak with entire sentences as if he would talk to a human being. Thereby, multiple parameters can be input at once and the dialog initiative switches frequently. Two different GUI design concepts were targeted to support the conversational dialog and to raise the level of naturalness. The first concept does not display the commands anymore but uses icons to suggest possible functions of the system to the driver. Based on the first GUI concept, the second concept contributes to a more conversational interaction by displaying additionally a humanlike character on the screen.

With the aid of the developed prototypes the different HMI concepts will be evaluated on usability and driving performance. The driving simulator experiments will be performed at DFKI in Saarbrücken. Based on the results of the experiments, the best HMI concept will be employed in the GetHomeSafe system and will be further improved.

Acknowledgements The research work described in this paper is performed in the context of the GetHomeSafe project which is conducted within the scope of the Seventh Framework Program of the European Commission. We would like to thank the European Commission for funding the GetHomeSafe project.

References

1. Bohlin, P., Bos, J., Larsson, S., Lewin, I., Matheson, C., Milward, D.: Survey of existing interactive systems. Deliverable 1.3, TRINDI (1999)
2. Devillers, L., Bonneau-Maynard, H.: Evaluation of dialog strategies for a tourist information retrieval system. In: Proceedings of International Conference on Spoken Language Processing, pp. 1187–1190 (1998)
3. Driver Focus-Telematics Working Group: Statement of Principles, Criteria and Verification Procedures on Driver Interactions with Advanced In-Vehicle Information and Communication Systems. Alliance of Automotive Manufacturers, Washington (2002)
4. Governors Highway Safety Association: Distracted driving: What research shows and what states can do. Technical report, U.S. Department of Transportation, 2011
5. Hofmann, H., Ehrlich, U., Berton, A., Minker, W.: Speech interaction with the Internet - a user study. In: Proceedings of Intelligent Environments, Guanajuato, Mexico (2012)
6. Hone, K.S., Graham, R.: Subjective assessment of speech-system interface usability. In: Proceedings of Eurospeech, pp. 2083–2086 (2001)
7. Kellar, M.: An examination of user behaviour during web information tasks. Ph.D. thesis, Dalhousie University, Halifax, Canada (2007)
8. Mattes, S.: The lane-change-task as a tool for driver distraction evaluation. Proceedings of IGfA, pp. 1–30 (2003)
9. Moniri, M., Mahr, A., Math, R., Feld, M., Müller, C.: The ConTRe (continuous tracking and reaction) task: A flexible approach for measuring driver distraction with high sensitivity. In: Adjunct Proceedings of the 4th International Conference on Automotive User Interfaces and Interactive Vehicular Applications, pp. 88–91 (2012)
10. Mutschler, H., Steffens, F., Korthauer, A.: Final report on multimodal experiments - part 1: Evaluation of the sammie system. d6.4. talk public deliverables. Technical Report (2007)
11. Pauzie, A.: Evaluating driver mental workload using the driving activity load index (DALI). In: Proceedings of European Conference on Human Interface Design for Intelligent Transport Systems, pp. 67–77 (2008)
12. Peissner, M., Doebler, V., Metze, F.: Can voice interaction help reducing the level of distraction and prevent accidents? Meta-study on driver distraction and voice interaction. Technical Report, Fraunhofer-Institute for Industrial Engineering (IAO) and Carnegie Mellon University, 2011
13. Walker, M.A., Litman, D.J., Kamm, C.A., Kamm, A.A., Abella, A.: PARADISE: A framework for evaluating spoken dialogue agents. In: Proceedings of the eighth conference on European chapter of the Association for Computational Linguistics, pp. 271–280 (1997)

Chapter 3
Real Users and Real Dialog Systems: The Hard Challenge for SDS

Alan W. Black and Maxine Eskenazi

Abstract Much of the research done in our community is based on developing spoken dialog systems and testing various techniques within those dialog systems. Because it makes it easier to deal with our experimental conditions, many of our tests and studies involve controlled (paid or volunteered) users. However, we have seen in a number of studies that these controlled users do not use the system in the same way as those for whom the system was actually designed. Sometimes the difference between the real user, who wants the information the spoken dialog system is providing or who wants to give information to it, and the controlled user, who is acting under some direction, is not that different. Certainly in some circumstances it is necessary to use the latter. But, since state-of-the-art systems have become increasingly reliant on large amounts of user data to train their models of behavior, it is critical that the user behavior we train on is real user behavior. This paper describes the issues that arise when building a spoken dialog system for real users. The goal is to provide both a service to the user and a realistic spoken dialog system (SDS) research platform.

3.1 Background

Many spoken dialog researchers have come to the realization that having real users test systems produces different results than testing with paid subjects. Ai et al. [1] showed several differences between real and paid subjects using the Let's Go system. This difference between paid and actual users also appeared in the results of the spoken dialog challenge 2010 (SDC 2010) [2]. In SDC 2010 the system with the best results for the control test with volunteers was different from the one that

A.W. Black (✉) • M. Eskenazi
Language Technologies Institute, Carnegie Mellon University,
Forbes Ave., Pittsburgh, PA, USA
e-mail: awb@cs.cmu.edu; max@cs.cmu.edu

J. Mariani et al. (eds.), *Natural Interaction with Robots, Knowbots and Smartphones: Putting Spoken Dialog Systems into Practice*, DOI 10.1007/978-1-4614-8280-2_3,

did best on the tests with real users. Paid/volunteer users have different goals from actual system users, so optimizations based on paid user data may not be valid when applied to a real system.

It should be noted that finding appropriate paid/volunteer users is, in itself, a major task. Recruiting takes time and money. While we do not advocate that all control tests on spoken dialog systems (SDS) should be considered invalid, since sometimes they are the only reasonable method of testing, we do believe that targeting tests using live systems whenever possible has the highest probability of providing more useful results.

Control tests have the advantage of targeting specific phenomena. Given an equal amount of real and controlled user dialogs, we may not find enough of the desired phenomena in the real data. So we need to collect many more live user dialogs in order to get statistically significant results. There is a tension between a requirement of smaller numbers of targeted controlled users and much larger numbers of real users.

We are however aware that simply stating that real users make better models for an SDS is somewhat obvious (and tautological). But in order to collect data from real users, building SDS that are capable of serving a significant number of real users is not simple.

Commercial dialog systems have significantly more users than research systems do (at least successful ones). But most commercial providers are reluctant to allow frontline research on their live systems (though if this is done properly, it would benefit the researcher, the commercial providers, and the users themselves). Finding research systems that can generate a significant user base is in fact very difficult. Most organizations wanting to acquire a telephone dialog system may not be interested in running a system provided by a research group, and most research groups do not have the experience necessary to provide real, reliable service.

We want to encourage others to create SDS that gather real users' data and so wish to relate some experience that can be useful to those who wish to try. We will discuss the task, system building and maintenance issues, and alternatives to creating one's own system.

3.2 Finding a Real Task

Finding an appropriate task for a successful SDS is a harder problem than it would seem, perhaps harder than developing the system itself. We need to find a task that will have a large number of users and will serve them in some useful way.

We often select a task for a SDS without considering the full consequences of that choice. If major issues come up once the system has been developed, it is usually too onerous to change the task. So, given the investment that we already have in it, we stick with the system we have created. The alternative to this is to move our dialog systems to a task that already has users.

There have been a number of successful real user dialog systems that have been platforms for researchers. We have been lucky to have built Let's Go [7], which provides bus information to the Pittsburgh community.

Let's Go has been running daily with live users since early 2005. At time of writing, it has had over 180,000 calls and continues to run. A fundamentally simple SDS, it provides the scheduled time for the requested buses. It is, however, complex enough to have allowed for substantial research projects to have been tested on it and to have been mentioned or used in over 185 published papers. As a result of our success in running studies on this platform, we have offered it as a research platform to others: using the live system and our real users. Where paid/volunteer testing might generate up to 100 calls, Let's Go Live typically offers some 1,300 calls a month.

Other systems with real users are starting to appear. Notably, we see museum guides [8, 10] and games [6]. Real users are attracted to an SDS that gives them something that they need. Usually they get information or some form of assistance or the enjoyment of playing a game. In essence, the real users have some defined personal need, not a script to follow.

From over 7 years of experience with the live system, we can identify some properties that have enabled it to become a platform that many callers use, although the exact requirements for such a system are still far from being as clear as we would like:

Regular Callers The task you choose should already, in some form, have many regular callers who almost all ask for the same type of information. A mixture of repeat and novel callers is an added advantage: these caller types may have quite different goals and require different optimal strategies. As far as numbers are concerned, Let's Go, in a city of around 300,000, rarely gets over 100 calls per night (it operates from 7 pm to 6 am on week nights, longer on weekends).

No Personal Information Due to human privacy concerns, it is desirable to have a system that does not require personal information. As we will see below, if the system has to take personal information, effort has to be devoted to removing it by putting it in a separate file or by detecting and erasing it automatically.

Simple is Good Enough Even a relatively small task, like getting three well-defined pieces of information from a caller, can still benefit from more research. There are many remaining issues worthy of investigation, for example, Let's Go has been a good test bed for belief-based architectures. Elaborate dialogs (and open-ended ones) can be interesting, but it is difficult to provide a real service that can benefit very large numbers of users.

When choosing an application, there are several considerations. The first is dialog type. Our group works on information giving and receiving systems: others may be interested in multi-agent, robotic, mixed initiative, etc. type SDS. Also, real-world applications bring other less scientific concerns into play. One is human subject clearance. We made a list of applications where people call to give or get information in which personal information was not needed. If there is an application that has other worthy qualities, but must contain some personal information, the latter can be recorded it in a separate file that the researchers do not keep or distribute (such as with our 311 pothole system where users leave contact information if they want to be called when the pothole is filled). Another possibility is to clean the personal information from the data, post acquisition. The latter solution is less desirable since it results in data with missing segments which may make it unsuitable for retraining models.

In general, we have found that it is better to move to where the users already are (a service that they are already using) rather than to get the users to come to a system that you have created. This implies that system creators must listen to the needs of the users and adapt to them. It also implies that if there is a partner that the real users already deal with on a regular basis, an agreement of some sort should to be drawn up. It may include answers to the following:

- How will the users be informed that they will be recorded (and implicit consent obtained)?
- What rights will you have to the data?
- How will you maintain user privacy?
- How will the partner and the users benefit from dealing with you? The Port Authority of Allegheny County (that runs the Pittsburgh bus service) had no one manning the phones at night, so they (and the Pittsburgh callers) gained a service that they would not have had otherwise.

In some cases, where the partner's benefit is high, you may want to ask for monetary participation in the project. This should not create a financial burden that would jeopardize the agreement. Our group has found that systems are more appreciated by management when they have had to pay something, even if it is a relatively small amount, for the service.

3.2.1 Where to House the System

The Let's Go system has run successfully for many years on a server, located at Carnegie Mellon linked to a direct telephone line. Our team has recently started to move the system from POTS to VOIP and from the server to a virtual machine in the cloud. The advantages of hardware-independence should afford less system downtime. The costs involved in this move can be shared with the partner. In the past, running the system on our own server, rather than moving it to the Port Authority, has had the advantage of giving us full access and control of the system. We have been able to update it and run studies when we need to. With the use of the cloud, both the partner and our team can equally have access to the system.

3.2.2 Maintaining the System

It is always important to know how real users will interact with an SDS. At first we used recordings of the human operators to understand how people ask for bus information. However after start-up, we spent significant resources bringing the system up to its core quality [7]. This included language and acoustic model retraining.

Long-term maintenance has also been a substantial part of our effort. Systems like this quickly deteriorate if they are not continually supported. In addition to regular bus schedule updates, the routes, bus numbers, and bus stop names have regularly changed over the past 7 years. We also have increased our coverage from the initial routes that passed the major universities to all buses in the city.

A real, live system is not one that sits unattended with no problems to deal with. Over the years many hardware, software, and administrative issues have affected the system. We have updated components, from information directly obtained from experiments run on the machine (e.g., error handling [3]) and from general updates to our speech components in the system. More recently we have changed the dialog manager itself to a belief-based system [5], but still keep the same core functionality for the users.

Although we would like to spend most of our engineering effort on core research components, a significant amount of effort goes to hardware maintenance updates, to the core computing systems, and to the obscure (changing the motherboard clock battery when the clock was not running smoothly). Operating system updates also generate significant problems. This takes effort and time away from core research, but it also provides us with core stability for a real research platform.

3.2.3 Knowing If It Is Working

Once the system has started gathering large amounts of data, it is not possible to listen to every dialog every day. Whether you have 50 or 5,000 dialogs a day, it is extremely costly to hand check everything. Thus it is important to build automated analysis tools that report on system performance.

Our developers receive a daily report that summarizes the previous day's calls. It includes number of calls, number of estimated successes (i.e., the system provided an actual bus time), average number of turns per call, and number of no-turn calls. Along with these absolute figures it provides the average statistics for calls for that day of the week. This enables someone who has become familiar with the report to easily see if something is not working correctly. The report also contains information related to specific failure mechanisms. The system also sends the developers email if certain modules fail.

We have noticed less obvious changes over time. A few years ago the estimated success rate gradually decreased over a few months. Listening to calls did not identify what the differences were. The callers were clearly less successful in using the system than before, but similar calls had always existed. One change that happened during that time was the release of a smartphone app by another group in the university (that also tracks real-time bus information). We assume that a group of our experienced users moved to their system. This left us with a less experienced user community. We changed our dialog strategies to make them more reliable for these users (e.g., removing a "how may I help you" type prompt in favor of a direct request "where are you leaving from?") and our success rate rose again. Such nonobvious external forces are costly to detect and correct.

We are aware that call analysis techniques are, in themselves, research topics. Having an automated system inform you of your system's performance and eventual changes in it (and their causes) is critical to running big live systems. Necessity has prompted us to build several such aids over the years.

Also it has been important to ensure that our users do actually get what they need with no decrease in the quality of service over time. The system must be efficient, accurate, and easy to use. The constant analysis of error types is an important part of our work, enabling us to fix the most serious errors rapidly with no apparent loss of service. Recently we have been ameliorating our response to shouting callers, first identifying them, and then changing our dialog strategy to stop the shouting [4].

3.3 Using Someone Else's Real System

Creating a full SDS with a significant set of real users is, as we have seen, onerous and rare. If this is not possible, an alternative to this is to use someone else's system.

Over the past years our team has offered Let's Go as a platform (with our real user base) that other researchers around the world can run their experiments on. After offering this within Carnegie Mellon, we opened it to researchers elsewhere [1, 9]. This open platform has been difficult to explain to others. The idea of obtaining the system code, making changes, having it tested, and then taking it live and reaping the data (logs and speech files) seems novel and should be well accepted by the community as it has been for the two SDC.

We employ a number of careful safeguards before taking a system live so that our users do not suffer any degradation in service.

In the SDC 2010 [2] and SDC 2011, we allowed complete external systems other than our own to serve our real users. In SDC 2010 three non-CMU systems were developed and then tested before being called directly by our users.

3.4 Crowdsourcing

A number of researchers have started looking at crowdsourcing as a method to get users for their systems. Although this is an excellent method to allow a user community to have access to a system, the same caveats apply to crowdsourced users as to any other users. If their task has been set by the requester, they will be optimizing that task (or maximizing their payments) rather than choosing to do the task itself.

One way to obtain surrogate data is to create a game (e.g., on Facebook). A game where the user tries to beat the clock to get some bus data from a dialog system is one way that data from a system like Let's Go might be augmented. This implies setting up a visually appealing game and insuring that the audio works in both directions. It also implies some quality control in case of malicious users.

Another way to obtain surrogates is to create crowdsourcing tasks on platforms such as Amazon mechanical turk (MTurk) or on your own. The reward of remuneration is valued slightly differently from remuneration of callers in the past. Some of the workers are truly interested in the success of the task that they work on and will try to provide nearly real data. The quality control issue remains, though. One way to control the dialogs is to have the same worker do the same dialog task twice in a set of dialogs and compare the two outputs, which should be nearly identical. Another control is to have a second set of workers listen to the dialogs produced by the first set and pass judgement on whether a real user participated in this dialog. The agreement of 3–5 workers on each dialog should be sufficient.

Following data acquisition, you should make statistical models of both the real and surrogate user data to determine whether the two datasets differ and, if they do, to understand the dimensions in which they differ. This can lead you to compensate for that difference when running a study or at least to explain the difference when examining your results.

To sum up, obtaining surrogate user data is possible, but it comes with an extra burden of work. As the search for subjects in a lab study is a long task, so is creating, gathering, and processing surrogate data.

3.5 Conclusions

Our aim of building SDS for real users seems an obvious one that many already feel strongly about. At some level this aim is due to the fact that SDS are now sufficiently accurate that it matters what user community the systems are tuned for. One could consider this aim for real user systems is perhaps due to the advancement of our technology. The target of real users is also aimed at ensuring that our systems are actually improving for the target audience and just for not some special paid/volunteer audience.

In spite of the apparently simple aim, we are acutely aware that finding real users and supporting real systems for them are hard and will unquestionably consume some of our resources, but the research that we can carry out on such systems will challenge us in novel ways and be more applicable to the real world.

References

1. Ai, H., Raux, A., Bohus, D., Eskenazi, M., Litman, D.: Comparing spoken dialog corpora collected with recruited subjects versus real users. In: SIGDial, Columbus, OH (2008)
2. Black, A., Burger, S., Conkie, A., Hastie, H., Keizer, S., Lemon, O., Merigaud, N., Parent, G., Schubiner, G., Thomson, B., Williams, J., Yu, K., Young, S., Eskenazi, M.: Spoken dialog challenge 2010: Comparison of live and control test results. In: SIGDial, Portland, OR (2012)
3. Bohus, D., Langner, B., Raux, A., Black, A., Eskenazi, M., Rudnicky, A.: Online supervised learning of non-understanding recovery policies. In: SLT-2006, Palm Beach, Aruba (2006)
4. Fandrianto, A., Eskenazi, M.: Prosodic entrainment in an information-driven dialog system. In: Interspeech, Portland, OR (2012)
5. Lee, S., Eskenazi, M.: An unsupervised approach to user simulation: Toward self-improving dialog systems. In: Proceedings of the 13th Annual Meeting of the Special Interest Group on Discourse and Dialogue, pp. 50–59. Association for Computational Linguistics, Seoul, South Korea (2012). URL http://aclweb.org/anthology-new/W/W12/W12-1606
6. McGraw, I., Gruenstein, A., Sutherland, A.: A self-labeling speech corpus: Collecting spoken words with an online educational game. In: Interspeech, Brighton, UK (2009)
7. Raux, A., Bohus, D., Langner, B., Black, A., Eskenazi, M.: Doing research on a deployed spoken dialogue system: one year of let's go experience. In: Interspeech, Pittsburgh, PA (2006)
8. Rojas-Barahona, L., Lorenzo, A., Gardent, C.: Building and exploiting a corpus of dialog interactions between french speaking virtual and human agents. In: LREC, Istanbul, Turkey (2012)
9. Stoyanchev, S., Stent, A.: Predicting concept types in user corrections in dialog. In: EACL Workshop on Semantic Representation of Spoken Language, Athens, Greece (2009)
10. Swartout, W., Traum, D.R., Artstein, R., oren, D.N., Debevec, P., Bronnenkant, K., Williams, J., Shrikanth Narayanan, A.L., Piepol, D., Lane, C., Morie, J., iti Aggarwal, P., Liewer, M., Chiang, J.Y., Gerten, J., hu, S.C., White, K.: Ada and grace: Toward realistic and engaging virtual museum guides. In: IVA 2010, Philadelphia (2010)

Chapter 4
A Multimodal Multi-device Discourse and Dialogue Infrastructure for Collaborative Decision-Making in Medicine

Daniel Sonntag and Christian Schulz

Abstract The dialogue components we developed provide the infrastructure of the disseminated industrial prototype RadSpeech—a semantic speech dialogue system for radiologists. The major contribution of this paper is the description of a new speech-based interaction scenario of RadSpeech where two radiologists use two independent but related mobile speech devices (iPad and iPhone) and collaborate via a connected large screen installation using related speech commands. With traditional user interfaces, users may browse or explore patient data, but little to no help is given when it comes to structuring the collaborative user input and annotate radiology images in real-time with ontology-based medical annotations. A distinctive feature is that the interaction design includes the screens of the mobile devices for touch screen interaction for more complex tasks rather than the simpler ones such as a mere remote control of the image display on the large screen.

4.1 Introduction

Over the last several years, the market for speech technology has seen significant developments [7] and powerful commercial off-the-shelf solutions for speech recognition (ASR) or speech synthesis (TTS). For industrial application tasks such medical radiology, we implemented a discourse and dialogue infrastructure for semantic access to structured and unstructured information repositories [13]. The infrastructure is based on the assumption that in order to support a rapid dialogue system engineering process for domain-specific dialogue applications, an ontology-based approach should be followed for all internal and external processing steps.

D. Sonntag (✉) • C. Schulz
German Research Center for AI (DFKI), Stuhlsatzenhausweg 3,
66123 Saarbruecken, Germany
e-mail: sonntag@dfki.de; chschulz@dfki.de

J. Mariani et al. (eds.), *Natural Interaction with Robots, Knowbots and Smartphones: Putting Spoken Dialog Systems into Practice*, DOI 10.1007/978-1-4614-8280-2_4,

The idea of semantic web data structures [1] has provided new opportunities for *semantically enabled user interfaces*. The explicit representation of the *meaning* of data allows us to (1) transcend traditional keyboard and mouse interaction metaphors and (2) provide representation structures for more complex, collaborative interaction scenarios that may even combine mobile and terminal-based interaction [11]. The collaborative speech-based interaction scenario in a multiparty setting for medical decision-making, namely in radiology, will be the focus of this paper. We relied on a semantic web toolbox for ontology-based dialogue engineering. In previous implementation work of this large-scale project (THESEUS[1]), we provided a technical solution for the two challenges of engineering ontological domain extensions and debugging functional modules [14].

In this paper, we basically provide two new contributions. First, we provide distinctive features of our new dialogue infrastructure for radiology and explain the first speech-based annotation system for this task. Second, we discuss the radiology interaction system in greater detail and explain the implemented dialogue sequences which constitute a running demo system at our partner hospital in Erlangen. Thereby we also focus on the special technical components and implementation aspects that are needed to convey the requirements of dialogical interaction in a medical application domain. With traditional user interfaces in the radiology domain (most of which are desktop-based monomodal keyboard input systems), users may browse or explore patient data, but little to no help is given when it comes to structuring the collaborative user input and annotate radiology images in real-time with ontology-based medical annotations. To meet these objectives, we implemented a distributed, ontology-based dialogue system architecture where every major component can be run on a different host (including the graphical interface and audio streaming on mobile devices). This increases the scalability of the overall system.

In earlier projects [8, 15] we integrated different sub-components into multi-modal interaction systems. Thereby, hub-and-spoke dialogue frameworks played a major role [9]. We also learned some lessons which we use as guidelines in the development of *semantic* dialogue systems [5]; the whole architecture can be found in [10]. Thereby, the dialogue system acts as the middleware between the clients and the backend services that hide complexity from the user by presenting aggregated ontological data. One of the resulting speech system, RadSpeech (http://www.youtube.com/watch?v=uBiN119_wvg), is the implementation of a multimodal dialogue system for structured radiology reports.

---

[1]This work is part of THESEUS-RadSpeech (see www.dfki.de/RadSpeech/) to implement dialogue applications for medical use case scenarios. It has been supported by the German Federal Ministry of Economics and Technology (01MQ07016).

4.2 Special Radiology Task Requirements and Implementation

In the MEDICO use case, we work on the direct industrial dissemination of a medical dialogue system prototype. Recently, structured reporting was introduced in radiology that allows radiologists to use predefined standardised forms for a limited but growing number of specific examinations. However, radiologists feel restricted by these standardised forms and fear a decrease in focus and eye dwell time on the images [2, 16]. As a result, the acceptance for structured reporting is still low among radiologists while referring physicians and hospital administrative staff are generally supportive of structured standardised reporting since it eases the communication with the radiologists and can be used more easily for further processing.

We implemented the first mobile dialogue system for radiology annotations, which is tuned for the standardised radiology reporting process. Our solution not only provides more robustness compared to speech-to-text systems (we use a rather small, dedicated, and context-based speech grammar which is also very robust to background noise), it also fits very well into new radiology reporting processes which will be established in Germany and the USA over the next several years: in structured reporting you directly have to create database entries of a special vocabulary (according to a medical ontology) instead of text. The semantic dialogue system presented by RadSpeech should be used to ask questions about the image annotations while engaging the clinician in a natural speech dialogue. Different semantic views of the same medical images (such as structural, functional, and disease aspects) can be explicitly stated, integrated, and asked for. This is the essential part of the knowledge acquisition process during the speech dialogue: the grammar of the ASR system only accepts the annotations of a specific grammar which stems from the used medical ontologies; this allows us to reject arbitrary annotations and recognitions with low probability which makes the system very reliable. Upon touching a region on the interaction device, the ASR is activated. After recognition, the speech and gesture modalities are fused into a complex annotation using a combination of medical ontologies. For disease annotations, for example, the complete RadLex (http://www.radlex.org) terminology can be used, but we also use an OWL version of ICD-10 [4] and FMA [3]. With this dedicated grammar, the annotation accuracy of single term annotations is above 96 %, whereby multi-term annotations (three annotations in one speech command) are difficult to handle (informal evaluation).

Another central requirement is the need for different graphical user interfaces and contents on the mobile devices and the screen. Currently, radiology working stations must feature an FDA clearing (http://www.fda.gov/) meaning that only cleared (mobile) devices can be used for active diagnostic purposes. Following this sub-requirement, we can use the FDA-cleared iPad (or iPhone) for diagnostic purposes and the big screen for non-diagnostic ones. As a result, the image series should only be manipulated and annotated on the mobile interaction devices, whereas key

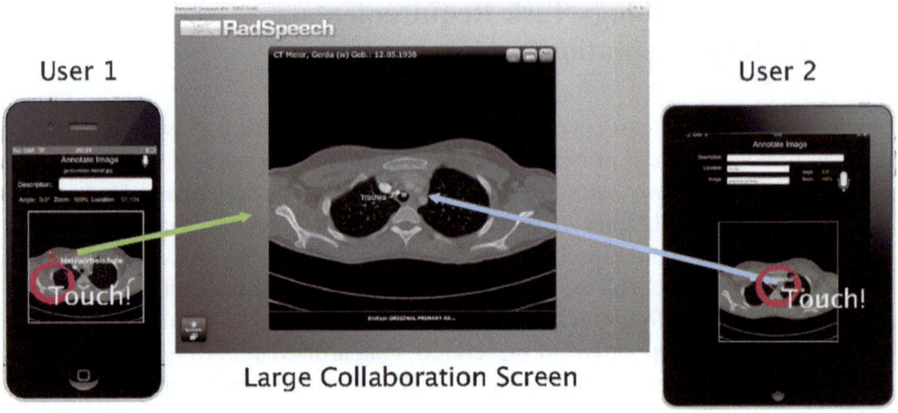

Fig. 4.1 Multimodal speech dialogue scenario with multiple input/output devices

images are displayed on the big screen, thereby allowing to synchronise individual annotations stemming from multiple FDA-cleared devices. A very nice feature of the resulting interaction scenario which takes on this special requirement is the effect that, on the mobile device, we can implement the multimodal setting with a mobile image series viewer which runs through the slices (see, e.g., the commercial DICOM app MIM, http://www.mimsoftware.com). The ASR activates upon touch, and the manipulation of the images can be done using touch instead of trying to do all of these things using speech and the big touch screen—thereby making a virtue of necessity (Fig. 4.1).

In addition to ASR, dialogue tasks include the interpretation of the speech signal and other input modalities, the context-based generation of multimedia presentations, and the modelling of discourse structures. According to the utility issues and medical user requirements we identified (system robustness/usability and processing transparency play the major roles), we provide for a special rule-based fusion engine of different input modalities such as speech and pointing gestures. We use a production-rule-based fusion and discourse engine which follows the implementation in [6]. Within the dialogue infrastructure, this component plays a major role since it provides basic and configurable dialogue processing capabilities that can be adapted to specific industrial application scenarios (e.g., the coordination of pointing gestures and ASR activation on the medical images). More processing robustness is achieved through the application of a special robust parsing feature in the context of RDF graphs as a result of the input parsing process. The domain-specific dialogue application is able to process the following medical multi-user-system dialogue on multiple devices (the cancer annotation is replaced by a simple anatomy annotation for illustration):

1 **U1:** "Show me the CTs, last examination, patient XY."
2 **S:** Shows corresponding patient CT studies as DICOM picture series and MR videos.
3 **U1:** "Show me the internal organs: lungs, liver, then spleen and colon."
4 **S:** Shows corresponding patient image data according to referral record on the iPad.
5 **U1:** "Annotate this picture with 'Heart' (+ pointing gesture on the iPad)."

6 **S:** "Picture has been annotated with 'Heart'."
7 **U1:** "Show it on screen."
8 **S:** "Shows patient XY on the large screen, automatically rendering the picture with the heart annotation in the foreground."
9 **U2:** "and 'Heart chamber' (+ pointing gesture on the iPhone)"
10 **S:** Adds the second annotation on screen.
11 **U1:** "Synchronise annotations with my iPad."
12 **S:** "Shows new annotation on the iPad."
13 **U2:** "Search for similar patients."
14 **S:** "The search obtained this list of patients with similar annotations including 'Heart' and 'Heart chamber'."
15 **U1:** "Okay."

Our system then switches to the comparative records to help the radiologist in the differential diagnosis of the suspicious case, before the next organ (e.g., liver) is examined in the collaborative session of the two doctors. The semantic search for similar cases is implemented by a SPARQL engine which computes semantic similarities between the ontology concepts on the images and the image series in the databases (see [12]).

4.3 Multimodal Interaction in the Multiparty Setting

For the collaborative scenario we need to be able to model the activity of each user that is connected to the infrastructure. The challenge in this setting is that, in our infrastructure, the input/output communication assigned to every individual user must be processed separately in one individual dialogue session. This architectural decision was made in the initial setting to cope with (deictic) dialogue references in the dialogue history and allow for a coherent representation of a specific session's working memory. In addition, we handle multiparty dialogue input by multiple devices. As a result, a single dialogue session has been restricted to a single user. Accordingly, a multisession operation is our answer to the new multi-user requirement (towards the direction that one user indicates something and the second can refer to it (future work)). In Fig. 4.2, the most relevant parts of the implementations concerning the multiparty scenario are displayed.

The ontology-based dialogue system (ODP) represents the central part in the architecture and handles the communication among the external device components through multiple channels (i.e., handshaking/messaging among clients, controlling the speech server to listen to audio streams, and the like). In addition, it provides the multisession infrastructure based on a rule engine in order to instantiate several dialogue system sessions in the desired multi-device setting. At this point, we want to emphasise the fact that all peripheral devices (our mobile devices such as iPhones or iPads) are associated with one session for one device, respectively, which is hold throughout the dialogue.

As a consequence, an event within one session will not directly affect the state of another session. In what follows, we will illustrate how we extend our infrastructure by implementing a multi-party-enabled Display Context Manager to meet the new requirements: to implement collaborative scenarios where actions on peripheral devices actually have an effect on other users (and corresponding dialogue sessions) connected to the dialogue system.

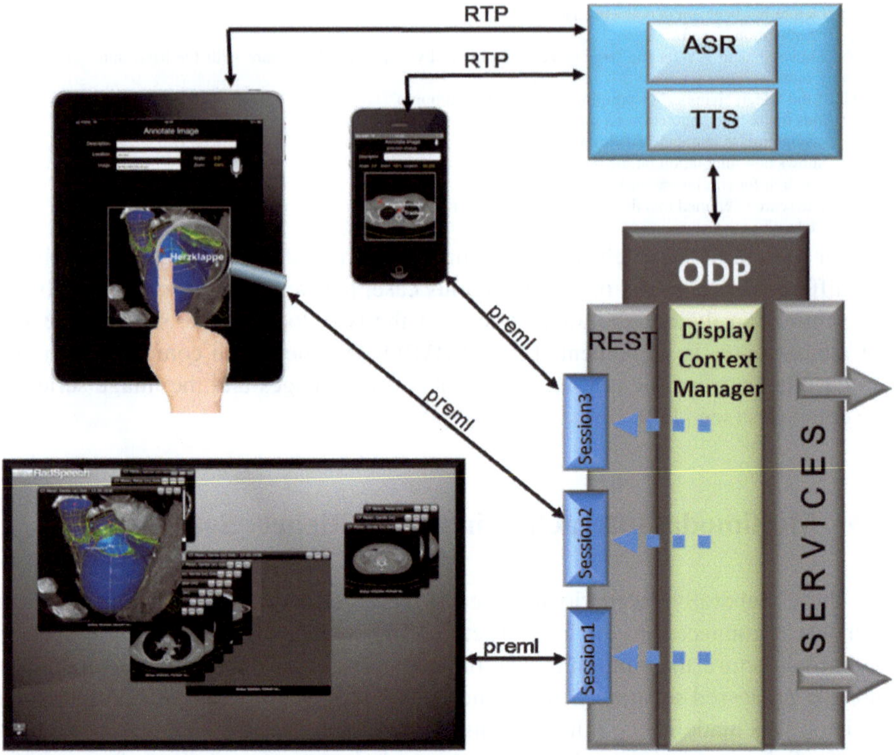

Fig. 4.2 The multiparty/multisession infrastructure: two active users on iPad and iPhone

The Display Context Manager is in charge of dispatching the command messages which are also ontological instances, with an internal representation as *typed feature structures* (TFS). The corresponding TFS is then handed over to proper operational components possessing exclusive access to write on medical data records. The medical data that are subject to the expert's analysis and manipulation are located inside a data container, maintaining so-called *spotlets* and *zones*. Spotlets are containers for meta-information of patient images (e.g., DICOM metadata about the image recording process in the hospital such as date, time, image modality, and the patient's name). Zones are containers administrating the annotations associated with the spotlets. Medical data inside the container are instantiated as soon as the user retrieves patient images at the backend service by using the dialogue engine. In this sense, the life cycle of the data in the working memory is determined by the image retrieval process and the length of a session. However, a user has the option to commit annotation results of his or her diagnostic analysis to dedicated servers as backend services at any point during a session.

Operations on data are categorised into different levels of intrusion. For instance, the deictic input on the user interface can be associated with a different operation than voice input which may contain the user's demand to attach an annotation to a

```xml
<object type="radspeech#ImageInputEvent">
  <slot name="odp#hasContent">
    <object type="medico#ImageAnnotation">
      <slot name="odp#isSelected"/>
      <slot name="medico#annotation"/>
    </object>
  </slot>
  <slot name="odp#action">
    <value type="String">
      <![CDATA[select_zone]]>
    </value>
  </slot>
  <slot name="radspeech#id">
    <value type="String">
      <![CDATA[1]]>
    </value>
  </slot>
  <slot name="comet#xCoordinate">
    <value type="Float">
      <![CDATA[252]]>
    </value>
  </slot>
  <slot name="comet#yCoordinate">
    <value type="Float">
      <![CDATA[190]]>
    </value>
  </slot>
</object>
```

```xml
<object type="medico#AnnotateTask">
  <slot name="odp#hasContent">
    <object type="medico#MedicoSpotlet"/>
  </slot>
  <slot name="odp#hasContent">
    <object type="medico#ImageAnnotation">
      <slot name="medico#annotation">
        <value type="String">
          <![CDATA[herzklappe]]>
        </value>
      </slot>
    </object>
  </slot>
  <slot name="medico#linked">
    <object type="medico#Modifier">
      <slot name="radspeech#modifier">
        <value type="String">
          <![CDATA[add_ann]]>
        </value>
      </slot>
    </object>
  </slot>
</object>
```

Fig. 4.3 TFS messages that represent different types of events which in turn invoke different classes of operations

medical image. In particular, the first operation is relevant to inform the Display Context about what the attentional focus of the user is (e.g., selecting medical images or performing image annotations), whereas the second operation performs data manipulation in a zone that belongs to some spotlet representing the selected medical image on the mobile device of the respective user.

Figure 4.3 (on the left) shows the corresponding TFS message that is transferred by a select gesture, while the TFS message (Fig. 4.3, on the right) encapsulates an annotation task triggered by voice. Please note, however, that the level of intrusion is independent of the input modality, a voice command may easily serve to change the attentional focus by saying "Open the patient's last image annotation," for example.

In order for the multisession scenario to use inputs from different users, we have implemented a class of operations that has the permission to make manipulations even on data which do not belong to the same session. As pointed out in the lower part of Fig. 4.4, each data container is assigned to a session ID.

Depending on the type of operation, the Display Context Manager identifies the corresponding session ID that is connected to the data container. In this way, we are able to model a process that a user is able to perform actions on an iPad whereupon the display content changes and displays further related results on a big screen. Table 4.1 shows an overview of the basic multisession interactions that support gesture and voice inputs for the setting where a mobile device propagates its contents to the big screen. For example, on the iPad the propagation of all manipulations of the images is only executable in the main view where all images are displayed. After manipulation, all annotation activities will be mirrored to and synchronised with the big screen. This refers to the actions in Table 4.1 that are indicated by "(both)."

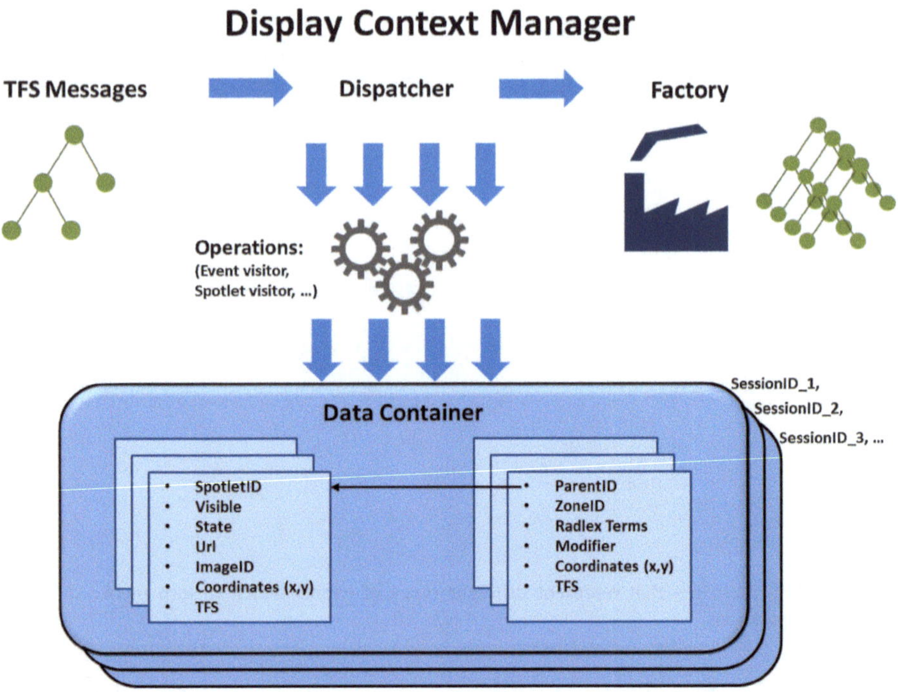

Fig. 4.4 The Display Context Manager and data container

The second user or additional passive user groups might then inspect the results of the dialogue-based annotation and diagnostic analysis in real-time on the big screen. In particular, the operations that are executed within the session dedicated to the iPad have access to not only the data container representing the display contents of the iPad but also the data container that is responsible for the display content of the big screen. The synchronisation of manipulation behaviour and TFS data between multiple data containers is achieved by an operation that enables instances of other sessions to obey to operations executed by the session in command. This means among other things that only the user who opens a session is allowed to make his or her actions shareable to other sessions.

Besides providing a mechanism to manipulate meta-information of data containers regardless of the device the command is issued from, we also have to make sure that the result reaches the correct recipient among the sessions. Again depending on the type of operation, the Display Context Manager detects the corresponding working memory being associated with a particular session/device on the basis of the session ID. After the operation has been executed on the data in terms of updating its internal state, the dispatching mechanism selects a factory method to produce the appropriate TFS result.

Table 4.1 Overview of the modelled collaborative interactions

Device type	Gesture/voice input	Multisession action
iPad	1-SwipeToRight(onImage)	Propagate image to screen
	1-SwipeToRight(onMainview)	Propagate all images to screen
	"Show the images on the screen"	Propagate all images to screen
	1-SwipeToLeft(onImage)	Remove image from screen
	"Synchronise with the screen"	Synchronise actions on screen
	"Stop synchronisation with screen"	Desynchronise actions on screen
	doubleTap(mainviewFooterCenter)	Close the patient file/images (both)
	longPress(ann)	Delete annotation (both)
	select(ann)+*"Delete annotation"*	Delete annotation (both)
	doubleTap(imageviewFooterRight)	Delete all annotations of image (both)
	drag(ann)	Repositioning the annotation (both)
iPhone	1-SwipeToRight(belowImage)	Synchronise actions on screen
	1-SwipeToLeft(belowImage)	Desynchronise actions on screen
	"Synchronise with the screen"	Synchronise actions on screen
	"Stop synchronisation with screen"	Desynchronise actions on screen
	longPress(annotation)	Delete annotation (both)
	select(ann)+*"Delete annotation"*	Delete annotation (both)
	drag(ann)	Repositioning the annotation (both)

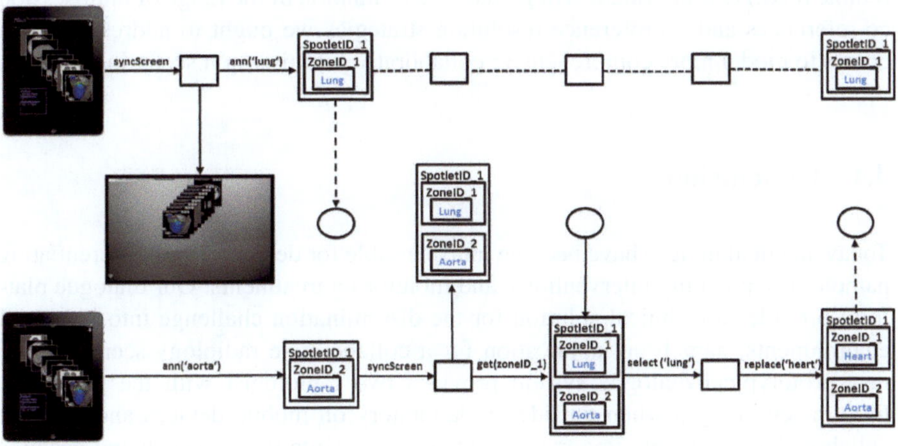

Fig. 4.5 Workflow of a collaborative scenario

Based on the identified working memory, the corresponding update rule inside the dialogue engine instance fires in response to the created TFS object that wraps the modified state of spotlets and zones.

The workflow of the collaborative scenario is shown in Fig. 4.5, where the behaviour of the multisession operations between multiple devices is outlined. The chart demonstrates a collaborative interaction example where an annotation of the first user is overwritten/corrected (or potentially specified in more detail) by another user while using the shared view on the big screen.

First, user 1 (upper iPad) propagates all relevant images of the patient being treated to the big screen. Then, user 1 annotates the zone with the id "ZoneID_1" of the image referring to "SpotletID_1" with the term *lung*.

Meanwhile, another user (user 2, lower iPad) annotates the same image with the term *aorta* but in another zone. The propagation of the annotation event by the second user allows the Display Context Manager to unify the annotations assigned to the same image and display them both on the screen.

Subsequently, the second user disagrees with the annotation of the first user for illustration. First she pulls the annotations of the image on the screen to her device (which is implemented as an update operation similar to subversion systems), namely the annotation she wants to correct. Only at the point when the second user obtains the annotation of the first user on her own device she is able to replace the annotation in question. In turn, this manipulation of the zone (replacing *lung* with *heart* by a voice command) will be reflected on the big screen. In this way, we obtained a clear "speech co-operation policy" and avoided too complex synchronisation behaviours, conflict solution strategies, and recovery mechanisms for unification failures. (Please note that the case with a remote client is slightly different; here the *syncScreen* function synchronises with the big screen and the remote iPad.) Our next steps will include the evaluation of the range of multisession co-references and co-reference resolution strategies we ought to address when it comes to model more comprehensive collaborative multisession scenarios.

4.4 Conclusion

Today, medical images have become indispensable for detecting and differentiating pathologies, planning interventions, and monitoring treatments. Our dialogue platform provides a technical solution for the dissemination challenge into industrial environments, namely an application for a collaborative radiology scenario. Our new prototypical dialogue system provides two radiologist with the ability to, first, review images when outside the laboratory on mobile devices and, second, collaboratively annotate important image regions while using speech and gestures on multiple mobile devices while co-operating in front of a large synchronised touch screen installation. Currently, the system is part of a larger clinical study about the acquisition of medical image semantics at Siemens Healthcare, the University Hospital in Erlangen, and the Imaging Science Institute (ISI). In future work, we will pursue the idea of multisession dialogue management in order to allow for more complex user interactions such as "What do you think about this lesion? + pointing gesture (user 1)" and user 2, "—it's a difficult case, but I think it's a subtype of non-Hodgkin lymphoma." Thereby, we would extend our first RadSpeech scenario (http://www.youtube.com/watch?v=uBiN119_wvg) not only to the collaboration described here but also to the highly desired multisession fusion scenario.

References

1. Fensel, D., Hendler, J.A., Lieberman, H., Wahlster, W. (eds.): Spinning the Semantic Web: Bringing the World Wide Web to Its Full Potential. MIT Press, Cambridge (2003)
2. Hall, F.M.: The radiology report of the future. Radiology **251**(2), 313–316 (2009)
3. Mejino, J.L., Rubin, D.L., Brinkley, J.F.: FMA-RadLex: An application ontology of radiological anatomy derived from the foundational model of anatomy reference ontology. In: Proceedings of AMIA Symposium, pp. 465–469, 2008. http://stanford.edu/~rubin/pubs/097.pdf
4. Möller, M., Ernst, P., Dengel, A., Sonntag, D.: Representing the international classification of diseases version 10 in OWL. In: Proceedings of the International Conference on Knowledge Engineering and Ontology Development (KEOD), Valencia, Spain, 2010
5. Oviatt, S.: Ten myths of multimodal interaction. Commun. ACM **42**(11), 74–81 (1999). citeseer.nj.nec.com/oviatt99ten.html
6. Pfleger, N.: FADE: an integrated approach to multimodal fusion and discourse processing. In: Proceedings of the Dotoral Spotlight at ICMI 2005, Trento, Italy, 2005
7. Pieraccini, R., Huerta, J.: Where do we go from here? research and commercial spoken dialog systems. In: Proceedings of the 6th SIGDdial Worlshop on Discourse and Dialogue, pp. 1–10, 2005
8. Reithinger, N., Fedeler, D., Kumar, A., Lauer, C., Pecourt, E., Romary, L.: MIAMM: A multimodal dialogue system using haptics. In: van Kuppevelt, J., Dybkjaer, L., Bernsen, N.O. (eds.) Advances in Natural Multimodal Dialogue Systems. Springer, Berlin (2005)
9. Reithinger, N., Sonntag, D.: An integration framework for a mobile multimodal dialogue system accessing the semantic web. In: Proceedings of Interspeech, Lisbon, Portugal, pp. 841–844 (2005)
10. Sonntag, D.: Ontologies and Adaptivity in Dialogue for Question Answering. AKA and IOS Press, Heidelberg (2010)
11. Sonntag, D., Deru, M., Bergweiler, S.: Design and implementation of combined mobile and touchscreen-based multimodal web 3.0 interfaces. In: Proceedings of the International Conference on Artificial Intelligence (ICAI), pp. 974–979 (2009)
12. Sonntag, D., Möller, M.: Unifying semantic annotation and querying in biomedical image repositories. In: Proceedings of International Conference on Knowledge Management and Information Sharing (KMIS) (2009)
13. Sonntag, D., Reithinger, N., Herzog, G., Becker, T.: a discourse and dialogue infrastructure for industrial dissemination. Proceedings of IWSDS—Spoken Dialogue Systems for Ambient Environment, Chapter. Lecture Notes in Artificial Intelligence, pp. 132–143. Springer, Berlin (2010)
14. Sonntag, D., Sonnenberg, G., Nesselrath, R., Herzog, G.: Supporting a rapid dialogue engineering process. In: Proceedings of the First International Workshop On Spoken Dialogue Systems Technology (IWSDS) (2009)
15. Wahlster, W.: SmartKom: symmetric multimodality in an adaptive and reusable dialogue shell. In: Krahl, R., Günther, D. (eds.) Proceedings of the Human Computer Interaction Status Conference 2003, pp. 47–62. DLR, Berlin, Germany (2003)
16. Weiss, D.L., Langlotz, C.: Structured reporting: Patient care enhancement or productivity nightmare? Radiology **249**(3), 739–747 (2008)

Part II
Spoken Dialog Prototypes and Products

Chapter 5
Yochina: Mobile Multimedia and Multimodal Crosslingual Dialogue System

Feiyu Xu, Sven Schmeier, Renlong Ai, and Hans Uszkoreit

Abstract Yochina is a mobile application for crosslingual and cross-cultural understanding. The core of the demonstrated app supports dialogues between English and Chinese and German and Chinese. The dialogue facility is connected with interactive language guides, culture guides and country guides. The app is based on a generic framework enabling such novel combinations of interactive assistance and reference for any language pair, travel region and culture. The framework integrates template-based translation, speech synthesis, finite-state models of crosslingual dialogues and multimedia sentence generation. Furthermore, it allows the interlinking between crosslingual communication and tourism-relevant content. A semantic search provides easy access to words, phrases, translations and information.

5.1 Introduction

The language barriers between Eastern and Western societies constitute big challenges for economic and cultural exchange. The dream of today's technophile travelers, educated by visionary science fiction, is to own a mobile speech-to-speech translation system functioning as a personal interpreter, allowing them to talk in their own language and to be understood in the language of the partner thanks to a combination automatic translation as well as speech recognition and synthesis. Recent technological breakthroughs incorporated into Apple's Siri and Google's Translate lend additional support to these expectations. However, as we know from experience and literature, both speech recognition and automatic translation are still far from being reliable [2]. Furthermore, the most reliable systems still suffer from

F. Xu (✉) • S. Schmeier • R. Ai • H. Uszkoreit
Yocoy Technologies GmbH and DFKI LT Lab, Berlin, Germany
e-mail: Feiyu.Xu@yocoy.com; feiyu@dfki.de; Sven.Schmeier@yocoy.com; schmeier@dfki.de; renlong.ai@dfki.de; uszkoreit@dfki.de

J. Mariani et al. (eds.), *Natural Interaction with Robots, Knowbots and Smartphones: Putting Spoken Dialog Systems into Practice*, DOI 10.1007/978-1-4614-8280-2_5,

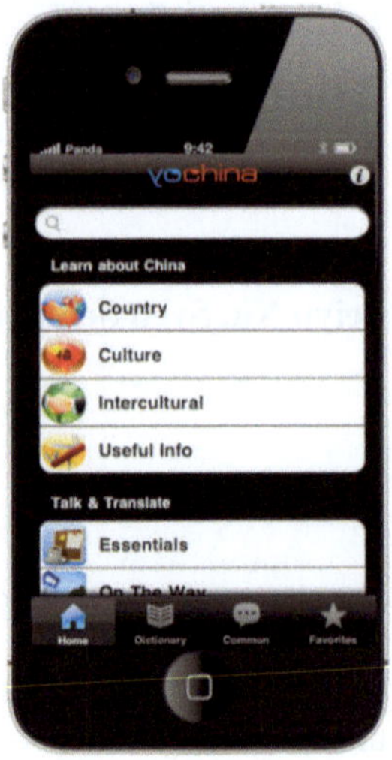

Fig. 5.1 Yochina: language, travel and culture guide for China

slow response times depending on input length, complexity and Internet access. Roaming costs still prevent travelers to enjoy the benefits of online translation and speech recognition services in most foreign countries such as China.

The Yochina crosslingual dialogue framework developed by Yocoy Technologies GmbH (Yocoy)[1] provides a realistic solution that helps foreigners to overcome language and communication barriers in countries such as China without depending on Internet connection. The pragmatic approach guarantees correct translations. Yochina incorporates various language technologies such as speech synthesis, template-based translation, dialogue and semantic search [3, 4]. The framework provides the following functions:

- Template-based and situation-based translation
- Crosslingual dialogue
- Multimodal dictionary
- Semantic search
- Interlinking of language and information
- Spoken output

Yochina is implemented as a mobile application available for two language pairs (English to Chinese and German to Chinese) at the Apple app store. Yochina contains three major components: language guide, country and travel guide and semantic search, as depicted in Fig. 5.1. The remainder of the paper is organized as follows. Section 5.2 describes the Yochina crosslingual dialogue system. Section 5.3 explains a novel strategy of linking provided knowledge with covered communication situations. Section 5.4 shows the search function and the visualization of the search results.

---

[1]http://www.yocoy.com

5.2 Crosslingual Spoken Dialogue System

The Yochina language guide aims to help foreigners to formulate their wishes, requests and questions in their own languages by providing phrases, phrase templates and multimedia phrases. The phrase templates provide slots for filling in words or phrases expressing numbers, currencies, time points, countries, languages, body parts, medical symptoms, etc. Figures 5.2 and 5.3 show the entire workflow

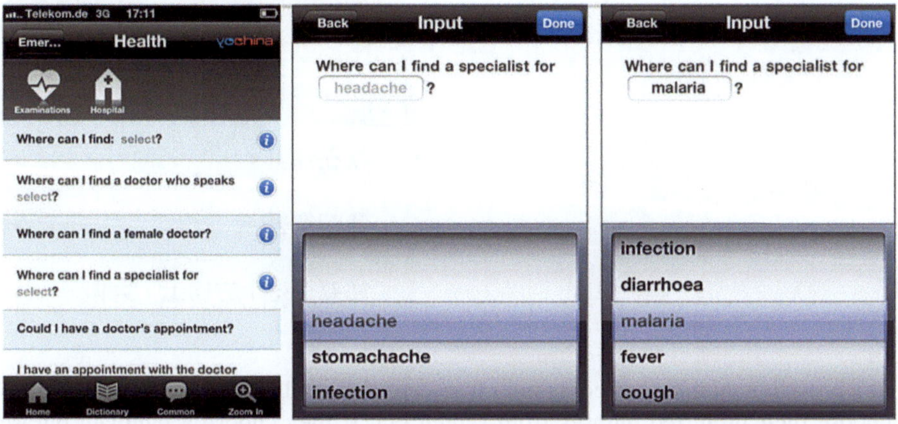

Fig. 5.2 Phrase templates with slots

Fig. 5.3 Crosslingual dialogue with bidirectional translations

Fig. 5.4 Multimedia phrase

of a crosslingual dialogue containing five steps depicted by five screens. Users can choose their preferred phrases from screen (1). If they choose a template phrase such as the one on screen (2), they can fill the slots either from a preselected list or by free input as shown on screen (3). Screen (4) displays the translation of the sentence from screen (3) together with options for responses to be shown to the Chinese conversation partner for selection. Screen (5) shows the translation of a selected answer into the language of the user.

Figure 5.2 shows an example in which a user asks for medical specialists in particular areas. Figure 5.3 exhibits the translation of the completed request given in screen (4) together with options for responses by the Chinese conversation partners. If the users touch the speech bubble, the system will read the Chinese sentence in the yellow area. The next screen displays the translation of the selected Chinese response into the language of the user.

Since most smart phones have a camera and can store pictures, we developed a new communication method allowing users to integrate pictures instead of referential phrases in their utterances. For example, the following phrase in Fig. 5.4 refers to the object in the picture which users want to buy. Users can insert a picture from their iPhone photo album into the slot just like filling a slot with a text snippet. This new function comes in handy when a picture can spare the user from describing a complex object.

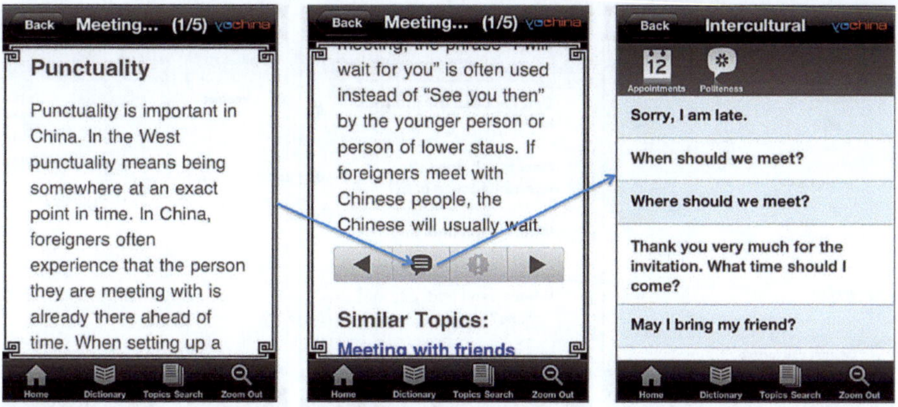

Fig. 5.5 Linking between communication and content

5.3 Interlinking of Communication and Knowledge

Traditionally language guides and travel guides are separate products. Thus there is no linking between descriptions of points of interests, historical events, cultural information or intercultural hints on the one side and useful related phrases on the other. However, in our real world, communication and information is tightly connected in many ways. During our conversations, we search for information and knowledge for understanding what we hear and better explaining what we mean. On the other hand, new information or content inspires and supports communication. In Yochina, we annotate content on country and culture with phrases which are useful in the context of the information. Figure 5.5 depicts one example of such linking, here between the intercultural issue of "punctuality" and related phrases. The linking is signaled by the language bubble symbol on the second screen.

5.4 Semantic Search and Expansion

In Yochina, all words, phrases, and content are indexed for free text search. Given a word as search query, users can find (1) the exact match and the related words with their translations and speech output, (2) phrases semantically related to the word, and (3) travel information mentioning the keyword. Figure 5.6 shows an example with the query *internet*. Given this query, the users are shown Internet-related words and phrases or sentences around Internet access and costs. Furthermore, Yochina explains the Internet usage situation in China.

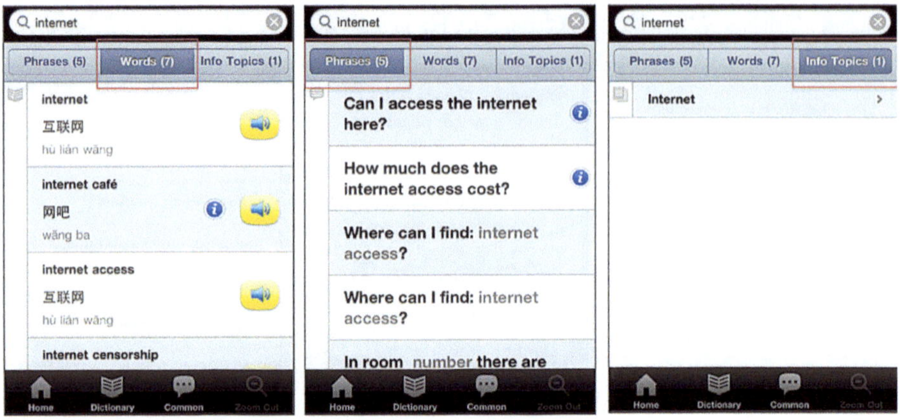

Fig. 5.6 Search

5.5 Robust Offline Application of Embedded ASR

Although the current Yochina applications available on the market do not feature speech recognition technologies, extensive research has been conducted with the aim to integrate embedded automatic speech recognition (ASR) technology. ASR is used to activate phrases and phrase templates available in the Yochina dialogue grammars. In [1], Yochina phrases and grammars have been adapted to various grammar formats of the corresponding ASR tools. Special methods have been developed to convert the template phrase slot features into features allowed by the corresponding grammar formats. An *n*-best recognition strategy has been applied to ensure the targeted robustness. Three ASR tools have been tested for this specific task in our experiments. Fonix performs a little better than SVOX and Nuance (overall recognition rate: Fonix 87.4%, SVOX 85.9% and Nuance 84.0%). But Nuance exhibits the best recognition for non-native speakers and in noisy open-air situations, while SVOX's result is best with female testers and native speakers. Although the recognition performance would not suffice for a sufficiently reliable speech-to-speech translation, it turned out that by restricting the vocabulary to the words needed for semantic access to the situation-relevant phrases a satisfactory recognition performance can be accomplished.

The upshot of the experiments was therefore that a restricted utilization of speech recognition for fast access will circumvent the problems of free speech input while still freeing the user from typing in the entire sentences and phrases to be translated.

5.6 Conclusion and Future Work

The demonstrated app exemplifies a thoughtful combination of various language technologies into a successful real-life application. We have argued for a creative novel pragmatic combination of matured technologies into a reliable product that avoids the pitfalls of imperfect leading-edge techniques such as completely free automatic translation and free speech recognition. Because of the modular design of the application framework, the modules for translation and speech input processing can be substituted at any time by improved MT and ASR technologies as soon as they reach the needed level of reliability. The continuous testing and gradual incorporation of maturing technologies into a modular application framework has proven an appropriate approach for securing maximal user benefits and technical competitiveness.

References

1. Ai, R.: Fuzzy and intelligent match between speech input and textual database. Master's thesis, Technical University of Berlin, Berlin, Germany (2010)
2. Feng, J., Ramabhadran, B., Hansen, J.H.L., Williams, J.D.: Trends in speech and language processing [in the spotlight]. IEEE Signal Process. Mag. **29**(1), 177–179 (2012)
3. Uszkoreit, H., Xu, F., Liu, W.: Challenges and Solutions of Multilingual and Translingual Information Service Systems (invited paper). In Proceedings of HCI International 2007, 12th International Conference on Human-Computer Interaction, Beijing (2007).
4. Uszkoreit, H., Xu, F., Liu, W., Steffen, J., Aslan, I., Liu, J., Müller, C., Holtkamp, B., Wojciechowski, M.: A Successful Field Test of a Mobile and Multilingual Information Service System COMPASS2008. In Proceedings of HCI International 2007, 12th International Conference on Human-Computer Interaction, Beijing (2007)

Chapter 6
Walk This Way: Spatial Grounding for City Exploration

Johan Boye, Morgan Fredriksson, Jana Götze, Joakim Gustafson, and Jürgen Königsmann

Abstract Recently there has been an interest in spatially aware systems for pedestrian routing and city exploration, due to the proliferation of smartphones with GPS receivers among the general public. Since GPS readings are noisy, giving good and well-timed route instructions to pedestrians is a challenging problem. This paper describes a spoken-dialogue prototype for pedestrian navigation in Stockholm that addresses this problem by using various grounding strategies.

6.1 Introduction

Recent years have seen an immense proliferation of smartphones among the general public. Smartphones feature an open computing platform and GPS satellite tracking facilities, and coupled with geographic databases they allow for the creation of spatially aware applications like routing a pedestrian from A to B (see, e.g., Google Navigation [8]) or providing information of sites and services in the immediate vicinity (see, e.g., Google Maps [9]). Though some services and experimental systems rely on spoken *output* ([1, 14]), so far no such spatially aware service has been based on spoken *dialogue* (e.g., the possibility for the user to intervene and ask "Should I turn left here?" or "What street am I walking on?"). Furthermore, the advantage of the spoken-dialogue approach over a map-based approach is that many people find interpreting maps on a small screen to be strenuous and confusing [16]. It is therefore safe to say that well-functioning spoken dialogue would be a valuable contribution to the plethora of spatially aware mobile applications.

J. Boye (✉) • J. Götze • J. Gustafson
KTH, School of Computer Science and Communication, Stockholm 100 44, Sweden
e-mail: jboye@kth.se; jagoetze@csc.kth.se; jocke@speech.kth.se

M. Fredriksson • J. Königsmann
Liquid Media, Hammarby allé 34, Stockholm 120 61, Sweden
e-mail: morgan@liquid.se; jurgen@liquid.se

J. Mariani et al. (eds.), *Natural Interaction with Robots, Knowbots and Smartphones: Putting Spoken Dialog Systems into Practice*, DOI 10.1007/978-1-4614-8280-2_6,

This paper describes an implemented dialogue system for helping a user explore the city of Stockholm. The system can either guide the user to a location of his choice ("I want to go to Odengatan") or to specific spots chosen by the system, like a statue or an interesting architectural detail on a particular building. The latter setting in particular is interesting as it allows us to investigate various methods for producing referring spatial expressions, in order to help the user find quite small objects in a complex city environment.

In general, the city exploration problem addressed here is challenging since it involves the interpretation and generation of utterances within a rapidly changing spatial context under uncertainty.

6.2 Background

Many researchers within cognitive psychology have investigated how people give route instructions to one another (see, e.g., [7]) and what the elements of a good route description are (see, e.g., [17, 23]). It is however not clear how these results transfer to computational models of route description generation. One finding is that a big portion of such dialogues are devoted to grounding, making sure that the dialogue partner actually sees and understands what is being referred to. Grounding is a well-studied phenomenon also in dialogue systems (see, e.g., [20, 24]).

The implemented systems for guiding pedestrians have mostly been based on spoken output from the system, with little or no possibility for the user to provide information ([1, 13, 14, 19, 26]). Spoken-dialogue systems in spatial domains have mostly focused on non-dynamic contexts where the user can ask questions about a static map (e.g., [5, 10, 25]), on virtual environments such as computer games (e.g., [3, 4, 21, 22]), in indoor environments [6], or on natural-language interfaces to robots ([12, 15, 18]). Few if any researchers have so far addressed the topic of spoken natural-language dialogue with a user in a real, dynamic city environment.

6.3 Uncertainty and Grounding

A recurring problem for any pedestrian routing system is to describe to the user how to get from his current position to the next node in the planned route. This has to be done reliably even though the user's position, speed, and direction are uncertain due to possible errors in GPS readings. Giving simple instructions like "turn left here" is therefore a risky strategy; such instructions might be nonsensical for the user if he is not quite where the system believes him to be. Furthermore, the interpretation of left and right is not always clear, for instance, in parks and open squares or when the user is standing still without the system knowing which way he is facing. Therefore, before giving directions, it is often preferable that the system first *grounds* the user's current position and orientation by means of reference landmarks in the near vicinity.

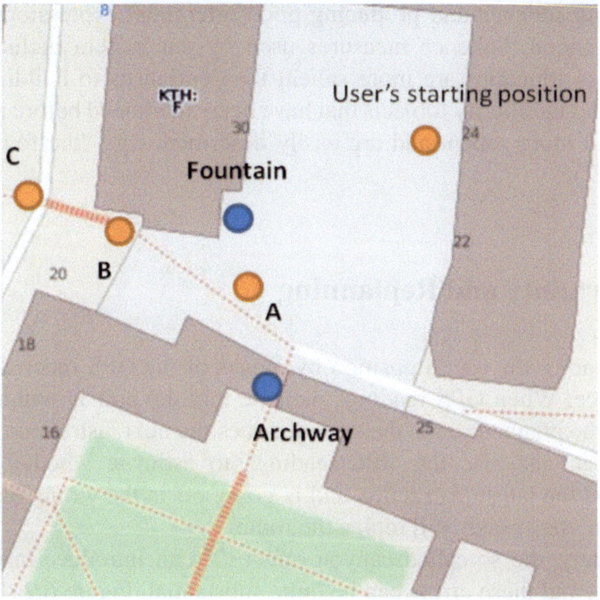

Fig. 6.1 Generating route instructions

Consider for instance the situation depicted in Fig. 6.1. Here the system seeks to describe the route given by the route planner, first to node A, then (when the user reaches A) to B, then down a flight of stairs to node C, then turn left, etc. Before giving instructions, our system first calculates if there is a clear line of sight from the user's assumed position to a number of reference landmarks. It then selects the most salient landmark, seeks to make the user aware of it, and describes the route relative to it. Here is a sample dialogue:

1. System: There is a fountain about 35 m from here. Can you see it?
2. User: Yes.
3. S: Good! Please walk to the left of the fountain. (*user walks*)
4. S: Please turn right and walk to the top of the stairs.
5. U: What?
6. S: There is a flight of stairs leading down about 25 m from here. Can you see it?

In utterance 1, since there is no good way of describing node A, the system cannot ask directly about it. Instead, the system calculates that there are two describable landmarks visible from the user's presumed position: a fountain and an archway, of which the fountain is considered most salient. When the user confirms (utterance 2), the system gives the next instruction with a reference to the fountain. If the user had answered in the negative, the system would have proceeded to ask about another visible landmark. If all possibilities are exhausted, the user is asked to simply start walking, so the system can adjust his course if needed.

Determining salience and producing good referential expressions is a difficult problem in general. Salience measures used by our system include rarity (rare objects such as fountains are more salient than entrances to buildings), distance, uniqueness, and familiarity (objects that have been mentioned before in the dialogue are considered more salient and are easily described, e.g., "the fountain that you passed before").

6.4 Uncertainty and Replanning

The system knows the user's position by means of the GPS receiver in the user's Android device. When GPS readings indicate that the user is within 20 m of the next node in the planned route, the system issues the next instruction. Furthermore, the system can also use the GPS readings to estimate whether the user has misunderstood the latest instruction and is going off in the wrong direction. In the latter situation, the system will replan the route.

Unfortunately, the so-called canyon effect [2] can introduce inaccuracies into GPS readings, and these errors can be quite substantial. Figure 6.2 shows a typical situation, in which the user is walking along the street (from left to right in the picture) and where the GPS readings (in red) are incorrect a large part of the time. These inaccuracies are a problem for two reasons.

Firstly, the user can appear to "miss" the 20-m circle around the next node and appear never to come sufficiently close. The result will be that no instruction is produced by the system at that node. Secondly, the user can appear to walk in the wrong direction when in fact he is not. Consider the situation 2 depicted in Fig. 6.2 below. The user has passed the next node A, but GPS errors have prevented the system from registering this. At 2, the user is getting further and further away from A, and since the system is still considering A to be the next node, it appears as if the user is going the wrong way. Clearly, it would be very misleading and confusing for the user if the system would say "Please turn around" at this point.

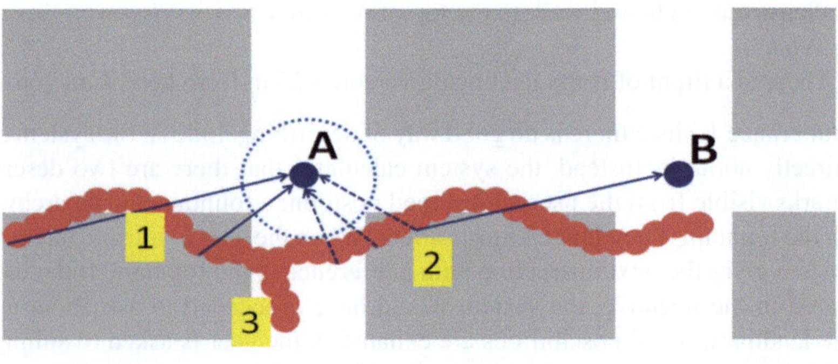

Fig. 6.2 The user appears to "miss" the expected next node due to GPS drift

The method we have adopted to address these problems is illustrated above. As long as the distance to the next node A is decreasing (situation 1), everything is fine. If the distance to the next node starts increasing (situation 2), the system checks the distance to the next-next node B as well. In situation 2, the distance to the next-next node B is *decreasing* while the distance to the expected next node A is *increasing*. If this pattern persists for 10 s, the system assumes that the user has passed the expected node A and is continuing in the correct direction.

Another possibility is when the distance is increasing both to the expected next node A and to the expected next-next node B (situation 3). If this pattern persists for 10 s, the system assumes that the user is walking in the wrong direction and will issue replanning.

6.5 Using Visibility Information

The system repeatedly performs visibility calculations to find out whether there is a free line of sight between two given points. Such visibility calculations are currently used for three purposes: Firstly, as mentioned in Sect. 6.3, they are used to find candidates for referring expressions (the objects of which have to be visible from the user's assumed position). Secondly, they are used to produce better route plans. The system currently gets its data from OpenStreetMap [11], and street objects in OpenStreetMap may contain many nodes very close to each other (in particular in roundabouts or curved streets). Consequently route plans can become very long. By iteratively weeding out any node visible from the preceding node, route plans become more suitable for narration. This process is depicted in Fig. 6.3 below.

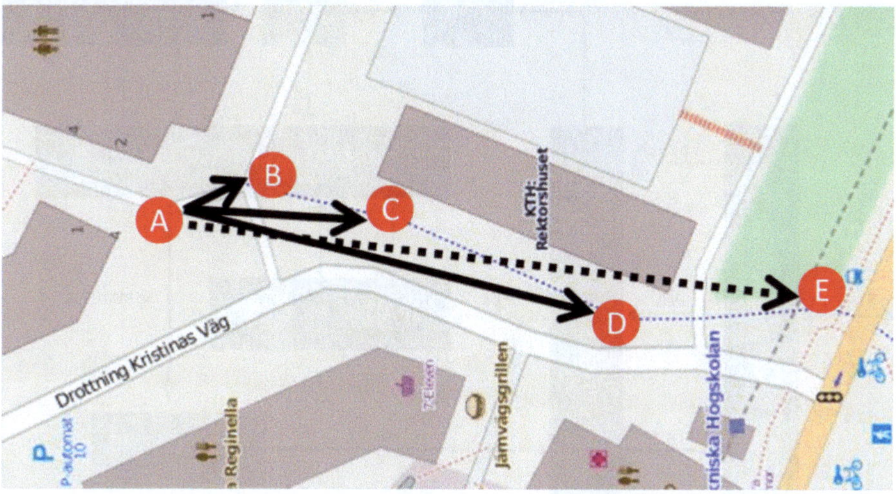

Fig. 6.3 Weeding out nodes in a route plan using visibility information

Here the produced route plan starts with the nodes A–E, in that order. The system checks visibility from A to B, from A to C, etc., and finds that D is the last visible node from A. It therefore concludes that B and C can be removed from the plan. The system then continues to check visibility from D to E, etc., until all unnecessary nodes are removed from the plan. However, no segment is allowed to be longer than 60 m.

A third use of the visibility information is to produce better route instructions. The visibility calculations also return information on the streets, parks, etc., that were intersected by the visibility vector. This allows for instructions like "Now cross X street."

6.6 System Architecture

The system is implemented to work speech-only and "eyes-free"—the user should not need to look down on a map on the screen, but rather be free to experience the city. The architecture described here is used both for the fully automatic system and for a Wizard-of-Oz data collection that preceded it. In the latter case, an operator GUI took the place of the dialogue manager. The operator GUI showed the user's position as a colored dot on a map and used Google street view to show an approximation of the user's visual context.

The user downloads a client app to his Android device which once started connects to a central phone server. The client app sends the sound stream from the microphone of the Android device to the phone server, and as soon as contact has

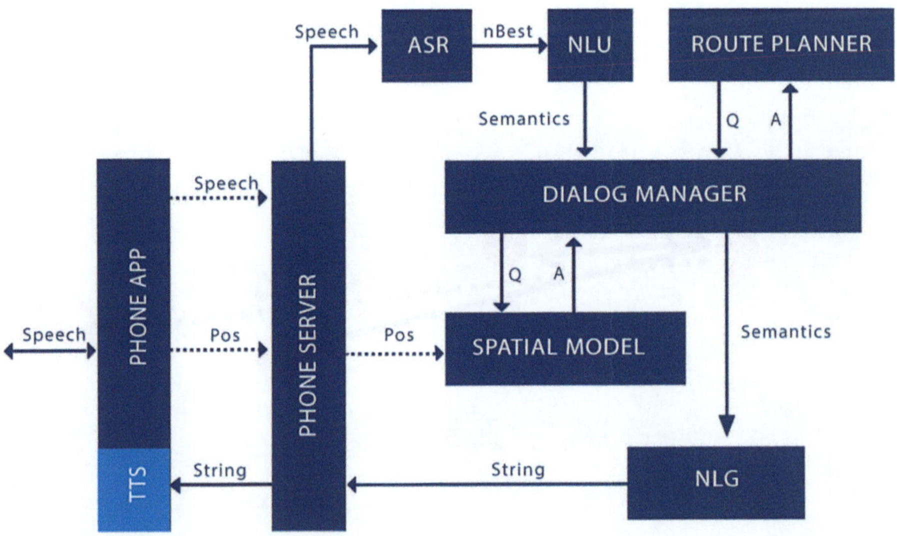

Fig. 6.4 System architecture

been established with the GPS satellites, it also starts sending the coordinates (i.e., latitude-longitude pairs) of its current location. The sound stream is sent to speech recognition and parsing, and a semantic expression representing the utterance is sent to the dialogue manager (DM). Coordinates are sent directly to the DM. The dialogue manager updates its context model based on the input and decides what to say to the user and when to say it. The DM may also call an external planner to compute a route between two points in the city. For user speech input we are currently using a commercial off-the-shelf speech recognizer with a handwritten language model. For speech output, we use the built-in speech synthesizer on the Android device. The architecture also supports the use of server-side speech synthesis streamed to the handset as well as the speech recognizer to be run on the handset. The latter feature would make it possible to maintain a dialogue in places where the 3G data connection fails.

Coordinates are also sent to the Spatial Model, which is a module that maintains the mapping from the logical representation of the city (in terms of buildings, streets, etc.) to the algebraic representation (in terms of polygons, lines, and coordinates). The Spatial Model also performs the visibility calculations described in Sect. 6.5. The polygon representation of the city is automatically generated from an export from OpenStreetMap, generated by indicating an area on the map. A minimal bounding rectangle is computed for each polygon in order to speed up visibility calculations, as it is faster to compute whether a line intersects a rectangle than an arbitrary polygon. If the dialogue manager needs to find out if B is visible from A, a request is sent to the Spatial Model, which first computes whether the line AB intersects any bounding rectangle in the entire city representation. If not, there is a clear line of sight from A to B. If the line intersects a bounding rectangle, a second more expensive calculation is carried out to check whether AB intersects the polygon inside the rectangle.

6.7 User Experiment

In order to evaluate the strategies described in Sects. 6.3, 6.4, 6.5, we performed a user test with eight subjects on four scenarios each. In each scenario, the user was guided to a specific spot in the city and asked to write down some inconspicuous detail (like the serial number on an electricity wiring box).

As a rough estimate of the success of the implemented strategies, we note that seven users managed to complete all four scenarios (one user only completed one scenario due to technical problems) and that, on average, the system had to replan 1.6 times per completed scenario (Table 6.1).

Table 6.1 User experiment

User #	Scenarios completed	#Instructions	Duration (min)	# Replannings
1	4	49	28.2	3
2	4	59	34.6	10
3	4	77	40.4	5
4	4	68	28.3	10
5	1	6	2.1	0
6	4	60	27.7	4
7	4	82	35.6	14
8	4	48	24.9	0

6.8 Concluding Remarks

The system presented here routes pedestrians to their destination, using spoken dialogue to first ground reference landmarks used in the routing instructions. Ongoing work includes further user tests in a part of Stockholm in order to assess and improve the implemented strategies.

Acknowledgements This paper is supported by the European Commission, project *SpaceBook*, grant no 270019.

References

1. Bartie, P., Mackaness, W.: Development of a speech-based augmented reality system to support exploration of cityscape. Trans. GIS **10**(1), 63–86 (2006)
2. Borriello, G., Chalmers, M., LaMarca, A., Nixon, P.: Delivering real-world ubiquitous location systems. Commun. ACM **8**(3), 36–41 (2005)
3. Boye, J., Gustafson, J.: How to do dialogue in a fairy-tale world. Proceedings of the 6th SIGDial Workshop on Discourse and Dialogue, Lisbon, Portugal (2005)
4. Boye, J., Gustafson, J., Wirén, M.: Robust spoken language understanding in a computer game. J. Speech Commun. **48**, 335–353 (2006)
5. Cai, G., Wang, H., MacEachren, A.: Communicating vague spatial concepts in human-gis interactions: a collaborative dialogue approach. In: Spatial Information Theory: Foundations of Geographic Information Science. Lecture Notes in Computer Science, vol. 2825, pp. 287–300. Springer, Berlin (2003)
6. Cuayáhuitl, H., Dethlefs, N.: Spatially-aware dialogue control using hierarchical reinforcement learning. ACM Trans. Speech Lang. Process (Special Issue on Machine Learning for Adaptive Spoken Dialogue Systems) **7**(3), 5:1–5:26 (2011)
7. Denis, M., Pazzaglia, F., Cornoldi, C., Bertolo, L.: Spatial discourse and navigation: an analysis of route directions in the city of Venice. Appl. Cognitive Psychol. **13**(2), 145–174 (1999)
8. Google Inc.: Google maps for mobile. http://www.google.com/mobile/maps (2012b)
9. Google Inc.: Google maps navigation. http://www.google.com/mobile/navigation/ (2012a)
10. Gustafson, J., Bell, L., Beskow, J., Boye, J., Carlson, R., Edlund, J., Granström, B., House, D., Wirén, M.: Adapt: a multimodal conversational dialogue system in an apartment domain. In: Proceedings of International Conference on Spoken Language Processing (ICSLP) (2000)

11. Haklay, M.: OpenStreetMap: user-generated street maps. Pervasive Comput. IEEE **7**(4), 12–18 (2008)
12. Johansson, M., Skantze, G., Gustafson, J.: Understanding route directions in human-robot dialogue. Proceedings of SemDial 2011: Proceedings of the 15th Workshop on the Semantics and Pragmatics of Dialogue, Los Angeles, pp. 19–27 (2011)
13. Jöst, M., Häussler, J., Merdes, M., Malaka, R.: Multimodal interaction for pedestrians: an evaluation study. In: IUI '05: Proceedings of the 10th International Conference on Intelligent User Interfaces, San Diego, pp. 59–66 (2005)
14. Krug, K., Mountain, D., Phan, D.: Webpark: location-based services for mobile users in protected areas. GeoInformatics **26**, 26–29 (2003)
15. Lemon, O., Bracy, A., Gruenstein, A., Peters, S.: A multi-modal dialogue system for human-robot conversation. In: Proceedings of North American Chapter of the Association for Computational Linguistics (NAACL), Pittsburgh (2001)
16. Looije, R., te Brake, G., Neerincx, M.: Usability engineering for mobile maps. In: Proceedings of Mobility'07, 4th International Conference on Mobile Technology, Applications, and Systems, Singapore (2007)
17. Lovelace, K., Hegarty, M., Montello, D.: Elements of good route descriptions in familiar and unfamiliar environments. In: Spatial Information Theory: Cognitive and Computational Foundations of Geographic Information Science. Lecture Notes in Computer Science, vol. 1661/1991. Springer, Berlin (1999)
18. MacMahon, M., Stankiewicz, B., Kuijpers, B.: Walk the talk: connecting language, knowledge, action in route instructions. National Conference on Artificial Intelligence (AAAI-06), Boston (2006)
19. Malaka, R., Zipf, A.: Deep map: challenging IT research in the framework of a tourist information system. In: Information and Communication Technologies in Tourism. Springer, New York (2000)
20. Skantze, G.: Making grounding decisions: Data-driven estimation of dialogue costs and confidence thresholds. Proceedings of SigDial, pp. 206–210. Antwerp, Belgium (2007)
21. Skantze, G., Edlund, J., Carlson, R.: Talking with higgins: Research challenges in a spoken dialogue system. In: Perception and Interactive Technologies, pp. 193–196. Springer, New York (2006)
22. Striegnitz, K., Denis, A., Gargett, A. Garoufi, K., Koller, A., Theune, M.: Report on the second challenge on generating instructions in virtual environments (GIVE-2.5). In: Proceedings of the Generation Challenges Session at the 13th European Workshop on Natural Language Generation (ENLG), Nancy (2011)
23. Tom, A., Denis, M.: Referring to landmark or street information in route directions: What difference does it make? In: Spatial Information Theory: Foundations of Geographic Science. Lecture Notes in Computer Science, vol. 2825, pp. 362–374. Springer, Berlin (2003)
24. Traum, D.: Computational models of grounding in collaborative systems. AAAI Technical Report FS-99-03 (1999)
25. Wang, H., Cai, G., MacEachren, A.: GeoDialogue: a software agent enabling collaborative dialogues between a user and a conversational GIS. In: 20th IEEE International Conference on Tools with Artificial Intelligence, 2008 ICTAI '08, Dayton (2008)
26. Zipf, A., Jöst, M.: Implementing adaptive mobile GI services based on ontologies: examples for pedestrian navigation support. Comput. Environ. Urban Syst. Special Issue on LBS and UbiGIS (2005)

Chapter 7
Multimodal Dialogue System for Interaction in AmI Environment by Means of File-Based Services

Nieves Ábalos, Gonzalo Espejo, Ramón López-Cózar, Francisco J. Ballesteros, Enrique Soriano, and Gorka Guardiola

Abstract This paper presents our ongoing work on the development of a multimodal dialogue system to enable user control of home appliances in an ambient intelligence environment. The physical interaction with the appliances is carried out by means of *Octopus*, a system developed in a previous study to ease communication with hardware devices by abstracting them as network files. To operate the appliances and get information about their state, the dialogue system writes and reads files using WebDAV. This architecture presents an important advantage since the appliances are considered as abstract objects, which notably simplifies dialogue system's interaction with them.

7.1 Introduction

Ambient intelligence (AmI) is an emerging research area that aims to ease users' everyday life by developing proactive, context- and user-awareness environments [3, 14]. A number of research projects concerned with this area can be found in the literature. For example, *ATRACO* works on the implementation of adaptive and trusted ambient ecologies considering activity spheres based on ontologies [18, 19]. Another example is [15], which aims to create adaptive AmI environments in which hardware devices can be easily set up in run-time. Many projects focus on agent-based context-awareness architectures to create dynamic and adaptive

N. Ábalos (✉) • G. Espejo • R. López-Cózar
Department of LSI, CITIC-UGR, University of Granada, Spain
e-mail: nayade@correo.ugr.es; gonzaep@correo.ugr.es; rlopezc@ugr.es

F.J. Ballesteros • E. Soriano • G. Guardiola
Laboratorio de Sistemas Universidad Rey Juan Carlos, Madrid, Spain
e-mail: nemo@lsub.org; esoriano@lsub.org; paurea@gmail.com

J. Mariani et al. (eds.), *Natural Interaction with Robots, Knowbots and Smartphones: Putting Spoken Dialog Systems into Practice*, DOI 10.1007/978-1-4614-8280-2_7,

environments by taking into account user preferences [4, 13, 16, 23]. For example, in the field of home assistance, AmI technology has been applied to help users carry out daily activities and save energy [2, 17]. Regarding healthcare, AmI systems have been applied to monitor user health from analyses of behavioural patterns [10, 27]. This technology has also been applied to provide assistance in public places, for example, [12] developed a guide that provides information to tourists while sightseeing, whereas [11, 26] presented systems to assist people during museum visits, suggesting routes and providing suitable information. Similarly, [20] presented a system to support people while shopping that takes into account user preferences and based on these compares products and prices.

The remainder of the paper is organised as follows. Section 7.2 discusses the *Mayordomo* dialogue system focusing on speech recognition and understanding, dialogue management and response generation. Section 7.3 discusses the *Octopus* system focusing on its file-based service access. Section 7.4 discuses the interaction between *Mayordomo* and *Octopus* in order to enable user operation of home appliances. Finally, Sect. 7.5 presents the conclusions and outlines possibilities for future work.

7.2 *Mayordomo*

Mayordomo is a multimodal dialogue system developed in a previous study to control home appliances using a multimodal interface [1]. The system has been designed to operate in an AmI environment in order to ease user interaction with the appliances. It can work in homes with any distribution of rooms and interact with any kind of appliances if their manufacturers have provided the necessary configuration files, which contain characteristics of the appliances as well as descriptions of the commands that can be operated with them, e.g., switching on and off. The system is multimodal as it provides several methods for the interaction with appliances, using spontaneous speech and a traditional GUI.

7.2.1 Speech Recognition and Understanding

Speech recognition in *Mayordomo* is carried out using the Microsoft ASR engine of Windows Vista. In order to interact with appliances, users can utter questions and commands. Questions are used for requesting information about the state of appliances, whereas commands are used for changing their state. Each appliance has associated a configuration file that allows the user to operate it by means of speech employing a grammar in SRGS (speech recognition grammar specification) format.

Table 7.1 Description of the "action" frame

Room	Room where the appliance is placed
Appliance	Particular appliance on which the command is executed
Attribute	Characteristic of the appliance that is affected by the command
Value	New value for the attribute

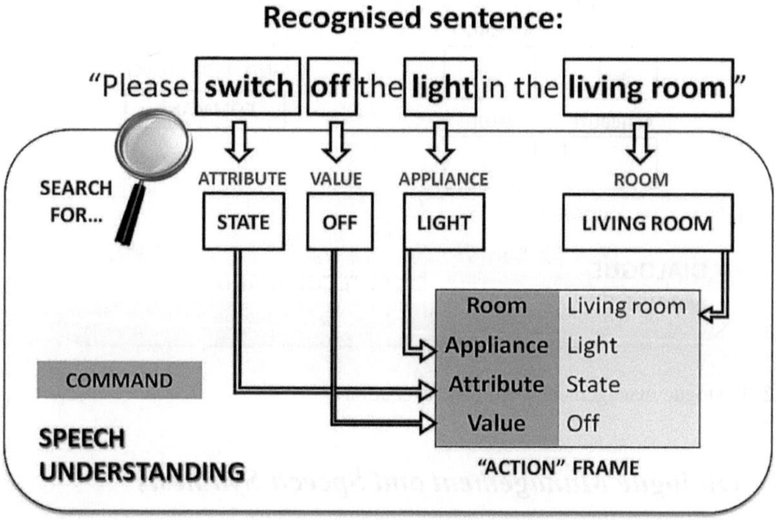

Fig. 7.1 Speech understanding from a recognised sentence

Speech understanding works on a frame structure that in this paper we call "action", which is comprised of four slots: *room*, *appliance*, *attribute* and *value* (see Table 7.1). Using this frame the system can execute commands on the appliances in the AmI environment as well as provide information about their state.

The system employs a method that analyses the recognised sentence and looks for names of rooms, appliances, attributes and possible values for the attributes. The obtained data is used to fill in the slots of the "action" frame, as shown in Fig. 7.1.

We have observed that sometimes users utter sentences omitting attributes whereas others they just pronounce values associated with the attributes. For example, in the command "Switch on the lights" the attribute is "state", but it does not appear explicitly in the sentence. Hence, to determine whether the user is asking a question or entering a command, the system proceeds as follows. If it finds a word beginning with "wh-" or the verb "to be" in present tense, it assumes the user has asked a question. If the word is "what" or "which", it assumes the user has asked about an attribute of an appliance. Finally, if the word is "where", the system assumes the user has requested information about the places where the appliances are located.

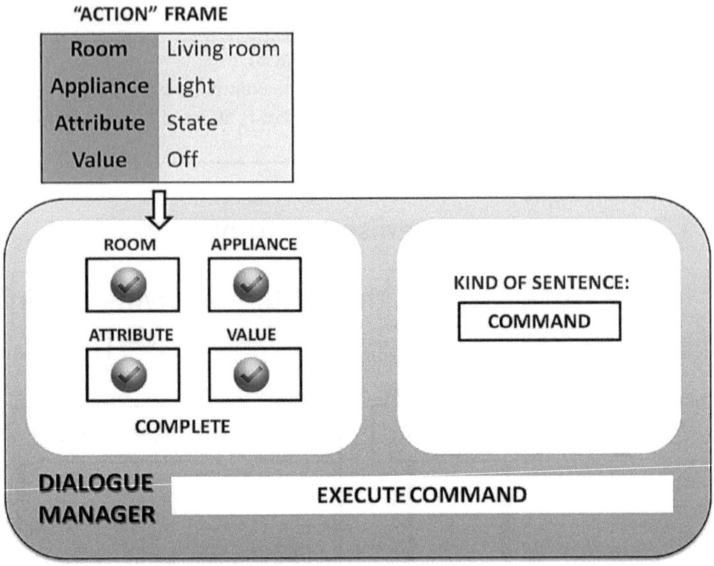

Fig. 7.2 Dialogue management: command execution

7.2.2 *Dialogue Management and Speech Synthesis*

After speech understanding, the system's dialogue manager checks the data stored in the "action" frame and determines whether to provide information to the user, prompt the user for additional data or execute a command (see Fig. 7.2).

If the user has requested for information, the dialogue manager calls a system's module that organises the information to be provided to the user into text sentences. If he has entered a command, the dialogue manager calls a system's module that executes the command. Whenever a command is executed, an entry is made in a log file that stores date and time of the execution. If the dialogue manager detects missing data in the recognised sentence, it decides the appropriate prompt to be generated in order to obtain that data from the user (see Fig. 7.3).

In order to execute commands, *Mayordomo* uses information about the number of rooms and the appliances installed in each room. This information is provided to the dialogue system through configuration files, one for the environment and another for each appliance. The former contains a generic description of the environment that includes the number of rooms and the room names.

The configuration file for each appliance contains information about the functionality of the appliance as well as details about attributes and possible values for them. For example, for interacting with the TV, the system uses a configuration file that contains two attributes: *volume* and *channel*, as well as possible values for these attributes. Values for the *volume* include "1", "2", "3", "4" and "5", whereas values for the *channel* include "tve1", "tve2", "antena3", "canalsur" and "tele5".

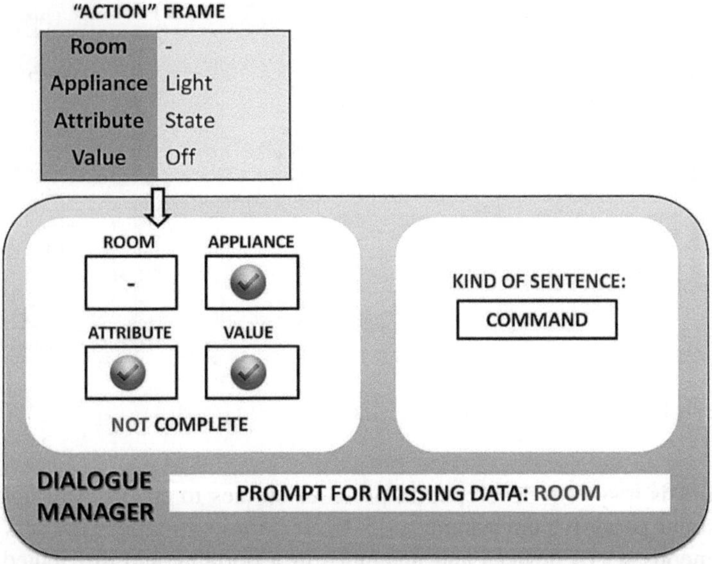

Fig. 7.3 Dialogue management: asking for missing data

To generate text sentences, *Mayordomo* uses a set of patterns that are instantiated with different values depending on the appliance, room, attribute and value involved. To transform these sentences into speech, the system uses the Microsoft TTS engine of Windows Vista.

7.3 File-Based Service Access: Octopus

Service-oriented architectures represent as a suitable solution for engineering an open distributed computing environment. Indeed, having clearly defined service concepts and interfaces makes it possible for providers to export resources in a structured fashion and enables clients to exploit them without caring about their internal implementation details.

Service access is usually described in the form of a procedural interface, e.g., a set of (remote) procedure or method calls [24]. An alternative option is to define a set of request-reply message pairs [25]. There is also work on capturing client-server interaction in a more elaborated way, e.g., as a dialogue following a formal grammar [21].

Taking a different viewpoint, services can be abstracted as file systems. In this case, the client perceives a service as a subdirectory within the file system namespace and interacts with it using the standard set of `open`, `read`, `write` and `close` operations.

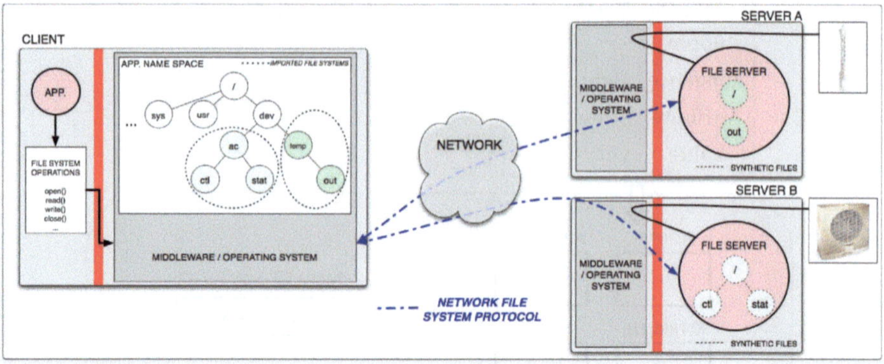

Fig. 7.4 File-based service abstraction

Octopus is a system developed in previous studies to enable AmI applications and personal pervasive environments [5–9]. It permits to access the services of a heterogeneous set of devices and machines in a portable and distributed manner. The key idea behind the system is to provide a unified namespace and a small set of operations to access resources represented by means of network files.

Using *Octopus*, system calls targeted to files are translated to one or more request messages which are sent to a server. The corresponding replies are used to deliver the result of the operations to the client that invoked them.[1] This interaction is transparent for the client, and the server may be running on a remote machine.

The file structure witnessed by the client can be *synthesised*. Put in other words, the server may report names and metadata for files that are not just a set of blocks in a storage device. The file server can handle file operations and process them according with its semantics. Every file operation can be associated with (implicit) processing functionality, e.g., controlling a resource, accessing it to push/retrieve data or performing a computation. This way, the file server provides the illusion of talking to a traditional file system.

Figure 7.4 illustrates this concept. It depicts a client with an air cooler service and a temperature device attached to file names `/devs/ac` and `/devs/temp`, respectively. These services are provided by different, possibly remote, processes. Each server furnishes a set of synthesised files that can be used to access the service. For instance, the client may read the `/devs/temp/out` file to retrieve the current temperature or write into the `/devs/ac/ctl` file to control the air cooler.

Octopus employs X10 services for implementing the connection with hardware devices. The file system provides a flat directory with a (synthesised on demand) file per X10 device. In this way, the state of a device can be accessed by reading its associated file and can be updated by writing to this file. This means that both users

[1]For details of such a file protocol and its mapping to the standard file operations, the interested reader may refer to the design and implementation of 9P/Plan9 and Styx/Inferno [22].

and programmers can rely on the well-understood file system interface to operate the device. The interface for the resulting service is very simple as each file contains either the value `on` or `off` depending on the state of the corresponding device.

7.4 Interaction Between *Mayordomo* and Home Appliances

Mayordomo interacts with home appliances by means of the network files handled by *Octopus*, which allows considering the appliances as "abstract objects" and thus getting rid of the problems concerned with the physical interaction with them. In order to access the files, we use WebDAV, a protocol that defines how basic file functions such as `copy`, `move`, `delete` and `create` are performed by using HTTP. In this way, when the system executes a command ordered by a user, for example, switching on of a light, it uses the statement `write()` to put the value `on` in the file associated with the light switch (see Fig. 7.5). To do this, the system executes the following instructions:

```
fd=open("/mnt/x10/hall/light", OWRITE);
write(fd, "on", 3);
close(fd);
```

As a result of the command execution there is a change in the state of the light, as can be observed in Fig. 7.5.

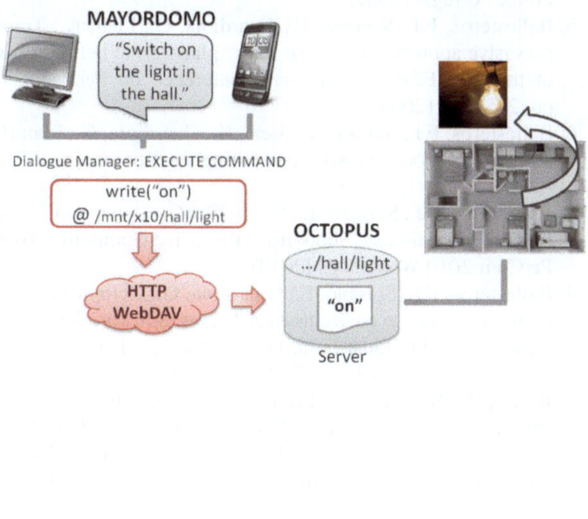

Fig. 7.5 Sample interaction between *Mayordomo* and *Octopus* to switch on a light

7.5 Conclusions and Future Work

In this paper we have presented the *Mayordomo* and *Octopus* systems focusing on their features and characteristics. Also, we have discussed our ongoing work on getting them to work together in an AmI environment for user operation of home appliances. Future work includes adding new devices to our experimental environment, such as RFID and motion sensors, in order to consider information about user location in the dialogue management, which would be very useful to make the dialogue system responses context-dependent. Also, we plan to carry out objective and subjective evaluations of this system.

Acknowledgements This research has been funded by the Spanish project ASIES TIN2010-17344.

References

1. Abalos, N., Espejo, G., Lopez-Cozar, R., Callejas, Z., Griol, D.: A multimodal dialogue system for an ambient intelligent application in home environments. In: Lecture Notes in Artificial Intelligence, vol. 6231/2010, pp. 491–498. Springer, Berlin (2010)
2. Allerding, F., Schmeck, H.: Organic smart home: architecture for energy management in intelligent buildings. Workshop Organic Computing as Part of ICAC, Karlsruhe, Germany (2011)
3. Augusto J.C.: Ambient intelligence: the confluence of pervasive computing and artificial intelligence. In: Schuster, A. (ed.) Intelligent Computing Everywhere, pp. 213–234. Springer, Berlin (2007)
4. Augusto, J.C., O'Donoghue, J.: Context-aware agents (the 6 ws architecture). In: Proceedings of the International Conference on Agents and Artificial Intelligence (ICAART). INSTICC, Porto, Portugal (2009)
5. Ballesteros, F.J., Soriano, E., Guardiola, G., Leal, K.: Traditional systems can work well for pervasive applications. a case study: plan 9 from bell labs becomes ubiquitous. In: Proceedings of the 3rd IEEE International Conference on Pervasive Computing and Communications, pp. 295–299 (2005)
6. Ballesteros, F.J., Soriano, E., Leal, K., Guardiola, G.: Plan B: using files instead of middleware abstractions for pervasive computing environments. IEEE Pervasive Comput. **6**(3), 58–65 (2007)
7. Ballesteros, F.J., Soriano, E., Guardiola, G.: Upperware: bringing resources back to the system. In: IEEE Middleware Support for Pervasive Computing Workshop 2010, in Proceedings of the PerCom 2010 Workshops (2010)
8. Ballesteros, F.J., Soriano, E., Guardiola, G.: Octopus: an upperware based system for building personal pervasive environments. J. Syst. Software **85**(7), 1637–1649 (2012)
9. Ballesteros, F.J., Guardiola, G., Soriano, E.: Personal pervasive environments: practice and experience. Sensors **12**(6), 7109–7125 (2012)
10. Barger, T., Brown, D., Alwan, M.: Health status monitoring through analysis of behavioral patterns. IEEE Trans. Syst. Man, and Cybernetics, Part A **35**, 22–27 (2005)
11. Costantini, S., Mostarda, L., Tocchio, A., Tsintza, P.: Dalica: agent-based ambient intelligence for cultural-heritage scenarios. IEEE Intell. Syst. **23**(2), 34–41 (2008)

12. Corchado, J.M., Pavón, J., Corchado, E., Castillo, L.F.: Development of CBR-BDI agents: a tourist guide application. Lecture Notes in Artificial Intelligence, vol. 3155, pp. 547–559. Springer, New York (2004)
13. Cook, D., Youngblood, M., Das, K..: A multi-agent approach to controlling a smart environment. Designing Smart Homes. The Role of Artificial Intelligence, pp. 165–182. Springer, Berlin (2006)
14. Cook, D., Augusto, J.C., Jakkula, V.: Ambient intelligence: Technologies, applications, and opportunities. J. Pervasive Mobile Comput. **5**(4), 277–298 (2009)
15. Gómez, J., Montoro, G., Haya, P.A., García-Herranz, M., Alamán, X.: Distributed schema-based middleware for ambient intelligence environments. In: Ubiquitous Developments in Ambient Computing and Intelligence: Human-Centered Applications, pp. 205–218. IGI Global, Pennsylvania (2011)
16. Hagras, H., Callaghan, V., Colley, M., Clarke, G., Pounds-Cornish, A., Duman, H.: Creating an ambientintelligence environment using embedded agents. IEEE Intell. Syst. **19**(6), 12–20 (2004)
17. Helal, A., Mann, W., El-Zabadani, H., King, J., Kaddoura, Y., Jansen, E.: The Gator tech smarthouse: a programmable pervasive space. IEEE Comput. **38**, 50–60 (2005)
18. Heinroth, T., Schmitt, T., Bertrand, G.: Enhancing speech dialogue technologies for ambient intelligent environments. In: 5th International Conference on Intelligent Environments (IE09). Ambient Intelligence and Smart Environments, vol. 2, pp. 42–49. IOS Press, Barcelona (2009)
19. Heinroth, T., Denich, D., Schmitt, A., Minker, W.: Efficient spoken dialogue domain representation and interpretation. In: Language Resources and Evaluation Conference 2010 is Organised by ELRA, Valletta, Malta (2010)
20. Keegan, S., O'Hare, G.M.P., O'Grady, M.J.: Easishop: Ambient intelligence assists everyday shopping. Inform. Sci. **178**(3), 588–611 (2008)
21. Lalis, S., Savidis, A., Karypidi, A., Gutknecht, J., Stephanides, C.: Towards dynamic and cooperative multi-device personal computing. In: The Disappearing Computer. Lecture Notes in Computer Science, vol. 4500. Springer, Berlin (2007)
22. Pike, R., Ritchie, D.M.: The styx architecture for distributed systems. Bell Syst. Tech. J. **4**(2), 146–152 (1999)
23. Richard, N., Yamada, S.: Context-awareness and user feedback for an adaptive reminding system. In: Augusto, J.C., Shapiro, D. (eds.) Proceedings of the 2nd Workshop on Artificial Intelligence Techniques for Ambient Intelligence (AITAmI'07), Hyderabad, India, pp. 57–61 (2007)
24. Sun's Java Remote Method Invocation Home: http://java.sun.com/javase/technologies/core/basic/rmi/index.jsp
25. W3C SOAP specifications: http://www.w3.org/TR/soap/
26. Wakkary, R., Evernden, D.: Museum as ecology: a case study analysis of an ambient intelligent museum guide. In: Trant, J., Bearman, D. (eds.) Museums and the Web 2005: Proceedings. Archives and Museum Informatics, Toronto, 31 Mar 2005
27. Xu, B., Ge, Y., Chen, J., Chen, Z., Ling, Y.: Elderly personal safety monitoring in smart home based on host space and travelling pattern identification. Inform. Tech. J. **11**(8), 1063–1069 (2012)

Chapter 8
Development of a Toolkit Handling Multiple Speech-Oriented Guidance Agents for Mobile Applications

Sunao Hara, Hiromichi Kawanami, Hiroshi Saruwatari, and Kiyohiro Shikano

Abstract In this study, we propose a novel toolkit to handle multiple speech-oriented guidance agents for mobile applications. The basic architecture of the toolkit is server-and-client architecture. We assumed the servers are located on a cloud-computing environment, and the clients are mobile phones, such as the iPhone. Huge amounts of servers exist on the cloud-computing environment, and each server can communicate with other servers. It is difficult to develop an omnipotent spoken dialog system, but it is easy to develop a spoken dialog agent that has limited but deep knowledge. If such limited agents could communicate with each other, a spoken dialog system with wide-ranging knowledge could be created. In this paper, we implemented speech-oriented guidance servers specialized to provide guide information for confined locations and implemented a mobile application that can get information from the servers.

8.1 Introduction

Spoken dialog systems based on server-client-type architecture have been widely investigated. In particular, many of the systems have been used on the World Wide Web (WWW) [2, 4, 8]. Some voice search applications focusing on mobile usage have been created by Google [9], Microsoft [6], and Yahoo, among others.

S. Hara (✉)
Graduate School of Natural Science and Technology, Okayama University, 3-1-1, Tsushima-naka, Kita-ku, Okayama, Japan
e-mail: hara@cs.okayama-u.ac.jp

H. Kawanami • H. Saruwatari • K. Shikano
Graduate School of Information Science, Nara Institute of Science and Technology, 8916-5, Takayama-cho, Ikoma, Nara, Japan
e-mail: kawanami@is.naist.jp; sawatari@is.naist.jp; shikano@is.naist.jp

J. Mariani et al. (eds.), *Natural Interaction with Robots, Knowbots and Smartphones: Putting Spoken Dialog Systems into Practice*, DOI 10.1007/978-1-4614-8280-2_8,

Currently, location information can easily be measured using GPS on a mobile phone, and this function is expected to be efficiently used for mobile applications. Among applications that marry automatic speech recognition (ASR) and location information, a speech-guidance system for local information is one of the most promising and useful. In this study, we developed a toolkit for multi-agent server-client spoken dialog systems. Each agent has only a small function as a guidance system in a confined location, e.g., restaurant, museum, community center, and railway station. However, if the agents can collaborate with each other, they can be viewed as a single dialog system with large-scale knowledge. In this paper, we propose a toolkit called "Tankred on rails (ToR)" and its client software called "iTakemaru." This software is based on "Takemaru-kun" [7] and its related systems [3]. Details and a demonstration are shown in later sections.

8.2 Dialog Management Toolkit for Mobile Applications

8.2.1 Speech-Oriented Guidance System for Community Center

The *Takemaru-kun* system has been implemented at the entrance hall of a community center since November 2002 [7]. Newer systems, *Kita-chan* and *Kita-robo*, have been in place at *Gakken Kita-ikoma* railway station since March 2006 [3]. The base system of *Takemaru-kun, Kita-chan*, and *Kita-robo* had a major alteration made by Cincarek [1]. He rewrote the whole system in Ruby and maintained it as "a Dialogue System Toolkit written in Ruby," which is called *Tankred.*[1]

This toolkit consisted of several modules: ASR, dialog management (DM), text-to-speech (TTS), Internet browser, and Computer Graphic Agent. The ASR module used Julius [5]. The DM module included functions of domain classification and response generation [10]. The TTS module used was a commercially available version. The toolkit was used on Debian GNU/Linux 4.0 (etch) with Ruby 1.8.

8.2.2 Speech-Guidance Information Service Software

We extended the "Tankred" system for WWW service software, i.e., Saas. Core modules such as ASR, DM, and TTS were implemented on the server-side, while interface modules such as display of the text and agent, speech input, and speech output were implemented on the client-side. The server system was implemented

[1]"Tankred" is an old German name meaning "the thinking adviser".

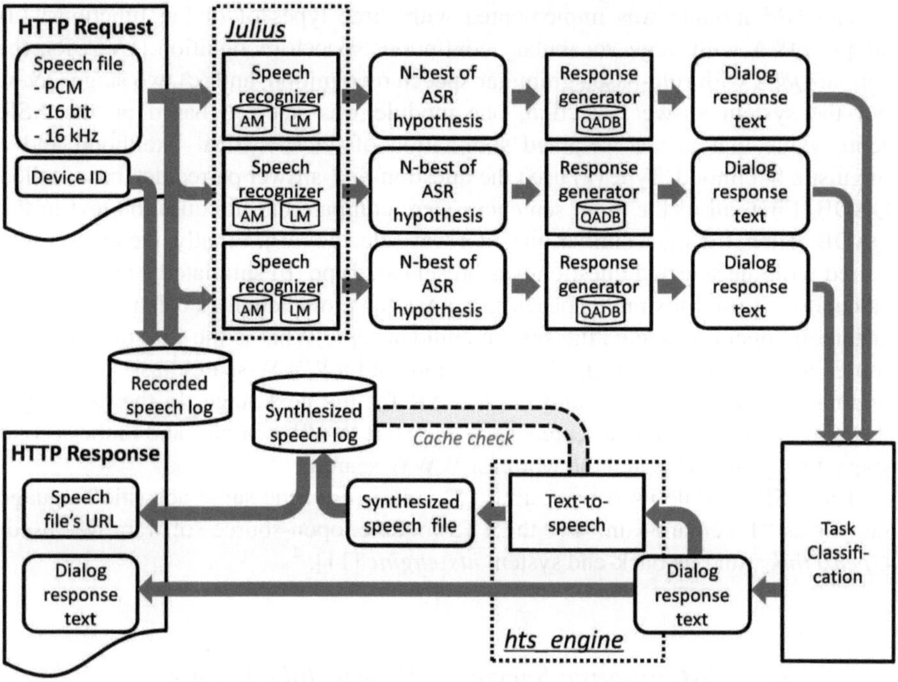

Fig. 8.1 Flowchart of server system

by *Ruby on Rails*,[2] which is one of the most popular frameworks for WWW development. We called it ToR.

A flowchart of the server system is shown in Fig. 8.1. The client system received the user's speech input through its microphone and converted the speech input to a speech file (PCM, 16 bit, 16 kHz, WAV-RIFF). Then, the speech file was sent to the server system as an HTTP POST request. The server system recognized the speech using Julius and generated the N-best recognition result. The recognition result was sent to the DM modules, and it selected an appropriate response from each DM module. The response was sent to the TTS module, which generated the speech response as a speech file with the file's URL. This response text and the speech file's URL were sent back to the client as an HTTP Response. The server recorded the speech file, recognition result, response result, an application ID generated in the client application, and its usage time.

The software used only HTTP protocols for the Internet. Therefore, the reversed HTTP proxy software, such as Apache httpd with mod_proxy, was used for the front-end of "ToR" to ensure high scalability.

[2]http://rubyonrails.org/.

The DM module was implemented with three types of dialog functions: (a) simple Q&A with large vocabulary continuous speech recognition (LVCSR), (b) simple Q&A with rule-based grammar speech recognition, and (c) two stage Q&A. For the system answer selection, one module was selected based on the ASR score value that is the weighted summation of an acoustical likelihood and a linguistic likelihood. Type (a) used the question-and-answer paired database, called QADB. First, all of the input sentences were compared to the question text in the QADB. Then, the most similar question was selected [10]. Finally, the answer text paired with the selected question was presented. Type (b) simulated a mathematics calculator. It used a simple speech grammar for a mathematics calculator of four arithmetic operations; e.g., the system could accept "What is the sum of three plus four?" and answered "seven." Type (c) simulated a WWW-search task. It used the grammar for the first stage and the LVCSR for the next stage. In the first stage, the system accepted magic sentences, e.g., "Start WWW search," and in the second stage, the system accepted the word for WWW search.

The ASR module used was Julius [5], and it used the same acoustic/language models as "Takemaru-kun." For the TTS module, open-source software was used: *Open JTalk* [3] and its back-end system *hts_engine* [11]. [4]

8.2.3 Speech-Guidance Service with Multiple Agents

The system was extended to handle multiple agents. A flowchart of the system is shown in Fig. 8.2. The whole system was constructed with one main agent and several subagents. The user of a client system talks with the main agent. At the same time, the main agent communicates with the subagents, which is implemented as server-to-server communication. Responses of the main agent include information about which subagent has been delegated a task from the main agent; therefore, the client can obtain more information from the subagent. Server-to-server communication is assumed to be faster than client-to-server communication; therefore, this architecture is expected to be efficient for mobile, which has a lot of limitations, such as calculation speed, amount of memory, and connection speed. This architecture also has high affinity with cloud computing. The user talks to just one agent but can get more information, as if he/she was talking with a huge amount of agents.

[3] http://open-jtalk.sourceforge.net/.

[4] http://hts-engine.sourceforge.net/.

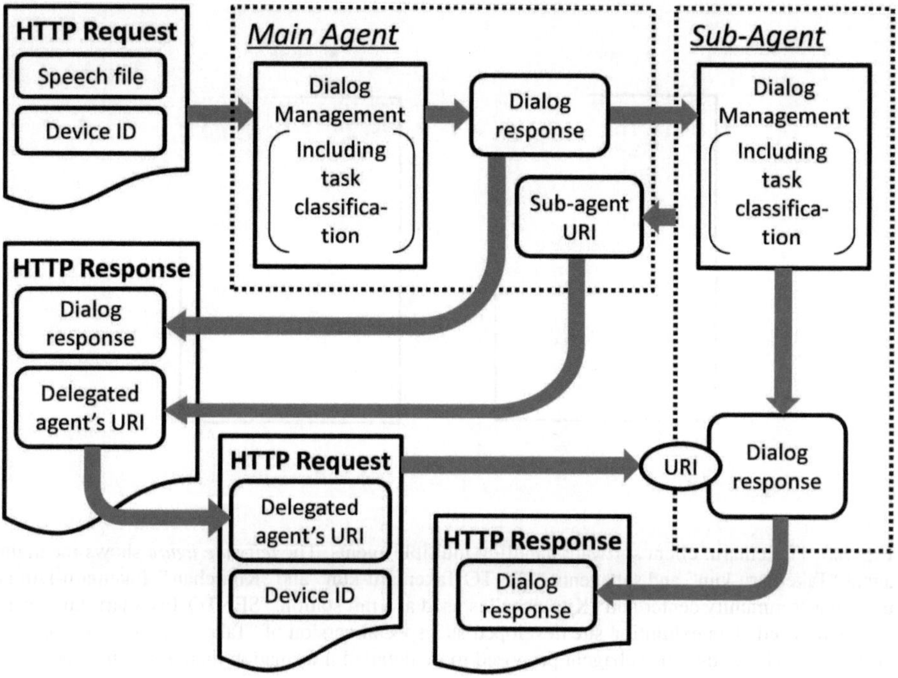

Fig. 8.2 Flowchart of server-to-server connection handling multiple agents

8.2.4 *iTakemaru: Client Software for Mobile Phone*

The example client system was implemented with Objective-C, and it could be used on the iPhone 4S. It could generate HTTP requests and interpret HTTP responses as described in Fig. 8.1, and it could guide users through Takemaru-kun, which is placed at the community center [7]. We called it "iTakemaru." A screenshot of the client system is shown in Fig. 8.3. The system accepted user's speech as Press-to-Talk architecture; that is, the speech was recorded while the user pushed buttons on the interface.

"iTakemaru" could handle one main agent and multiple subagents. In the current implementation, the agents were selected by users before using the system. It is more convenient to select the appropriate agents automatically based on their usage location, but this implementation remains as future work.

In the left side of Fig. 8.3, the user used "Takemaru-kun" as the main agent and he/she mainly talked with "Takemaru-kun". The user selected two subagents, and the main agent was connected to the subagents in their background server systems. One is "Kita-chan [3]," which is a guidance system at *Gakken Kita-ikoma* railway station in Nara, Japan. The other is "SENTO Takemaru-kun," which was previously used as a guidance system for the "Heijo-Sento 1,300th anniversary," which was held in Nara from July to October 2010.

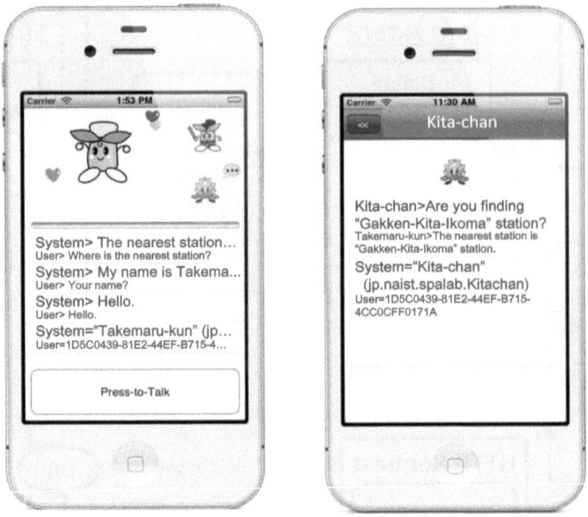

Fig. 8.3 iTakemaru: client software handling multiple agents. The *left-side figure* shows the main agent "Takemaru-kun" and subagents "SENTO Takemaru-kun" and "Kita-chan." Takemaru-kun is used at a community center and "Kita-chan" is used at a rail station. "SENTO Takemaru-kun" was originally used at an exhibition site developed as a special version of "Takemaru-kun." The *image on the right* shows that the subagent proposed more detailed information than the main agent

In the dialog example of the figure, "Takemaru-kun" knows "Gakken Kita-ikoma railway station is the nearest station from his location" but does not know details about the station. The subagent "Kita-chan" is a guidance system of the station and of course knows details about the station. "Kita-chan" always monitored the conversation between the user and "Takemaru-kun," and he notified the user by balloon icon that he could give more detailed information about their conversation topic than "Takemaru-kun." The right side of Fig. 8.3 is the screen for "Kita-chan," and the user could get more information from him about the railway station.

8.3 Conclusion

In this study, we developed a speech service software for speech-oriented guidance systems with multiple agents. This software will be used for the spoken dialog service of a mobile phone and will also be distributed as a developer's toolkit. We hope that a lot of dialog agents will be produced by nonprofessionals, and they will be widely used as one of methods for transmitting information.

If the software is to be widely used in a lot of organizations, we should consider a sharing method for the speech, text, and other resources between the organizations. For example, investigation of the training method of the acoustic model for the distributed resources is an important issue.

Acknowledgements This work was partially supported by CREST (Core Research for Evolutional Science and Technology) at the Japan Science and Technology Agency (JST).

References

1. Cincarek, T.: Selective training for cost-effective development of real-environment speech recognition applications. Ph.D. thesis, Nara Institute of Science and Technology (2008)
2. Gruenstein, A., McGraw, I., Badr, I.: The WAMI toolkit for developing, deploying, and evaluating web-accessible multimodal interfaces. In: Proceedings of International Conference on Multimodal Interaction 2008, pp. 141–148 (2008)
3. Kawanami, H., Takeuchi, S., Torres, R., Saruwatari, H., Shikano, K.: Development and operation of speech-oriented information guidance systems, Kita-chan and Kita-robo. In: Proceedings of APSIPA Annual Summit and Conference (2011)
4. Lau, R. et al.: WebGALAXY: integrating spoken language and hypertext navigation. In: Proceedings of Eurospeech-97, pp. 883–886 (1997)
5. Lee, A., Kawahara, T.: Recent development of open-source speech recognition engine julius. In: Proceedings of APSIPA-ASC 2009, pp. 131–137 (2009)
6. Levit, M., Chang, S., Buntschuh, B., Kibre, N.: End-to-end speech recognition accuraty metric for voice-search tasks. In: Proceedings of International Conference on Acoustics, Speech, and Signal Processing (ICASSP) 2012, pp. 5141–5144 (2012)
7. Nisimura, R., Lee, A., Saruwatari, H., Shikano, K.: Public speech-oriented guidance system with adult and child discrimination capability. In: Proceedings of International Conference on Acoustics, Speech, and Signal Processing (ICASSP) 2004, pp. I-433–I-436 (2004)
8. Nisimura, R., Miyake, J., Kawahara, H., Irino, T.: Development of speech input method for interactive voiceweb systems. In: Proceedings of Human-Computer Interaction, Part II, pp. 710–719. Springer, New York (2009)
9. Schalkwyk, J., Beeferman, D., Beaufays, F., Byrne, B., Chelba, C., Cohen, M., Kamvar, M., Strope, B.: Google search by voice: A case study. In: Neustein, A. (ed.) Advances in Speech Recognition: Mobile Environments, Call Centers and Clinics, chap. 4, pp. 61–90. Springer, New York (2010)
10. Takeuchi, S., Cincarek, T., Kawanami, H., Saruwatari, H., Shikano, K.: Question and answer database optimization using speech recognition results. In: Proceedings of Interspeech 2008, pp. 451–454 (2008)
11. Zen, H., Nose, T., Yamagishi, J., Sako, S., Masuko, T., Black, A.W., Tokuda, K.: The HMM-based speech synthesis system (HTS) version 2.0. In: Proceedings of ISCA SSW6, pp. 294–299, Bonn, Germany (2007)

Chapter 9
Providing Interactive and User-Adapted E-City Services by Means of Voice Portals

David Griol, María García-Jiménez, Zoraida Callejas, and Ramón López-Cózar

Abstract Digital cities offer new ways to provide information and services of a town or a region in an integrated form, favoring citizen participation and the use of services in previously unavailable ways. In addition, Speech Technologies and Language Processing have made possible the development of a number of new applications which are based on spoken dialog systems. One of them is voice portals, which facilitate spoken interaction with the Internet to provide their users with specific information or web services. In this chapter, we describe a voice portal developed to provide municipal information. The different functionalities provided by the system include to query information about the City Council, access city information, carry out several steps and procedures, complete surveys, access the citizen's mailbox to leave messages for suggestions and complaints, and to be transferred to the City Council to be attended by a teleoperator. This way, the voice portal improves the support of public services by increasing their availability, flexibility, and control while reducing costs and missed calls.

9.1 Introduction

Technological advances currently reached by computers and mobile devices allow their use to access information and applications on the Internet from almost everywhere and at anytime. In addition, users want to access these services in a natural, intuitive, and efficient way. Speech access is then as a solution to the shrinking

D. Griol (✉) • M. García-Jiménez
Computer Science Department, Carlos III University of Madrid, Madrid, Spain
e-mail: dgriol@inf.uc3m.es; 100025080@alumnos.uc3m.es

Z. Callejas • R. López-Cózar
Department of Languages and Computer Systems, University of Granada, Granada, Spain
e-mail: zoraida@ugr.es; rlopezc@ugr.es

J. Mariani et al. (eds.), *Natural Interaction with Robots, Knowbots and Smartphones: Putting Spoken Dialog Systems into Practice*, DOI 10.1007/978-1-4614-8280-2_9,

size of mobile devices (both keyboards to provide information and displays to see the results) [7, 10]. In addition, speech interfaces facilitate the communication in environments where this access is not possible using traditional input interfaces (e.g., keyboard and mouse). It also facilitates information access for people with visual or motor disabilities. In addition, the use of mobile technologies has been currently defined as one of the main indicators of the evolution of information technologies.

Speech and natural language technologies also allow users to access applications in which traditional input interfaces cannot be used (e.g., in-car applications, access for disabled persons). Also speech-based interfaces work seamlessly with small devices and allow users to easily invoke local applications or access remote information. For this reason, conversational agents are becoming a strong alternative to traditional graphical interfaces which might not be appropriate for all users and/or applications [10].

Spoken interaction can be the only way to access information in some cases, like for example when the screen is too small to display information (e.g., handheld devices) or when the eyes of the user are busy in other tasks (e.g., driving) [12]. It is also useful for remote control of devices and robots, especially in smart environments [9]. One of the most widespread applications is information retrieval. Some sample applications are tourist and travel information [5], weather forecast over the phone [13], speech-controlled telephone banking systems [8], conference help [3], etc. They have also been used for education and training, particularly in improving phonetic and linguistic skills: assistance and guidance to F18 aircraft personnel during maintenance tasks [2] and dialog applications for computer-aided speech therapy with different language pathologies [11]. Finally, one of the most demanding applications for fully natural and understandable dialogs is embodied conversational agents and companions [1, 4].

In addition, Spanish Law on electronic access to public services[1] defines the development of multichannel access to information as one of the main obligations of the municipalities. This law also recognizes explicitly the right of citizens to interact electronically with public administration. However, a detailed study of the current situation in the provision of public services by local councils provides the main conclusion of a lack of electronic public services that meet the functions of providing information, guidance and advice, manage suggestions and complaints, as well as consult official announcements and procedures. Therefore, public voice portals not only are compliant with this law but also constitute an efficient tool to provide speech access public services.

In this chapter we describe a voice portal that integrates different technologies such as the VoiceXML standard, databases, web and speech servers, and several programming languages (SQL, PHP, HTML), which make it more dynamic and flexible and increase its quality, efficiency, and adaptation to the users' specific preferences and needs. The functionalities of the system are to consult information

[1]http://www.boe.es/aeboe/consultas/bases_datos/doc.php?id=BOE-A-2007-12352.

about the City Council (government team, councils, etc.) and the city (history, geographic and demographic data, access to the city, yellow pages, movie show times, news, events, weather, etc.), carry out several steps and procedures (checklists and personal files, book municipal facilities, or make an appointment), complete surveys, access the citizen's mailbox to leave messages for suggestions and complaints, and be transferred to the City Council to be attended by a teleoperator.

Thanks to the developed application, speech services are automatically provided, allowing a 24 h a day access that facilitates the access for people with visual or motor disabilities, helping them to eliminate access barriers and enabling a more accessible technology world.

9.2 Interactive Voice Portal to Provide Municipal Information

The voice portal has been developed following the client-server paradigm with the architecture described in Fig. 9.1. The architecture is based on two main components: an IVR (interactive voice response) server and a set of web servers. The IVR provides users with pages following the VoiceXML standard,[2] the ASR and TTS interfaces, and VoIP and telephony technologies. Different web servers connected to the IVR via Internet provide dialog management facilities, grammars and system prompts, and the access to the information and different web services.

Regarding the VoiceXML server, it receives the users' calls and interprets the documents to provide the required services. The interpreter also requests the required resources for the application, defines the logic of the services, and stores users' session state to interact accordingly. To carry out these actions, the VoiceXML interpreter includes different systems to deal with users' calls, manage the communication with the servers and access the required resources, play audio files, convert text to speech, collect user data, perform voice recording, and manage sessions and events.

There are currently many VoiceXML interpreters. One of the most important ones, given the number of functionalities provided, is Voxeo Evolution.[3] Voxeo allows creating VoiceXML applications and access them by means of a local phone number and/or a Skype number. Voxeo also allows to track calls in real time, as well as automatically create log files. These files are very useful for debugging and optimizing the application. In addition, the Voxeo platform provides a fast and efficient support system, which includes forums, support tickets, and very complete documentation. Finally, Voxeo also provides the VoiceXML interpreter and the ASR and TTS components required for the voice portal. Our system uses the Prophecy

[2] http://www.w3.org/TR/voicexml21/.

[3] http://evolution.voxeo.com/.

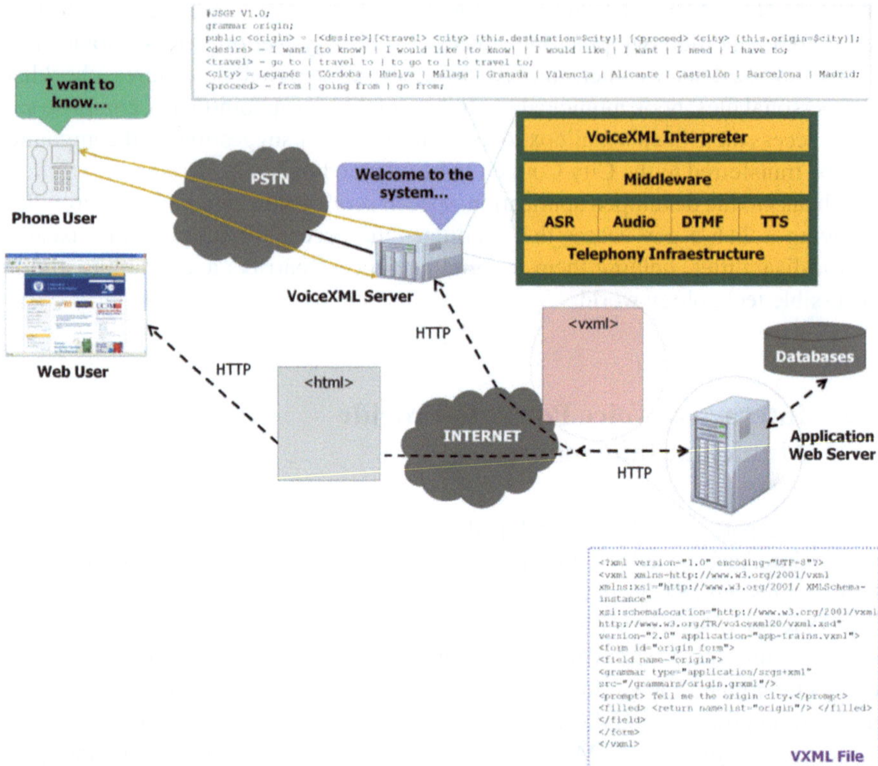

Fig. 9.1 Architecture designed for the interactive voice portal

11 Multi-Language VXML implementation, which facilitates the interaction with the application in different languages.

Regarding the web servers, PHP and VXML files are used to implement each service provided by the voice portal, in addition to access MySQL databases containing the specific information. The different functionalities and corresponding files allow users to complete more than one action in each call.

The *Home module* implements the first dialog that is provided to the user. The options that the system provides to users are divided into five well-differentiated modules that take into account the type of interaction and data that is facilitated: information, procedures and formalities, surveys, mailbox of the citizen, and human operator. Therefore, this module facilitates the access to the rest of functionalities provided by the portal.

The *Information module* provides specific information about the city. This information has been divided into six categories and classified so that users can easily access each functionality provided by this module:

- *City Council*: This module provides specific information related to the local government, local governing entities, teams, and areas.

- *City*: By means of this module it is possible to access information related to the city (history, access, and yellow pages with bars, cafes, restaurants, shops, hostels, hotels, and entertainment guides for cinemas and theaters).
- *Thematic Areas*: This module integrates the rest of information that users can consult by means of the voice portal. This information has been divided into 15 areas, which respectively provide general information, competencies, and contact details. Additional functionalities can be perfectly incorporated by simply adding more static information in the corresponding database of the application.
- *News*: This section provides current news of the municipality, including the date, title, and corresponding description of each one of them.
- *Events*: This section reproduces the list of events of the municipality, providing its specific area, title, date, location, and detailed description.
- *Weather Information*: Users can get the current weather information from the municipality and a forecast for the next two days.

An example of a dialog related to a query about a specific thematic area is shown below (dialog translation from Spanish to English).

```
S: Please select your area of interest:
   1-Social Welfare
   2-Circulation and Transportation
   3-Culture and Leisure
   4-Sports
   5-Administration
   6-Education
   7-Employment
   8-Environment
   9-Parks and Gardens
   10-Health and Consumer
   11-Security
   12-Planning
   13-Housing
   14-Finance and Equity
   15-Cooperation.
U: Social Welfare.
S: You have selected the Social Welfare area. Please select a
   specific topic:
   1-Area Expertise.
   2-General Information.
   3-Contact Information.
U: Contact.
S: The contact information of Social Welfare is:

   Social Services Centre "VALDERAS".
   Parque del Teide, 1. Alcorcón 28924.
   MetroSur: Puerta del Sur. Bus: 514, 511.
   Phone: 916 106 611 - 916 106 797 - 916 426 024.
   Fax: 916 425 311.

   Social Services Centre "EL TEJAR".
   Tiled Square, 4. Alcorcón 28921.
   MetroSur and rail: Central Alcorcón.
```

```
   Phone: 916 648 503 - 916 648 504.
   Fax: 916 425 311.

S: What do you want now?
   1 - Additional information of this area.
   2 - Menu of subject areas.
   3 - Information Menu.
   4 - Main menu.
   5 - Exit the application.
U: Exit.
S: You have chosen to exit the application. Thanks for using
   our voice portal. See you soon!
```

The *Procedures and Formalities* module provides information about procedures and steps, such as verifying the correct incorporation into specific lists, checking the status of dossiers and documents, booking a municipal facility, or making an appointment related to a specific municipal service. An example of a dialog for a user asking about a specific procedure is shown below (dialog translation from Spanish to English).

```
S: Here you can access and complete procedures and formalities.
   To verify whether you are included in a list or not, say
   list or press 1. To check the status of your dossier, say
   dossier or press 2. To make a reservation of a sport
   facility, say reservation or press 3. If you want to
   schedule an appointment, say appointment or press 4.
U: Dossier.
S: Please tell us your ID saying or dialing the 8 digits one
   by one.
U: 47452060.
S: You have provided 47452060. The status of your dossier is
   as follows: Your certificate of registration has been
   requested.
S: What do you want to do now?
   To return to the main menu of the voice portal, say Alcorcón
   or press 1. To access or complete other procedures, say
   procedures or press 2. If you want to exit the application,
   say exit or press 3.
U: Exit.
S: You have chosen to exit the application. Thanks for using
   our voice portal. See you soon!
```

The *Surveys* module allows the City Council to easily and quickly know citizens' opinion about specific issues and queries. Surveys in the application have been designed to be completely anonymous, so the voice portal does not require the users' identification. Once users have finished answering a survey, they are given the options to listen to the answers they have provided. An example of a dialog for a user answering a specific survey is shown below (dialog translation from Spanish to English).

```
S: Then you can complete a municipal survey and know the
   preliminary results. I will ask you a question and
```

```
   provide you 4 possible answers.
   Which political party are you going to vote?
   1: Partido Socialista.
   2: Partido Popular.
   3: Izquierda Unida.
   4: Los Verdes.
U: Los Verdes.
S: You have selected: Los Verdes.
S: Do you want to know the results of the survey?
   Say yes or press 1, or say no or press 2.
U: Yes
S: The results of the survey "Which political party are you
   going to vote?" are as follows:
   Answer 1: Partido Socialista, 25%.
   Answer 2: Partido Popular, 25%.
   Answer 3: Izquierda Unida, 25%.
   Answer 4: Los Verdes, 25%.
   Total votes: 8.
S: What do you want to do now?
   1 - Return to the main menu of the voice portal.
   2 - Exit the application.
U: Exit.
S: You have chosen to exit the application. Thanks for using
   our voice portal. See you soon!
```

The *Citizen's mailbox* implements the functionality of recording a user' speech message and store it for further processing. Thus, citizens can provide their requests, complaints, claims, or comments at anytime and anywhere. The Citizen's mailbox is then managed by a specific office of the City Hall. In addition, if users provide their contact information (telephone, mobile phone, or email), this office would contact them to provide a personalized response to their request. An example of a dialog for a user accessing the Citizen's mailbox is shown below (translated from Spanish to English). An example of a dialog for a user accessing the Citizen's mailbox is shown below (dialog translation from Spanish to English).

```
S: Welcome to the Citizen mailbox. You can contact us and make
   your suggestions, complaints and other comments according to
   the subjects of our voice portal.
S: You can record your message after the tone. Please provide
   first your name and phone so we can respond appropriately,
   and then provide your message. Finally, please keep waiting
   to confirm your recording. Thank you.
S: 'beep'
U: Hello, I am Juan Pérez and I want to congratulate you for
   the new voice portal service in Alcorcón. Thanks.
S: Your message is as follows: "Hello, I am Juan Pérez and I
   want to congratulate you for the new voice portal service
   in Alcorcón. Thanks".
S: Say yes or press one for sending this message. Say no or
   press 2 if you want to re-record your message.
U: Yes
```

```
S: We have saved your message. Your recording will be now
   managed by our staff, and then we will contact you. Thank
   you for collaborating with the City of Alcorcón.
S: What do you want to do now?
   To return to the main menu of the voice portal say Alcorcón
   or press 1. If you want to exit the application say exit or
   press 2.
U: Exit.
S: You have chosen to exit the application. Thanks for using
   our voice portal. See you soon!
```

Finally, the *Teleoperator* module transfers the user's call to a human operator.

The information provided by the voice portal can be classified into static and dynamic information. Static information has been collected from web pages, stored and classified in the databases of the application. Each time users request this information, the system accesses the database and returns this information encapsulated into a VoiceXML file. Examples of this type of information include the history of the city, access information, or contact information of hotels and main offices in the city. Dynamic information includes local news and events, weather information, surveys, and entertainment guides for cinemas and theaters. This information is automatically updated in the application by means of a PHP-based procedure that accesses the required web pages, carries out a syntax processing of this information, and stores the updated information the database. Each time the user requires this type of information, the system only has to access the database and return it.

All the application dialogs use voice grammars and DTMF, which means that users can access menus by speech or using the phone keys, making the application more accessible. Grammars are encoded following the XML standard format defined by the W3C and, therefore, supported by any VoiceXML platform. In addition, this format allows greater flexibility in terms of grammar structure and debugging. Static grammars deal with information that does not vary over time, including a small number of options to choose from. These grammars are coded in the same file where they are used. Dynamic grammars include information that varies with time and often deal with large amounts of data. These grammars are automatically created using PHP files to manage their contents (creation, obtain contents, modify and update information).

9.2.1 Added-Value Functionalities

One of the main aspects in the development of the voice portal was the introduction of different functionalities that allow the adaptation of the system taking into account the current state of the dialog as long as the characteristics of each user. On the one hand, we captured the different VoiceXML events considering different messages for the main events: *noinput* (the user does not answer in a certain time interval or it was not sensed by the recognizer), *nomatch* (the input did not match the recognition grammar or was misrecognized) and *help* (the user explicitly asks for help).

Additionally, VoiceXML provides the *property* element to establish the value of a property that affects the behavior of the platform. These properties may be defined for the whole application, for the document, or for a certain element in a form or menu. For the implementation of the voice portal we have tuned the following properties: *Confidencelevel*, *Sensitivity*, *Documentfetchhint* y *Grammarfetchhint*. The property *Confidencelevel* allows to adjust the accuracy of recognition in order to be accepted. The *Sensitivity* allows to adjust the sensitivity of the recognizer. The properties *Documentfetchhint* and *Grammarfetchhint* allow to adjust the usage of the cache to make searches either safer or faster. In our voice portal all these properties are adjusted dynamically depending on the analysis of the generated events and the history of the dialog, following the procedure described in [6].

On the other hand, the voice portal adapts to specific characteristics of the users. It can be used in different languages (Spanish, English, French, German, and Italian), as the speech recognition is tuned using the property *xml:lang* and the prompts have been stored in the different languages using different encodings in the database. Also the voice portal stores the telephone numbers from which the users access the system in order to compute which are the most frequent queries and predict the user preferences which can be directly accessed by the user in the next calls in order to save time and provide a better user experience.

9.3 Evaluation

The assessment, performance study, and usability analysis of conversational agents are procedures to minimize costs and optimize results in applications in which these agents are integrated. The evaluation of this voice portal has been carried out through quality assessments. To do this, a questionnaire has been designed to evaluate users' subjective opinion and satisfaction with the developed voice portal, thus obtaining a qualitative assessment of users' perception of the system.

This assessment is focused on how users appreciate that they are understood by the system and how they understand the messages generated by the system, the perceived interaction rate, the presence of errors, users understanding about next actions required by the system, the similarity between the developed system and a human operator, and the overall satisfaction with the system. In addition, additional information from users about their knowledge level about new technologies and previous use of dialog systems were considered as an estimator of the users' profile. The questionnaire developed for this purpose consists of the following ten questions: (1) *Q1*: State on a scale from 1 to 5 your previous knowledge about new

Table 9.1 Results of the subjective evaluation of the voice portal

	Q1	Q2	Q3	Q4	Q5	Q6	Q7	Q8	Q9	Q10
Average value	3.3	3.0	2.9	4.1	4.5	3.1	4.2	4.1	2.9	4.3
Minimum value	1	2	2	3	4	2	3	3	2	3
Maximum value	5	4	4	5	5	4	5	5	4	5
Standard deviation	1.2	0.6	0.7	0.6	0.5	0.5	0.5	0.8	0.5	0.6

technologies; (2) *Q2*: State on a scale from 1 to 5 your previous uses of speech-based interfaces; (3) *Q3*: State on a scale from 1 to 5 your previous uses of voice portals; (4) *Q4*: Did the system correctly understand you during the interaction?; (5) *Q5*: Did you understand correctly the messages of the system?; (6) *Q6*: Do you think that the interaction rate was adequate?; (7) *Q7*: Was it simple to obtain the requested information?; (8) *Q8*: *Set the difficulty level of the system for you.*; (9) *Q9*: Do you think that the system behaved in a similar way as human being?; (10) *Q10*: In general terms, are you satisfied with the performance of the system? The possible answers to the complete set questions were the same: *Never*, *Rarely*, *Sometimes Usually*, and *Always*.

The assessment test was completed by 20 professors and students of our university who were introduced on the main functionalities of the voice portal and required to complete the questionnaire once finished their interaction. Users freely chose to perform actions and select between the different functionalities, modules, and submodules. Table 9.1 shows the results for each one of the questions in the subjective evaluation of the application.

From the analysis of the results of the evaluation, it can be observed that users' knowledge about new technologies and use of dialog systems is varied. Most of the users found that the interaction with the system was very easy and the interaction rate is considered as suitable. They also considered that the system correctly understood their messages. The same fact was considered regarding the messages generated by the application. Related to the similarity between the system and a human operator, most users believed this characteristic has to be improved. Finally, users considered that they can easily obtain the required information and they are very globally satisfied with the system.

In addition, we completed an objective evaluation considering the following measures:

- *Successful Dialogs*. This is the percentage of successfully completed tasks. In each scenario, the user has to obtain one or several items of information, and the dialog success depends on whether the system provides correct data (according to the aims of the scenario) or incorrect data to the user.
- *Average Number of Turns per Dialog* (nT).
- *Confirmation Rate*. It was computed as the ratio between the number of explicit confirmations turns (nCT) and the number of turns in the dialog (nCT/nT).

Table 9.2 Results of the objective evaluation of the voice portal

Successful dialogs	nT	Confirmation rate	ECR	nCE	nNCE
94%	8.4	29%	9%	0.88	0.09

- *Average Number of Corrected Errors per Dialog* (nCE). This is the average of errors detected and corrected by the dialog manager. We have considered only those errors that modify the values of the attributes and that could cause the failure of the dialog.
- *Average Number of Uncorrected Errors per Dialog* (nNCE). This is the average of errors not corrected by the dialog manager. Again, only errors that modify the values of the attributes are considered.
- *Error Correction Rate* (ECR). The percentage of corrected errors, computed as nCE/ (nCE + nNCE).

Table 9.2 presents the results of the objective evaluation. These results also show that the developed system could interact correctly with the users in most cases. However, the statistical system obtained a higher success rate, improving the initial results by 9 % absolute. The average number of required turns has been reduced to 8.4 with only a 29 % of confirmation turns. The nCE and the value obtained for the ECR show how the system is able to detect and correct most of the errors made by the speech recognition and understanding modules.

9.4 Conclusions

This chapter describes a voice portal implemented using the VoiceXML standard that constitutes a useful tool to access information of a specific city, carry out procedures and formalities, complete surveys, use a mailbox to make their complaints and requests, and be transferred to a PBX (private branch exchange). The different services offered by the municipal voice portal are divided into modules which are accessed taking into account user's decisions during each dialog. These modules are interconnected so that users can complete more than one action.

To our knowledge, there are no voice portals in the Spanish city halls offering the functionalities described. This way, it creates a new communication channel which is useful, efficient, easy to use, and accessible. In addition, the voice portal improves the support of public services by increasing the availability, flexibility, control, and reducing costs and missed calls. The system is also easily adaptable to the requirements of each municipality. Future works include the development of additional functionalities for the described modules and the adaptation of the voice portal to each user or group of users by taking into account additional languages, previous dialogs, and preferred functionalities.

References

1. Bailly, G., Raidt, S., Elisei, F.: Gaze, conversational agents and face-to-face communication. Speech Commun. **52**(6), 598–612 (2010)
2. Bohus, D., Rudnicky, A.: LARRI: a language-based maintenance and repair assistant. In: Proceedings of Multi-Modal Dialogue in Mobile Environments Conference (IDS'02). Kloster Irsee, Germany (2002)
3. Bohus, D., Grau, S., Huggins-Daines, D., Keri, V., Krishna, G., Kumar, R., Raux, A., Tomko, S.: Conquest: an open-source dialog system for conferences. In: Proceedings of 7th Meeting of the North American Chapter of the Association for Computational Linguistics (HLT/NAACL'07), pp. 9–12, Rochester, USA (2007)
4. Brahnam, S.: Building character for artificial conversational agents: ethos, ethics, believability, and credibility. Psychol. J. **7**(1), 9–47 (2009)
5. Glass, J., Flammia, G., Goodine, D., Phillips, M., Polifroni, J., Sakai, S., Seneff, S., Zue, V.: Multilingual spoken-language understanding in the MIT Voyager system. Speech Commun. **17**, 1–18 (1995)
6. Griol, D., Callejas, Z., López-Cózar, R.: Statistical dialog management methodologies for real applications. In: Proceedings of the 11th SIGdial Meeting, pp. 124–131 (2010)
7. López-Cózar, R., Araki, M.: Spoken, Multilingual and Multimodal Dialogue Systems. Wiley, Chichester (2005)
8. Melin, H., Sandell, A., Ihse, M.: CTT-bank: A speech controlled telephone banking system - an initial evaluation. In: TMH Quarterly Progress and Status Report (TMH-QPSR), vol. 1, pp. 1–27 (2001)
9. Menezes, P., Lerasle, F., Dias, J., Germa, T.: Humanoid Robots, Human-like Machines, chap. Towards an Interactive Humanoid Companion with Visual Tracking Modalities, pp. 48–78. Advanced Robotic Systems Int. and I-Tech Education and Publishing, Vienna (2007)
10. Pieraccini, R., Rabiner, L.: The Voice in the Machine: Building Computers that Understand Speech. The MIT Press, USA (2012)
11. Vaquero, C., Saz, O., Lleida, E., Marcos, J., Canals, C.: VOCALIZA: an application for computer-aided speech therapy in spanish language. In: Proceedings of IV Jornadas en Tecnología del Habla, pp. 321–326, Zaragoza, Spain (2006)
12. Weng, F., Varges, S., Raghunathan, B., Ratiu, F., Pon-Barry, H., Lathrop, B., Zhang, Q., Scheideck, T., Bratt, H., Xu, K., Purver, M., Mishra, R., Raya, M., Peters, S., Meng, Y., Cavedon, L., Shriberg, L.: CHAT: a conversational helper for automotive tasks. In: Proceedings of the 9th International Conference on Spoken Language Processing (Interspeech/ICSLP), pp. 1061–1064, Pittsburgh, USA (2006)
13. Zue, V., Seneff, S., Glass, J., Polifroni, J., Pao, C., Hazen, T., Hetherington, L.: JUPITER: a telephone-based conversational interface for weather information. IEEE Trans. Speech Audio Process. **8**(1), 85–96 (2000)

Part III
Multi-domain, Crosslingual Spoken Dialog Systems

Chapter 10
Efficient Language Model Construction for Spoken Dialog Systems by Inducting Language Resources of Different Languages

Teruhisa Misu, Shigeki Matsuda, Etsuo Mizukami, Hideki Kashioka, and Haizhou Li

Abstract Since the quality of the language model directly affects the performance of the spoken dialog system (SDS), we should use a statistical language model (LM) trained with a large amount of data that is matched to the task domain. When porting a SDS to another language, however, it is costly to re-collect a large amount of user utterances in the target language. We thus use the language resources in a source language by utilizing statistical machine translation. The main challenge in this work is to induct automatic speech recognition results collected using a speech-input system that differs from the target SDS both in the task and the target language. To select appropriate sentences to be included in the training data for the LM, we induct a spoken language understanding module of the dialog system in the source language. Experimental construction using over three million user utterances showed that it is vital to conduct a selection from the translation results.

10.1 Introduction

The quality of the language model (LM) directly affects the performance of the spoken dialog system (SDS). It is desirable to use a statistical LM trained with a large amount of data that matches the task domain. When constructing a new SDS, however, it is almost impossible to prepare a large amount of user utterances. Thus, an initial LM is made by handcrafting grammar or conducting a Wizard-of-Oz data collection. Since such an approach is costly and often unreliable, methods have

T. Misu (✉) • S. Matsuda • E. Mizukami • H. Kashioka
National Institute of Information and Communications Technology (NICT), Kyoto, Japan
e-mail: teruhisa.misu@gmail.com

H. Li
Institute for Infocomm Research, Singapore, Singapore

J. Mariani et al. (eds.), *Natural Interaction with Robots, Knowbots and Smartphones: Putting Spoken Dialog Systems into Practice*, DOI 10.1007/978-1-4614-8280-2_10,

been studied to reduce the cost of collecting training data. In particular, there is a compelling need for such methods when porting a SDS to a new target language.

In this work we construct a LM for the dialog system in a target language by inducting language resources from different languages. Recently, such language portability has successfully led to the portability of spoken language understanding (SLU) systems [5,9,12,16], showing that using the machine translation results from the corpus in the source language can reduce the cost of system construction. In this work, we apply this procedure to the portability of a LM by constructing a LM of the target language by using the translation results of several resources from the source language. We here assume that the SLU component in the source language functions properly. This assumption is reasonable when extending it to the target language of a currently running SDS. Furthermore, recent advances in machine translation (MT) techniques have enabled us to easily access MT software (e.g., Google translation[1]).

In addition, we adopt resources that do not necessarily match the task or the language itself. Specifically, we tackle the challenge of using automatic speech recognition (ASR) results collected from a running speech-to-speech translation system in the source language, which has been tested publicly, and a training language model for a SDS in the target language.

The major problems in this trial are that they contain sentences that do not match the target task[2] and they may include erroneous results due to ASR and statistical machine translation (SMT) errors. Using such data for training a LM would often result in a worse model than when domain-matched utterances are available (e.g., [11]). Consequently, using all of the MT results may degrade the performance of the LM.

In this paper, we propose a method to select appropriate texts from SMT results that are suitable for inclusion in the training data for the LM in the target language. The paper is organized as follows: Sect. 10.2 overviews the proposed method to select appropriate sentences. Section 10.3 describes our statistical SLU module used for the selection. Section 10.4 explains our SMT module. Section 10.5 shows the details of the experimental results in a tourist information domain. Section 10.6 reviews related works, and Sect. 10.7 concludes the paper.

10.2 Strategies to Achieve Language Portability of LM

With the progress in corpus-based statistical machine learning methods, ASR and MT performances have been improving remarkably. In this work, we consider an attempt to obtain training data for the target language by combining these techniques. The ASR results collected from a running system with speech input (in the source language) are translated into the target language.

---

[1] http://translate.google.com/.

[2] Although the speech-to-speech translation system is oriented to a travel domain, the log data contains many irrelevant inputs.

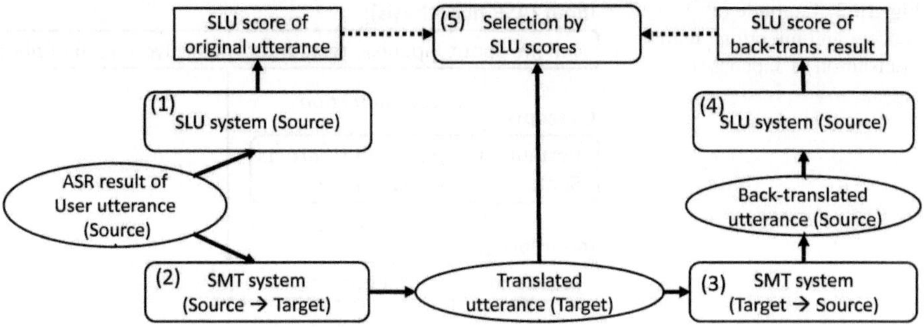

Fig. 10.1 Overview of proposed sentence selection scheme

From the translated ASR results, we eliminate (1) those that are irrelevant to the target domain and (2) those containing ASR and SMT errors. To handle domain-mismatched text (1), we use the SLU module in the source language as a selection criterion. By calculating the degree of matching, even though it is measured in the source language, we expect to eliminate irrelevant training data. In addition, to handle erroneous translation results (2), we consider SLU scores for back-translation results. Usually, SMT and SLU systems are trained using a corpus without ASR error, and unnatural word sequences caused by ASR errors are expected to result in unnatural translation results, which mismatch the SLU module. We expect the translation module to work as a filter to remove erroneous ASR results.

The flow of the selection method is illustrated in Fig. 10.1 and summarized as follows.

For each ASR result of user utterance collected from the running system (in the source language), we perform the following:

1. Annotate the utterance using the SLU module to obtain SLU score.
2. Translate it into the target language.
3. Back-translate the SMT result into the source language.
4. Annotate tags to the back-translation result using the SLU module.
5a. Accept 3. If the SLU score (the likelihood of the in-domain class with the largest score) exceeds a certain threshold θ.
5b. Reject 3. Otherwise.

In this work, we adopted a tourist information task as target domain and constructed a LM with Japanese as the source language and English as the target language. We use the log data collected from our speech-to-speech translation system called "VoiceTra[3]," which is now under public test.

[3]http://mastar.jp/translation/voicetra-en.html.

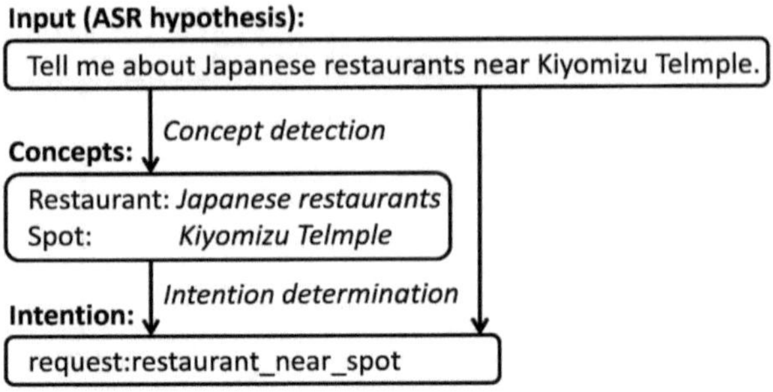

Fig. 10.2 Example of concept and intention tagging (translation of Japanese)

10.3 Configuration of SLU Module

We used our statistical SLU module to annotate the utterances. Our SLU module consists of a concept-detection (NE detection) part and an intention-determination (dialog act tagging) part (Fig. 10.2). Note that the intention-determination score is used for the text selection.

In the concept-detection part, concepts that correspond to the slot values used in the subsequent dialog manager are detected from an input ASR hypothesis. We train linear-chain CRF as a model to predict the sequence of concepts and label the tags using CRF++ toolkit.[4] The utterance features for the prediction consist of the word surface, the part-of-speech, and their 2-g information.

In the intention-detection part, the user's intention, which is associated with the system actions of the dialog system, is determined. We train a multi-class classifier using logistic linear regression with LIBLINEAR [6][5] toolkit and defined 83 user-intention classes. We also trained a binary classifier that determines if the utterance is within the target domain or out-of-domain (OOD). We annotated 82 classes as in-domain and just one as OOD and labeled the former as 1 and the latter as 0. This binary classifier is based on logistic linear regression. The SLU score range from 0 to 1. The utterance features used for the prediction consist of the number of times that word surface, part-of-speech, concepts, and 2-g each appear in the input.

10.4 Translation Module

We used our state-of-the-art phrase-based SMT system CleopATRa[6] [7] which is comprised of a beam search decoder based on a log-linear model, a language model, a translation model, and a distortion model. The models were trained

---

[4]http://crfpp.sourceforge.net.

[5]http://www.csie.ntu.edu.tw/~cjlin/liblinear/.

[6]This translation system has been used in our speech-to-speech translation application called "VoiceTra," http://mastar.jp/translation/index-en.html.

using our basic travel expression corpus (BTEC) which is comprised of 700 K Japanese-English parallel sentences. The parallel corpus covers tourism-related conversational sentences similar to those usually found in phrase books for tourists traveling abroad. The BLEU score of the SMT system for in-domain inputs is 0.46 in Japanese to English (J-E) and 0.50 in English to Japanese (E-J).

10.5 Evaluation

10.5.1 Training Data

As training data, we used the machine translation output from about 3.8 M ASR results of the user queries collected by the VoiceTra system in the source language (VoiceTra (w/o selection)). Note that the data have no manual transcription, and the transcription is given by ASR in Japanese and translated into English. Among these translated user utterances, we selected suitable utterances to include in the training data using the selection methods and prepared another set of training data. Because our SDS explains about 100 sightseeing spots in natural language, we used the system knowledge base (KB) employed by the system.

Our test set consists of 2,537 manual translation results of user queries collected from our SDS in the source language (Test set). We asked 20 (ten males and ten females) native English speakers to read the test set (= 127 utterances per speaker). The specifications of these data are summarized in Table 10.1.

We trained a 3-g language model using the MITLM toolkit [8], and the modified Kneser-Ney smoothing was used for all our models in this work. The lexicon of the language model was fixed to 42 K words that were selected based on the KB and the frequency in the VoiceTra corpus. For speech recognition, we used our WFST-based decoder SprinTra [4] and speaker-independent triphone acoustic models trained on feature vectors that were comprised of 12 MFCCs with their deltas and delta power.

The word error rates (WERs) obtained by the baseline LM trained on the KB only, or on the KB and the translated VoiceTra data (KB+(VoiceTra w/o selection)), were 55.43 % and 29.16 %, respectively.

Table 10.1 Specification of training and test data

	# Sentences	# Words
VoiceTra (w/o selection) (translation of Japanese)	3,767,081	13,388,082
Knowledge base (KB)	1,539	38,588
Test set	2,539	16,885

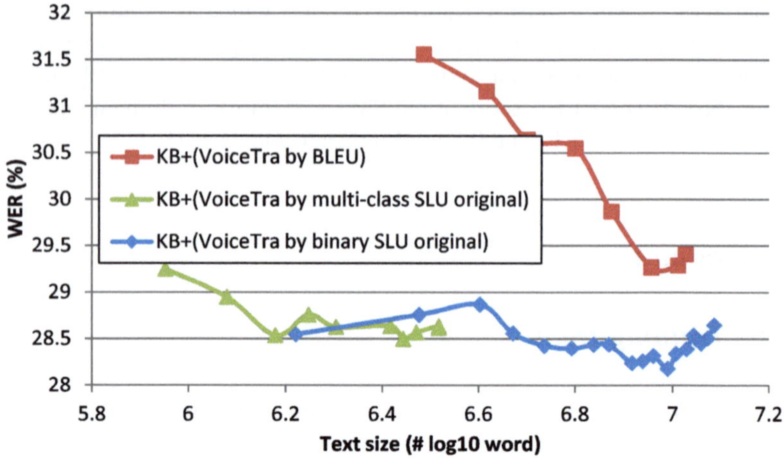

Fig. 10.3 Comparison of selection criteria

10.5.2 Comparison of Selection Criteria

We first compared the selection criteria. The proposed method using SLU module-based selection was marked as (KB+(VoiceTra by SLU multi-class) and (KB+(VoiceTra by SLU binary)) in Fig. 10.3. In this experiment, we calculated the SLU score against the original ASR results of VoiceTra data (not the back-translation results). In the figure, we plotted the WERs against the common logarithm of the amount of data used for language model training by changing threshold θ used for the VoiceTra text selection. For comparison, we also plotted the results when the BLEU score was used for selection (KB+(VoiceTra by BLEU)) as a typical measure of the correctness of machine translation. We calculated the similarity of the original and back-translated sentences, which were included in the training data when the similarity exceeded a certain threshold. This result is also illustrated in Fig. 10.3.

Unfortunately, the model constructed using BLEU-based selection failed to outperform the baseline method using all the texts (29.16 %). In contrast, the models using the proposed SLU-based selection methods outperformed the baseline method, although the improvement was not large. These results indicate that it is important to select texts based on the domain, and using the SLU system is effective as a selection criterion, even when calculated in the source language. In addition, we selected more texts using the binary classifier than texts using multi-class classifier and achieved better performance. This result suggests that using multi-class classifier is too strict because many sentences were annotated as OOD.

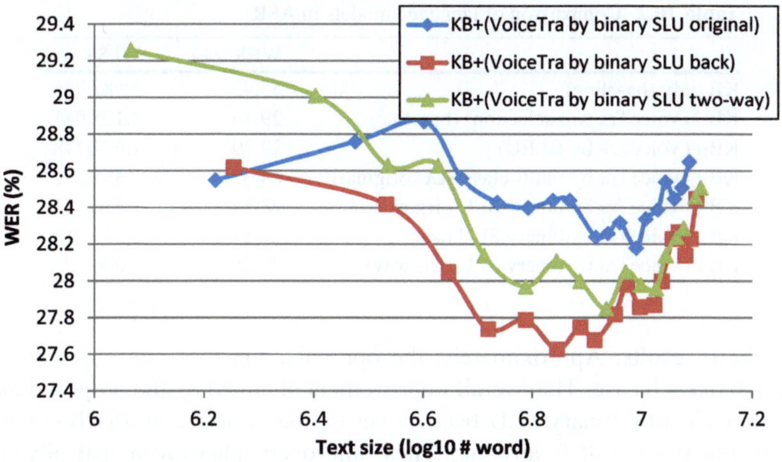

Fig. 10.4 Comparison of SLU score calculation target

10.5.3 Comparison of Calculation Targets of SLU Score

There are two choices in the evaluation targets with which the SLU scores were calculated: the original texts and the back-translation results. We tried three conditions. The first is the case where the original ASR results were used to calculate the SLU score, and the training data were selected using the score (KB+(VoiceTra by binary SLU original)). In the second condition, the back-translation results of the original ASR results were used to calculate SLU scores (KB+(VoiceTra by binary SLU back)). In the third condition, the average of the above two SLU scores was used for the selection (KB+(VoiceTra by binary SLU two-way)). The results are illustrated in Fig. 10.4.

All of these conditions achieved better performance (29.16 %) than the baseline method without selection. By adopting the back-translation results as the target of the SLU score calculation, we still achieved a better result than the case where the original ASR results were used for the calculation. These results suggest that it is important to select texts considering errors in machine translation, although more detailed analysis of the selected texts is needed.

10.5.4 Evaluation Based on Cross Validation

We then determined the value of threshold θ for text selection by twofold cross validation by splitting the test set into sets 1 and 2. Set 1 was used as a development set to estimate threshold θ for the evaluation of set 2, and vice versa. Table 10.2

Table 10.2 Comparison of language models in ASR

	WER (%)	WFST arcs
KB only (baseline)	55.43	87,755
KB+(VoiceTra w/o selection) (baseline)	29.16	3,105,041
KB+(VoiceTra by BLEU)	29.29	1,974,778
KB+(VoiceTra by multi-class SLU original)	28.62	855,295
KB+(VoiceTra by binary SLU original)	28.34	1,736,121
KB+(VoiceTra by binary SLU back)	27.84	1,379,710
KB+(VoiceTra by binary SLU two-way)	28.26	1,046,175

shows these results. Approximately, the optimal point was chosen by the cross validation in each case. The overall improvement obtained by the proposed method (KB+(VoiceTra by binary SLU back)) over the baseline method (KB+(VoiceTra w/o selection)) was 1.45 % absolute. These improvements were statistically significant ($p < 0.01$).

The numbers of arcs in the optimized WFST[7] for recognition in the best operating point are also listed in Table 10.2. The number of arcs, which indicates the maximum size of the search space during speech recognition, is related to the decoding speech; more arcs usually indicate larger decoding time. The number of the arcs in the WFST of (KB+(VoiceTra by binary SLU back)) was less than half of the without selection case . This indicates that the ASR model is more compact, and shorter decoding times can be expected.

10.6 Related Works

Because porting an ASR system to a new target language requires enormous resources to match the target domain in the target language, many works have been conducted on them.

Most previous studies addressed the portability of acoustic models. For example, Schultz and Waibel [15] proposed a method to construct a language-independent acoustic model based on the GlobalPhone database. Liu et al. [3] and Abe et al. [1] borrowed resources from a resource-rich language to build acoustic models for a resource-poor language. In contrast, few works have realized the language portability of a LM for SDSs due to dependence on the domain, although several works have adapted the model using the translation results from the speech recognition of read speech of news articles [10, 13].

For rapid prototyping of language models, several works have used external resources. For example, Zhu et al. [17] used the n-g count in Web search results to

[7]The lexicon, the acoustic model, and language model WFSTs are composed to generate a large WFST for recognition.

estimate unreliable trigram probabilities in a spoken document retrieval task. Bulyko et al.[2] used n-g entries that frequently appeared in the Switchboard corpus as a search query to retrieve relevant texts from the Web. Sarikaya et al.[14] adopted a similar approach for SDSs to enhance the training data. They made search queries with a small amount of domain-specific data and filtered the retrieved texts by considering the similarity with the domain-specific data using BLEU scores. Misu and Kawahara [11] selected text from the Web by considering the domain and the utterance style of the target dialog system.

The main point of our work is to realize language portability of the LM using language resources that do not match the target recognition task either as in the target language or as in the domain.

10.7 Conclusion

We proposed a method of constructing a statistical LM for a new SDS by translating and selecting sentences from the ASR results collected from a speech-based system in a different language. To select suitable sentences that match the target domain, we used SLU modules in the source language. The effectiveness of the proposed method was confirmed by constructing a LM for a sightseeing guidance dialog system with resources collected from a speech-to-speech translation system. Our future works include a combination of the selection methods that use the confidence measures of ASR and SMT, which we could not use due to the limitations of the ASR and SMT decoders. We also expect to apply our proposed method to the construction of SLU systems and evaluate our proposed method with a SLU task.

References

1. Abe, K., Sakti, S., Isotani, R., Kawai, H., Nakamura, S.: Brazilian portuguese acoustic model training based on data borrowing from other language. In: Proceedings of Interspeech, pp. 861–864 (2010)
2. Bulyko, I., Ostendorf, M., Stolcke, A.: Getting more mileage fromweb text sources for conversational speech language modeling using class-dependent mixtures. In: Proceedings of Human Language Technology (HLT), vol. 2, pp. 7–9 (2003)
3. C Liu, L.M.: Training acoustic models with speech data from different languages. In: Proceedings of Multilingual Speech and Language Processing (2006)
4. Dixon, P., Finch, A., Hori, C., Kashioka, H.: Investigation on the effects of asr tuning on speech translation performance. In: Proceedings of The International Workshop on Spoken Language Translation (IWSLT), pp. 167–174 (2011)
5. Lefèvre, F., Mairesse, F., Young, S.: Cross-lingual spoken language understanding from unaligned data using discriminative classification models and machine translation. In: Proceedings of Interspeech, pp. 78–81 (2010)
6. Fan, R., Chang, K., Hsieh, C., Wang, X., Lin, C.: LIBLINEAR: a library for large linear classification. J. Machine Learn. Res. **9**, 1871–1874 (2008)

7. Goh, C., Watanabe, T., Paul, M., Finch, A., Sumita, E.: The NICT translation system for IWSLT 2010. In: Proceedings of The International Workshop on Spoken Language Translation (IWSLT), pp. 139–146 (2010)
8. Hsu, B., Glass, J.: Iterative language model estimation: efficient data structure and algorithms. In: Proceedings of Interspeech, pp. 841–844 (2008)
9. Jabaian, B., Besacier, L., Lefèvre, F.: Combination of stochastic understanding and machine translation systems for language portability of dialogue systems. In: Proceedings of International Conference on Acoustics, Speech and Signal Processing (ICASSP), pp. 5612–5615 (2011)
10. Kim, W., Khudanpur, S.: Lexical triggers and latent semantic analysis for cross-lingual language model adaptation. ACM Trans. Asian Lang. Inf. Process. **3**(2), 94–112 (2004)
11. Misu, T., Kawahara, T.: A bootstrapping approach for developing language model of new spoken dialogue systems by selecting web texts. In: Proceedings of Interspeech, pp. 9–12 (2006)
12. Misu, T., Mizukami, E., Kashioka, H., Nakamura, S., Li, H.: A bootstrapping approach for SLU portability to a new language by inducting unannotated user queries. In: Proceedings of International Conference on Acoustics, Speech and Signal Processing (ICASSP) (2012)
13. Nanjo, H., Oku, Y., Yoshimi, T.: Automatic speech recognition framework for multilingual audio contents. In: Proceedings of Interspeech, pp. 1445–1448 (2007)
14. Sarikaya, R., Gravano, A., Gao, Y.: Rapid language model development using external resources for new spoken dialog domains. In: Proceedings of International Conference on Acoustics, Speech and Signal Processing (ICASSP), vol. 1, pp. 573–576 (2005)
15. Schultz, T., Waibel, A.: Language-independent and language-adaptive acoustic modeling for speech recognition. Speech Commun. **35**(1–2), 31–51 (2001)
16. Servan, C., Camelin, N., Raymond, C., Béchet, F., De Mori, R.: On the use of machine translation for spoken language understanding portability. In: Proceedings of International Conference on Acoustics, Speech and Signal Processing (ICASSP), pp. 5330–5333 (2010)
17. Zhu, X., Rosenfeld, R.: Improving trigram language modeling with the world wide web. In: Proceedings of International Conference on Acoustics, Speech and Signal Processing (ICASSP), vol. 1, pp. 533–536 (2001)

Chapter 11
Towards Online Planning for Dialogue Management with Rich Domain Knowledge

Pierre Lison

Abstract Most approaches to dialogue management have so far concentrated on offline optimisation techniques, where a dialogue policy is precomputed for all possible situations and then plugged into the dialogue system. This development strategy has however some limitations in terms of domain scalability and adaptivity, since these policies are essentially static and cannot readily accommodate runtime changes in the environment or task dynamics. In this paper, we follow an alternative approach based on online planning. To ensure that the planning algorithm remains tractable over longer horizons, the presented method relies on probabilistic models expressed via *probabilistic rules* that capture the internal structure of the domain using high-level representations. We describe in this paper the generic planning algorithm, ongoing implementation efforts and directions for future work.

11.1 Introduction

Dialogue management is at its core a decision-making operation: given a specific conversational situation the agent finds itself in, the objective of the dialogue manager is to find the optimal action to perform at that stage, depending on some performance criteria. In order to search for this optimal action, the agent often needs to take into account not only the local effect of the action, but also how it might influence future states and actions. An action might therefore be locally suboptimal but still be selected if it contributes to a higher objective within the interaction. A typical example of these lookahead strategies pertains to clarification requests.

P. Lison (✉)
Department of Informatics, Language Technology Group, University of Oslo, Oslo, Norway
e-mail: plison@ifi.uio.no

J. Mariani et al. (eds.), *Natural Interaction with Robots, Knowbots and Smartphones: Putting Spoken Dialog Systems into Practice*, DOI 10.1007/978-1-4614-8280-2_11,

These requests might indeed have a slightly negative effect in the short term but are often beneficial on the longer term, since they can help reducing the state uncertainty and therefore lead to more successful dialogues [15].

This observation has led many researchers to cast dialogue management as a decision-theoretic *planning problem* [4, 22, 23]. Such formalisation relies on the specification of rewards associated to particular (state, action) pairs. In this setting, the role of the dialogue manager is to select the action which yields the highest return (cumulative discounted reward) over the course of the interaction. One major benefit of this formalisation is that it enables the system designer to flexibly encode the trade-offs between the various sometimes conflicting objectives of the interaction. Dialogue strategies can then be automatically optimised for the domain at hand, leading to conversational behaviours which are often more flexible and natural than handcrafted strategies [8].

So far, most approaches to dialogue management relying on decision-theoretic planning perform this optimisation entirely *offline*, by precomputing a dialogue policy mapping every possible state of the dialogue to the optimal action to perform at that state. While this approach is attractive in terms of runtime computational savings, it also presents a number of limitations. The first challenge is that policy optimisation becomes increasingly hard as the size of the dialogue state grows, since the policy must be calculated for every hypothetical situation which might be encountered by the agent. The problem is especially critical when the dialogue state is partially observable and therefore expressed in a high-dimensional, continuous space. The second challenge is the difficulty of adapting or refining the policy once it has been calculated. Precomputed policies must be recalculated every time the domain-specific models are modified or extended. This is an important drawback for application areas such as human-robot interaction, cognitive assistants or tutoring systems, since these domains often rely on environment or task models whose dynamics can vary at runtime, while the interaction unfolds (for instance in order to adapt to shifting user preferences).

An alternative approach that has recently gained popularity in the partially observable Markov decision process (POMDP) planning literature is to perform planning *online*, at execution time [17]. Compared to offline policies, the major advantage of online planning is that the agent only needs to consider the current state to plan, instead of enumerating all possible ones. It can also more easily adapt to changes in the environment or task models. The available planning time is however more limited, since planning is in this case interleaved with dialogue system execution and must therefore meet real-time constraints.[1]

---

[1]Interestingly, offline and online approaches to planning are not mutually exclusive, but can be combined together to offer "the best of both worlds." The idea is to perform offline planning to precompute a rough policy and use this policy as a heuristic approximation to guide the search of an online planner [17]. These heuristic approximations can for instance be used to provide lower and upper bounds on the value function, which can be exploited to prune the lookahead tree.

To address these performance constraints, we investigate in this paper the use of prior domain knowledge to filter the space of possible actions and transitions that need to be considered by the planning algorithm. The key idea is to structure the probability and reward models used for planning in terms of high-level, probabilistic *rules*. These rules can notably express which actions are deemed to be relevant for a given dialogue state and thus filter the space of actions to consider at planning time. The main intuition is that by exploiting the internal structure of the domain and integrating it in our models, we can develop algorithms which are significantly more efficient than approaches relying on unstructured models.

The structure of this paper is as follows. We start by briefly reviewing the mathematical foundations of our work and then describe our representation formalism, based on the concept of a probabilistic rule. We then detail our planning algorithm, which relies on these rules to find the optimal action sequence to perform at a given state. We then describe our current implementation efforts to develop an efficient planner based on this algorithm. Finally, we compare our approach to related work in the field and conclude the paper.

11.2 Background

11.2.1 Dialogue State

Virtually all dialogue management frameworks rely on the notion of a *dialogue state*, which can take various forms, depending on the chosen level of expressivity and account of uncertainty. In this work, we encode the dialogue state as a *Bayesian Network* [21], where each node represents a distinct state variable deemed to be relevant for decision-making, such as the hypothesised user intention or contextual features. These variables might be conditionally dependent on each other, which is easily encoded in a Bayesian Network via directed edges.

Formally, let $X_1 \dots X_n$ denote a set of random variables. Each variable X_i is associated with a range of mutually exclusive values. A Bayesian Network defines the joint probability distribution $P(X_1 \dots X_n)$ via conditional dependencies between variables using a directed graph where each node corresponds to a variable X_i. Each edge $X_i \rightarrow X_j$ denotes a conditional dependence between the two nodes, in which case X_i is said to be a *parent* of X_j. A conditional probability distribution $P(X_i|Parents(X_i))$ is associated with each node X_i, where $Parents(X_i)$ denotes the parents of X_i.

A Bayesian Network can be straightforwardly extended for capturing utility-related information. In practice, this is realised by adding *utility* and *decision* nodes to the network. A utility node encodes the utility associated with a particular set of dependent variables. Typically, at least one of these dependent variables is a decision node, which describes a set of possible actions that the agent can perform. Figure 11.1 illustrates a Bayesian Network extended with such nodes.

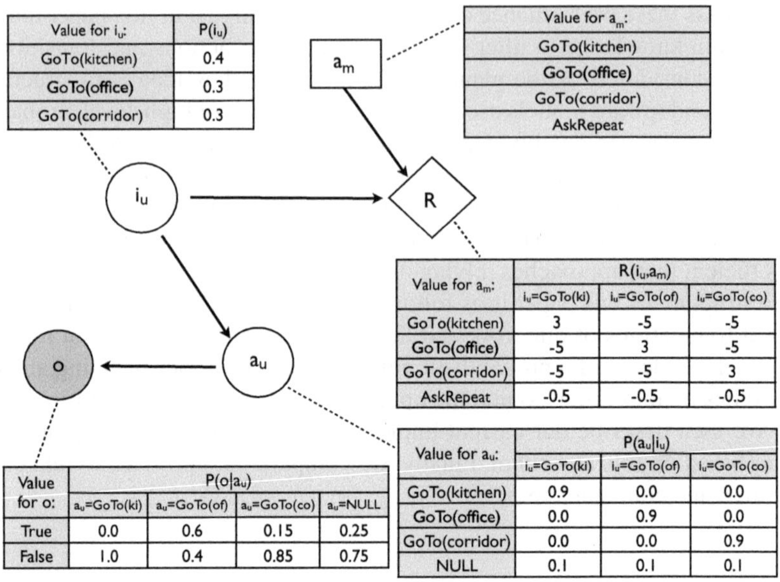

Fig. 11.1 Example of Bayesian Network extended with utility nodes (shown as *diamonds*) and decision nodes (shown as *rectangles*). The network represents a domain for a robot that is able to perform four actions (going to the kitchen, the office, the corridor, or asking the user to repeat). The node i_u represents the user intention and the node a_u the last dialogue act from the user. The latter is connected to an evidence node o which represents the actually observed N-best list of user acts (e.g. coming from speech recognition). Depending on the user intention, the execution of a specific action will yield different values for the reward R. In our case, we can easily calculate that $R(a_m = \text{GoTo(office)} \,|\, o = \text{True}) = 0.96$, $R(a_m = \text{GoTo(corridor)} \,|\, o = \text{True}) = -3.3$, $R(a_m = \text{GoTo(kitchen)} \,|\, o = \text{True}) = -4.65$ and finally $R(a_m = \text{AskRepeat} \,|\, o = \text{True}) = -0.5$

11.2.2 Decision-Theoretic Planning

In dialogue management, we are interested in the action which has the highest expected discounted cumulative reward for a given horizon. In order to find this optimal action, the dialogue manager must be able to perform some form of forward planning to estimate the expected future rewards of every action.

To this end, let us assume that we have a dialogue state (also called belief state) b, with $b(s) = P(s)$ being a joint probability distribution over the possible state values. This joint probability distribution can for instance be described as a Bayesian Network of state variables (cf. previous section). In addition, we will also assume that we can structure our models in the generic form of a *POMDP*, with a reward model $R(s,a)$ describing the rewards associated with particular actions, a transition model $P(s'|s,a)$ describing the state following the execution of action a in state s and an observation model $P(o|s,a)$ encoding the expected observations (in our case, an N-best list of user dialogue acts) for a given state s after action a.

The expected cumulative reward of a state-action sequence $\langle b_0, a_0, b_1, a_1, \ldots, b_n, a_n \rangle$ with a discount factor γ is then defined as:

$$Q[(b_0, a_0, b_1, a_1, \ldots b_n, a_n]) = \sum_{t=0}^{n} \gamma^t R(b_t, a_t) = \sum_{t=0}^{n} \gamma^t \sum_{s \in S} b_t(s) R(s, a_t) \tag{11.1}$$

where the dialogue state b_{t+1} is defined as an update from b_t:

$$b_{t+1}(s') = P(s'|o_{t+1}, a_t, b^t) = \alpha\, P(o_{t+1}|s', a_t) \sum_{s \in S} P(s'|s, a_t)\, b^t(s) \tag{11.2}$$

with α being a normalisation constant. Of course, the observations o^t are not known in advance, so the planning strategy must take into account the range of possible observations which might follow from the execution of a given action.

Using the fixed point of Bellman's equation [2], we know that the expected return for the optimal policy can be written with the following recursive form:

$$Q(b,a) = R(b,a) + \sum_{o \in O} P(o|b,a) max_{a'} Q(b', a') \tag{11.3}$$

where b' is the updated dialogue state following the execution of action a and the observation of o, as in Eq. (11.2). Furthermore, for notational convenience, we used $R(b,a) = \sum_{s \in S} R(s,a) b(s)$ and $P(o|b,a) = \sum_{s \in S} P(o|s,a) b(s)$.

Extracting an optimal policy for such POMDP is known to be a hard problem, with intractable exact solutions. Fortunately, many good approximations exist, often based on sampling a limited number of trajectories [13, 19]. The online planning algorithm we present in the next section makes use of such sampling techniques.

11.3 Approach

The general architecture of our approach revolves around a shared dialogue state, which is read and written asynchronously by a collection of modules (for dialogue understanding, interpretation, decision-making, generation, etc.). Each module can update the current state with new information. We have already described in our previous work the general dialogue system workflow [10] and will not repeat it here. Instead, we will concentrate on the dialogue manager module and in particular on its internal, domain-specific models.

As we will shortly describe, our planning algorithm relies on rich domain knowledge to speed up the action selection process. Our starting point is the observation that the probability and reward models used in dialogue management usually contain quite a lot of *internal structure* that can be readily exploited to yield more efficient algorithms. For instance, we can see in Fig. 11.1 that the probability $P(a_u = \mathsf{NULL}|i_u)$ does not actually depend on the specific value of i_u. Similarly, the reward $Q(a_m = \mathsf{AskRepeat}, i_u)$ does not depend on i_u either, since it is equal to -0.5 for all possible values of i_u. Generally speaking, we can often group the enumeration of possible values for the dependent variables into a set of distinct, mutually exclusive *partitions* that yield similar outcomes.

11.3.1 Probabilistic Rules

Probabilistic rules are a generic description formalism to capture such structure. They take the form of *if... then... else* cases mapping a list of *conditions* on input variables to specific *effects* on output variables. At runtime, these rules are then directly applied on the dialogue state, thereby extending the Bayesian Network with new nodes and conditional dependencies. This Bayesian Network can then be directly used for inference, e.g. to compute the marginal distribution of a particular variable or the utility of a given action. The probabilistic rules thus function as high-level *templates* for the incremental construction of a classical probabilistic model.

11.3.1.1 Probability Models

For probabilistic models of the form $P(X_1, \ldots X_n | Y_1, \ldots Y_m)$, a rule is formally expressed as an ordered list $\langle c_1, \ldots c_n \rangle$, where each case c_i is associated with a condition ϕ_i and a distribution over stochastic effects $\{(\psi_i^1, p_i^1), \ldots, (\psi_i^k, p_i^k)\}$, where ψ_i^j is a stochastic effect and probability $p_i^j = P(\psi_i^j | \phi_i)$, where $p_i^{1 \ldots m}$ satisfy the usual probability axioms. The rule reads as such:

if (ϕ_1) **then**

$$\{[P(\psi_1^1) = p_1^1], \ \ldots \ [P(\psi_1^k) = p_1^k]\}$$

elseif (ϕ_2) **then**

$$\{[P(\psi_2^1) = p_2^1], \ \ldots \ [P(\psi_2^l) = p_2^l]\}$$

...

elseif (ϕ_n) **then**

$$\{[P(\psi_n^1) = p_n^1], \ \ldots \ [P(\psi_n^m) = p_n^m]\}$$

A final **else** case is implicitly added to the bottom of the list and holds if no other condition applies. If not overridden, the default effect associated to this last case is void—i.e. it causes no changes to the distribution over the output variables.

The rule conditions ϕ_i are expressed as logical formulae grounded in the dependent variables. They can be arbitrarily complex formulae connected by conjunctive, disjunctive and negation operators. Formally speaking, a condition is therefore a function mapping state variable assignments to a boolean value. The conditions on the input variables can be seen as providing a compact partition of the state space to mitigate the dimensionality curse. Without this partitioning in alternative conditions, a rule ranging over m variables each of size n would need to enumerate n^m possible assignments. The partitioning with conditions reduces this number to p mutually exclusive partitions, where p is usually small.

The rule effects ψ_i^j are similarly defined: given a condition holding on a set of input variables, the associated effects define specific *value assignments* for the output variables. The effects can be limited to a single variable or range over several output variables. Each effect is assigned a probability, and several alternative stochastic effects can be defined for the same case. The effect probabilities are parameters which can be hand-coded or estimated from data .

As an illustrative example, consider the probability model $P(a_u|i_u)$ from Fig. 11.1. This model can be encoded with the rule r_1:

$$\textbf{Rule r}_\mathbf{1}: \textbf{ if } (i_u = \mathsf{GoTo(X)}) \textbf{ then}$$
$$\{[P(a_u = \mathsf{GoTo(X)}) = 0.9], \ldots\ [P(a_u = \mathsf{NULL}) = 0.1]\}$$

The rule simply expresses that the value of a_u should be identical to the one in i_u with probability 0.9 and equal to NULL with probability 0.1. The exact value for the argument X will be filled at runtime given the instantiation in i_u.

11.3.1.2 Reward Models

Rules can also be applied to describe reward models, with minor notational changes. Assume a reward model $R(X_1,\ldots X_n,A_1,\ldots A_m)$, where $X_1 \ldots X_n$ are random variables and $A_1,\ldots A_m$ decision variables. The rule is defined as an ordered list $\langle c_1,\ldots c_n\rangle$, where each case c_i has a condition ϕ_i ranging over the random variables $X_1,\ldots X_n$ and a set of reward values $\{(\psi_i^1, r_i^1),\ldots,(\psi_i^k, r_i^k)\}$, where ψ_i^j is an assignment of values for the decision nodes and $r_i^k = R(\psi_i^k, \phi_i)$. The rule reads similarly:

$$\textbf{if } (\phi_1) \textbf{ then}$$
$$\{R(\psi_1^1) = r_1^1,\ \ldots\ R(\psi_1^k) = r_1^k\}$$
$$\ldots$$
$$\textbf{elseif } (\phi_n) \textbf{ then}$$
$$\{R(\psi_n^1) = r_n^1,\ \ldots\ R(\psi_n^m) = r_n^m\}$$

By convention, the reward of an action value which is not explicitly expressed in the effect is assumed to be 0. The conditions ϕ_i are defined in the exact same

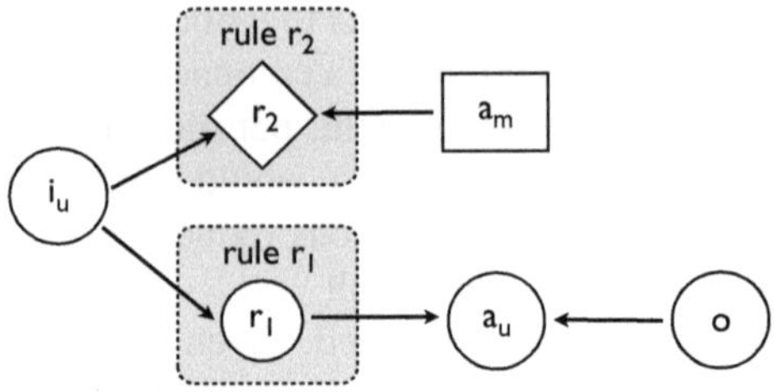

Fig. 11.2 Example of Bayesian Network generated by the instantiation of rules r_1 and r_2, producing a distribution similar to Fig. 11.1. The nodes r_1 and r_2 are rule nodes expressing which of the rule effects hold given the value of the i_u variable. The node r_1 represents the probabilistic effect on a_u given i_u, while r_2 encodes the utility of a_m given i_u. Finally, the distribution $P(a_u|r_1)$ is a simple deterministic distribution following the value assignment ascribed in the effect

way as for probability models—that is, as functions mapping assignments of values for $X_1 \ldots X_n$ to a boolean value. The effects express assignments for the decision variables $A_1, \ldots A_m$.

To illustrate the use of such rules to capture the model structure, consider the reward model $R(i_u, a_m)$ from Fig. 11.1, which can be encoded with the rule r_2:

$$\textbf{Rule r}_\mathbf{2}: \textbf{ if } (i_u = \mathsf{GoTo(X)}) \textbf{ then}$$
$$\{[R(a_m = \mathsf{GoTo(X)}) = 3], [R(a_m = \mathsf{AskRepeat}) = -0.5],$$
$$[R(a_m = \mathsf{GoTo(Y)} \wedge Y \neq X) = -5]\}$$

The general rule structure is provided by the system designer, while their parameters (probabilities or utilities) can be estimated from data, as shown in [11].

11.3.1.3 Rule Instantiation

At runtime, the rules are *instantiated* on the current dialogue state by creating new nodes and dependencies, thereby converting ("grounding") the rules into a standard probabilistic model which can then be straightforwardly used for inference based on standard algorithms such as variable elimination or importance sampling. The outlined procedure is an instance of *ground inference* [6], since the rule structure is grounded in a standard Bayesian Network (Fig. 11.2).

Practically, this instantiation is realised by creating one node for each rule:

- For probability models, the rule node is a chance node that is conditionally dependent on the input variables and expresses the effect which is likely to hold given their values. This rule node also has outward dependencies to the set of output variables it determines—for instance, a_u for rule r_1.
- For utility models, the rule node is a utility node dependent on the input variables and expresses the reward associated with specific action values given the inputs.

Algorithm 1: PLAN (**b**)

Require: **b**: Current dialogue state (expressed as a Bayesian Network)
Require: *nbTrajectories*: Number of trajectories to sample
Require: γ: Discount factor
Require: *horizon*: Planning horizon

$sequences \leftarrow \emptyset$
for $i = 0 \rightarrow nbTrajectories$ **do**
  $\mathbf{b}' \leftarrow$ copy of $\mathbf{b}$
  $a_m \leftarrow$ sample system action
  $trajectory \leftarrow [a_m]$
  $Q \leftarrow \Sigma_{s \in S} R(s, a_m)\mathbf{b}'(s)$
  for $t = 0 \rightarrow horizon$ **do**
    $a_u \leftarrow$ sample next user action given $\mathbf{b}'$
    $\mathbf{b}' \leftarrow BeliefUpdate(\mathbf{b}' \cup a_u)$
    $a_m \leftarrow$ sample system action
    $trajectory \leftarrow trajectory \cup [a_m]$
    $Q \leftarrow Q + \gamma^t \Sigma_{s \in S} R(s, a_m)\mathbf{b}'(s)$
  end for
  if $trajectory \in sequences$ **then**
    $sequences[trajectory] \leftarrow sequences[trajectory] \cup [Q]$
  else
    $sequences[trajectory] \leftarrow [Q]$
  end if
end for
$sequence^* \leftarrow arg\,max_i \frac{\Sigma_{Q \in sequences[i]} Q}{|sequences[i]|}$
return $sequence^*$

11.3.2 Planning Algorithm

The pseudocode of the planning algorithm we are currently developing is given in Algorithm 1. It works by sampling a set of *trajectories* starting from the current dialogue state until a given horizon limit is reached. One benefit of this sampling-based strategy is the ability to work in anytime mode, which means that at any point in time, the procedure is able to deliver a solution. The quality of the solution will of course depend on the number of trajectories that are sampled—more trajectories leading to a more accurate plan, but at a higher computational cost. The anytime nature of the algorithm is important since the planner operates online and must thus satisfy real-time constraints.

The trajectories are sampled by repeatedly selecting system actions and subsequent user actions until the horizon limit is reached. The system actions can be sampled uniformly or following heuristic distributions if some are available to guide the search towards high-utility regions [7]. For the user actions, the sampling relies on a (also rule-structured) user action model $P(a_u|s, a_m)$ that predicts the next user action given the current state and last system action, modulo some standard noise inserted into the distribution to simulate speech recognition errors.

In order to navigate through the trajectories, the algorithm needs to update its dialogue state after the selection of a user action. The belief update algorithm is described in detail in [10]. For each trajectory, the algorithm records the rewards accumulated after each action. Once enough trajectories have been sampled, the algorithm computes the average expected return for each possible sequence and selects the sequence with the highest score. This sequence will correspond to the optimal plan, and the only remaining step for the dialogue system is to execute the first action of this plan.

The originality of our approach lies in the use of probabilistic rules for updating the dialogue state and determining the possible actions available at that state. More specifically, the reward rules will determine a set of actions which can be performed if their conditions hold and which reward is expected from their execution. The major benefit is that instead of having to search through the whole space of possible actions, the planning algorithm can be limited to consider only a subset of *relevant* actions. The reward rules can therefore be seen as providing a high-level filter on the action space [9], and we expect them to significantly speed up the search for the optimal action.

We are currently in the process of implementing the details of this online planner and integrating it in the dialogue system architecture described in [10]. Unsurprisingly, the main bottleneck we currently encounter remains the runtime performance, which doesn't yet scale to real-time requirements for more than trivial domains. We hope to solve these tractability issues soon and be able to report empirical results on the performance of this online planner for a human-robot interaction domain similar to the one described in [11].

11.4 Related Work

Online planning has a long history in dialogue systems [1, 20], but it has usually been confined to classical planning, relying on a clear-cut set of goal states instead of relative utilities and with no account of observation uncertainty. Decision-theoretic approaches on the other hand have mostly focused on offline optimisations of MDPs [12, 16] or POMDPs [23] via reinforcement learning.

The field of POMDP planning has recently witnessed a surge of interest for online methods [17, 19]. As mentioned in the introduction, online planning offers clear advantage in terms of scalability to large domains and adaptivity to dynamic changes in the environment or task models. Moreover, it can be combined with offline methods, the precomputed policy being in this case employed as a heuristic approximation to guide the online search for the optimal action. Another related development is the use of online planning for model-based Bayesian reinforcement learning [18]. These approaches rely on the inclusion of model uncertainty as part of the state space. Due to the increased size and continuous nature of the resulting state space, direct policy optimisation is not feasible, and online planning based on sampling methods is the only viable alternative.

Online reinforcement learning is another alternative for adapting the system behaviour to its context. Most work so far has concentrated on model-free learning algorithms such as Gaussian Process SARSA or Kalman Temporal Differences [3, 5]. Model-free reinforcement learning seeks to directly derive an optimal policy through interaction with the environment. The approach we take in this work is closer to (Bayesian) model-based approaches to reinforcement learning [14, 18], where the learner first seeks to estimate a model of the environment and then uses this model to plan an optimal behaviour. The model estimation is often done by incorporating model uncertainty into the state space. Model-based approaches are able to directly incorporate prior knowledge and constraints into their models, which make them attractive for dialogue management tasks.

The use of high-level representations such as probabilistic rules in planning has been explored in previous work [7, 24]. The common intuition behind most of these approaches is that capturing the inner structure of the domain via high-level representations can yield models which are:

- Easier to learn and generalise better to unseen data, since they depend on a greatly reduced number of parameters (as shown in [11])
- More efficient to use, since the model structure can be exploited by the inference algorithms (as we have tried to show in this paper)

11.5 Conclusions

We have presented in this paper a general approach to online planning for dialogue management, based on the use of high-level probabilistic rules. These rules enable the system designer to provide important domain knowledge which can help filtering the space of possible actions to consider at a given time point.

Our underlying hypothesis is that online or hybrid planning can be beneficial for dialogue management, especially for open-ended domains such as human-robot interaction or tutoring systems. The environment's dynamics of these domains is rarely static and is likely to change over time due to shifting user preferences or contextual factors. Online planning strategies are able to naturally cope with such changes without having to recompile policies.

The major bottleneck for online planning remains however the runtime performance. We still need to find ways to make online planning tractable for real domains. In this respect, we would like to investigate the use of precomputed policies as heuristic approximation to guide the lookahead search.

Another interesting venue for future work is the combination of online planning with reinforcement learning. As described in [18], online planning can be ideally combined with Bayesian approaches to reinforcement learning, where uncertainty in the model parameters is directly captured in terms of additional variables in the state space. Given the high-level representations provided by the probabilistic rules, the rule parameters could potentially be estimated from limited amounts of (raw) interaction data and be continuously refined as more interaction experience becomes available, either from real users or from simulation.

References

1. Allen, J., Byron, D., Dzikovska, M., Ferguson, G., Galescu, L., Stent, A.: An architecture for a generic dialogue shell. Nat. Lang. Eng. **6**, 213–228 (2000)
2. Bellman, R.: Dynamic Programming. Princeton University Press, Princeton (1957)
3. Daubigney, L., Geist, M., Pietquin, O.: Off-policy learning in large-scale pomdp-based dialogue systems. In: 2012 IEEE International Conference on Acoustics, Speech and Signal Processing (ICASSP), pp. 4989–4992 (2012)
4. Frampton, M., Lemon, O.: Recent research advances in reinforcement learning in spoken dialogue systems. Knowl. Eng. Rev. **24**(4), 375–408 (2009)
5. Gasic, M., Jurcicek, F., Thomson, B., Yu, K., Young, S.: On-line policy optimisation of spoken dialogue systems via live interaction with human subjects. In: 2011 IEEE Workshop on Automatic Speech Recognition and Understanding (ASRU), pp. 312–317 (2011)
6. Getoor, L., Taskar, B.: Introduction to Statistical Relational Learning. The MIT Press, Cambridge (2007)
7. Lang, T., Toussaint, M.: Planning with noisy probabilistic relational rules. J. Artif. Intell. Res. **39**, 1–49 (2010)
8. Lemon, O., Pietquin, O.: Machine learning for spoken dialogue systems. In: Proceedings of the 10th European Conference on Speech Communication and Technologies (Interspeech'07), pp. 2685–2688 (2007)
9. Lison, P.: Towards relational POMDPs for adaptive dialogue management. In: Proceeding of the Student Research Workshop of the 48th Annual Meeting of the Association for Computational Linguistics. Association for Computational Linguistics (2010)
10. Lison, P.: Declarative design of spoken dialogue systems with probabilistic rules. In: Proceedings of the 16th Workshop on the Semantics and Pragmatics of Dialogue (2012)
11. Lison, P.: Probabilistic dialogue models with prior domain knowledge. In: Proceedings of the SIGDIAL 2012 Conference, pp. 179–188. Seoul, South Korea (2012)
12. Pietquin, O.: Optimising spoken dialogue strategies within the reinforcement learning paradigm. In: Cornelius Weber, M.E., Mayer, N.M. (eds.) Reinforcement Learning, Theory and Applications, pp. 239–256. I-Tech Education and Publishing, Vienna (2008)
13. Pineau, J., Gordon, G., Thrun, S.: Point-based value iteration: An anytime algorithm for POMDPs. In: International Joint Conference on Artificial Intelligence (IJCAI), pp. 1025–1032 (2003)
14. Poupart, P., Vlassis, N.A.: Model-based bayesian reinforcement learning in partially observable domains. In: International Symposium on Artificial Intelligence and Mathematics (ISAIM) (2008)
15. Purver, M.: The theory and use of clarification requests in dialogue. Ph.D. Thesis (2004)
16. Rieser, V., Lemon, O.: Learning human multimodal dialogue strategies. Nat. Lang. Eng. **16**, 3–23 (2010)
17. Ross, S., Pineau, J., Paquet, S., Chaib-Draa, B.: Online planning algorithms for POMDPs. J. Artif. Intell. Res. **32**, 663–704 (2008)
18. Ross, S., Pineau, J., Chaib-draa, B., Kreitmann, P.: A Bayesian approach for learning and planning in partially observable markov decision processes. J. Mach. Learn. Res. **12**, 1729–1770 (2011)
19. Silver, D., Veness, J.: Monte-carlo planning in large POMDPs. In: Lafferty, J., Williams, C.K.I., Shawe-Taylor, J., Zemel, R., Culotta, A. (eds.) Advances in Neural Information Processing Systems, vol. 23, pp. 2164–2172 (2010)

20. Steedman, M., Petrick, R.P.A.: Planning dialog actions. In: Proceedings of the 8th SIGDIAL Meeting on Discourse and Dialogue, pp. 265–272. Antwerp, Belgium (2007)
21. Thomson, V., Young, S.: Bayesian update of dialogue state: A POMDP framework for spoken dialogue systems. Comput. Speech Lang. **24**, 562–588 (2010)
22. Williams, J.: A case study of applying decision theory in the real world: POMDPs and spoken dialog systems. In: Sucar, L., Morales, E., Hoey, J. (eds.) Decision Theory Models for Applications in Artificial Intelligence: Concepts and Solutions, pp. 315–342. IGI Global, Pennsylvania (2012)
23. Young, S., Gašić, M., Keizer, S., Mairesse, F., Schatzmann, J., Thomson, B., Yu, K.: The hidden information state model: A practical framework for POMDP-based spoken dialogue management. Comput. Speech Lang. **24**, 150–174 (2010)
24. Zettlemoyer, L.S., Pasula, H.M., Kaelblin, L.P.: Learning planning rules in noisy stochastic worlds. In: Proceedings of the 20th AAAI Conference on Artificial Intelligence, pp. 911–918. AAAI Press (2005)

Chapter 12
A Two-Step Approach for Efficient Domain Selection in Multi-Domain Dialog Systems

Injae Lee, Seokhwan Kim, Kyungduk Kim, Donghyeon Lee, Junhwi Choi, Seonghan Ryu, and Gary Geunbae Lee

Abstract This paper discusses a domain selection method for multi-domain dialog systems to generate the most appropriate system utterance in response to a user utterance. We present a two-step approach for efficient domain selection. In our proposed approach, the domain candidates are listed in descending order of scores and then each domain is verified by content-based filtering. When we applied our method, the accuracy increased and the time cost decreased compared to baseline methods.

12.1 Introduction

Recently, spoken dialog systems (SDSs) have been applied in various tasks including city tour guiding, telematics, home networking, and entertaining. Although most previous work has been focused on the systems dealing with a single task, the communications are often not bounded in a single task in the real world as people require increasingly more functionalities in the system. Depending on the dialog flows and contextual environment, the communication can span more than one task. Thus, the systems should be able to handle multiple tasks rather than only a single task.

The *distributed architecture* is a well-known approach to the multi-task dialog systems [1–3]. In distributed architectures, a dialog system employs a separated

I. Lee • K. Kim • D. Lee • J. Choi • S. Ryu • G.G. Lee (✉)
Department of Computer Science and Engineering, Pohang University of Science and Technology, Pohang, South Korea
e-mail: lij1984@postech.ac.kr; getta@postech.ac.kr; semko@postech.ac.kr; chasunee@postech.ac.kr; ryush@postech.ac.kr; gblee@postech.ac.kr

S. Kim
Human Language Technology Department, Institute for Infocomm Research, Singapore
e-mail: kims@i2r.a-star.edu.sg

J. Mariani et al. (eds.), *Natural Interaction with Robots, Knowbots and Smartphones: Putting Spoken Dialog Systems into Practice*, DOI 10.1007/978-1-4614-8280-2_12,

dialog manager for each task. Each task is a unit of service that the users want the system to provide. A single domain dialog system processes a set of tasks related to a specific topic, accessing a common database to satisfy the user's requirements. In contrast, a multi-domain dialog system considers more than one domain through a single interface.

Selecting the most appropriate domain is one of the key technologies to develop a multi-domain dialog system.

Domain selection can be performed in two ways: preselection or post-selection. A preselection approach determines the most appropriate domains by considering the features extracted from the input utterance and then executes the dialog manager for the selected domain [4]. Although this approach is efficient in execution time, it has a limitation in incorporating domain-specific knowledge to improve the selection performances. The other way of domain selection is operated based on the results of dialog managers for all target domains [5]. This post-selection approach has the advantage in performance by considering rich domain-specific features. However, executing all the dialog managers on the unrelated domains can be a waste of time especially when the number of domains increases.

In this paper, we propose a two-step domain selection approach for multi-domain dialog systems. Although our approach is basically based on the post-selection framework utilizing domain-specific features from multiple dialog managers, both accuracy and efficiency have been improved with domain filtering constrained by the results of preselection.

12.2 Method

The goal of our proposed method (Fig. 12.1) is to select the domain expert, which can generate the system response to the user intention in multi-domain dialog systems. To obtain the features for selecting the appropriate domain, each user utterance is analyzed by a series of domain-independent analyzers including linguistic analysis, generic spoken language understanding (SLU) analysis, and keyword analysis. Based on the analyzed results, domain selection is performed by two-step approaches: domain ordering and domain filtering. In domain-ordering step, the domain candidates are listed in descending order of scores computed by a preselection model. Then, content-based domain filtering is performed for each domain in order to determine the final domain.

12.2.1 Input Sentence Analysis

Each input sentence should be analyzed to obtain a set of features that are used for domain selection. First, linguistic analysis extracts uni-gram and bi-gram identity features from the speech recognition results. Generic SLU analysis is to extract dialog act (DA) features from the user utterance regardless of the domain. The DA

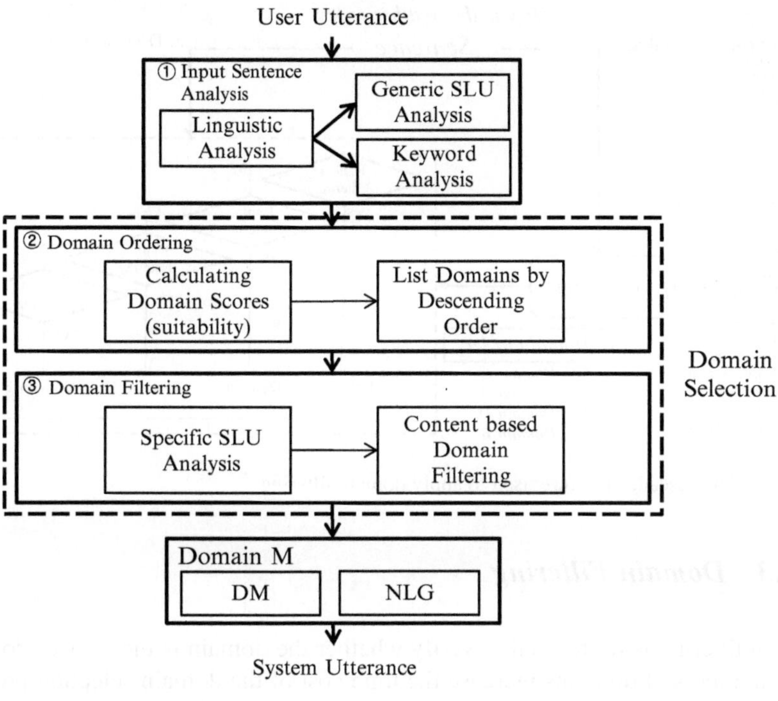

Fig. 12.1 The overall architecture

of an utterance is a domain-independent label of the speaker's intention [4]. The role of keyword analysis is to extract keywords which are the most informative words for domain selection in a sentence. The importance of each word is calculated based on the TF*IDF algorithm, which weighs the words for each domain in terms of term frequency and inversed document frequency. We selected the words with the n-best TF*IDF values as the n-best keywords.

12.2.2 Domain Ordering

We calculated a *domain score* for each domain based on the extracted information. The domain score is the probability that the input utterance is suitable for each domain. We used the maximum entropy (ME) classifier as a preselection model to compute the domain score. The model was trained using the following features: word, word-bigram, DA, and keyword features. The ordering is organized by listing the domains in descending order according to the domain scores.

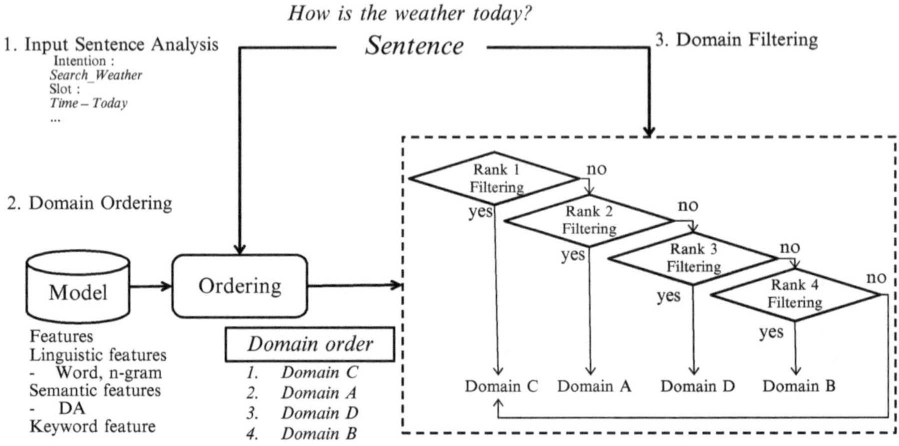

Fig. 12.2 The cascade structure used to apply domain filtering

12.2.3 Domain Filtering

Domain filtering is performed to verify whether the domain is the answer domain. Large numbers of domains increase the time cost of the domain selection process. We constructed a cascading structure to efficiently apply the domain filtering. The cascading structure invokes an iterative procedure, which first applies the domain filtering of the highest rank (Fig. 12.2). The domain filtering of each domain either confirms a domain selection or invokes an additional iteration for the next-ranked domain. The iteration continues until a domain is selected or until every candidate domain is tried. This structure decreases the computation cost because a fewer number of domain filtering is invoked.

When the system relies on a multi-domain mixed-dialog corpus, there are two problems in building a multi-domain dialog system; the first problem involves building a natural multi-domain mixed-dialog corpus for training the system which is time consuming and labor intensive, and the second problem involves adding and removing specific domains once the system is stabilized. This problem arises because of diverse dialog flows according to different users and/or environments. Introducing domain-shifting results in a high level of diversity that is hardly covered by a given fixed multi-domain mixed-dialog corpus.

To alleviate these difficulties, instead of using multi-domain mixed-dialog corpus, we applied domain filtering to each domain. Each domain filtering is constructed by each domain corpus and the domain knowledge. Domain-specific features, such as historical features, are considered in the domain filtering, avoiding the cost of building a multi-domain dialog corpus.

12.2.3.1 Domain-Specific SLU Analysis

The domain-specific user intention must be identified by analyzing the user utterance. For this purpose, domain-specific SLU extracts main action (MA) and named entities (NEs) slots for given domain. The domain-specific SLU results are used as a feature of the domain filtering. The MA indicates the user goal of an utterance. NEs are employed in the content-based domain-filtering phase.

12.2.3.2 Content-Based Domain Filtering

The fundamental idea of this phase is that the system selects the answer output domain when it includes content retrieval results. The domain's nonempty content retrieval result indicates that the domain expert of the domain has contents to suggest to the user to complete the task corresponding to the user intention. The NE extraction process determines whether a phrase is semantically related to a domain. A nonempty content retrieval result implies that the phrase is a valid NE in the domain. This result provides strong evidence that the input belongs to the domain; thus, it is reasonable to directly select the domain. We used extracted NE slots to retrieve the contents from the database.

NE slots are accumulated to retrieve content from the database even when NE slots are not extracted from the current user utterance. The accumulated NEs consider the dialog history. The final domain is chosen when a nonempty content retrieval result is returned. If the database search does not provide any content retrieval results from the accumulated slots, the current slots are used and the contents are retrieved again.

When none of the domain filtering can confirm the user by filtering, the top-ranked domain, which has the highest domain score, is chosen.

12.3 Experiments

12.3.1 Experimental Setup

We collected four task-oriented information-seeking dialog corpora: a navigation domain (NAVI) is a direction guide domain; a Pohang Institute of Intelligent Robotics (PIRO) domain is a building guide domain; a TV domain is a TV-controlling and TV-content-seeking domain; and a WEATHER domain is a weather-information-seeking domain (Table 12.1). We used each dialog corpus separately for the training data and constructed an arbitrary multi-domain mixed corpus for the test data.

Based on these corpora, we implemented three domain selection modules: one is based on our proposed two-step approach and the others are baseline systems

Table 12.1 Multi-domain corpora

	Training		Test	
Domains	Dialog	Utterance	Dialog	Utterance
Navi	122	525	14	61
PIRO	383	1,581	43	183
TV guide	123	500	14	62
Weather	89	455	10	48

Table 12.2 Domain selection accuracy and time cost

Methods	Accuracy (%)	Number of visiting SLU
Baseline (preselection)	328/354 92.65	354
Baseline (post-selection)	330/354 93.22	1,614
Proposed method	339/354 95.76	693

Table 12.3 Accuracy of distribution of answer domains in ordering

Selected position	Answer distribution in ordering	Number of correct answer in proposed method
1-best	328	321
2-best	22	14
3-best	2	2
4-best	2	2
Total	354	339

for comparisons. The first baseline uses only the preselection classifier without any domain-specific features, and the other baseline determines the domain by post-selection after extracting the domain-specific features for all domains.

12.3.2 Experimental Results

Table 12.2 shows the domain selection accuracy with three approaches. When our proposed two-step approach was considered, the accuracy increased by 3.11% from the first baseline with preselection. Since preselection approach considers 1-best domain only, it is impossible to rescue the lower-ranked domains. As shown in Table 12.3, 7.35% of evaluation datasets have to be assigned to the domain which is not ranked as the best by preselection classifier. Even though all the answer domains ordered in the 1-best position are not selected as the final domain, the answers ordered in the lower-ranked domains are finally selected as the final domain; the total accuracy increases. The performance improvements compared to the preselection baseline were achieved by considering these low-ranked domains in our proposed method.

Moreover, our proposed approach also achieved better accuracy than post-selection baseline. This result shows that the domain ordering with preselection constraints helped to improve the performance of domain selection in comparison to the post-selection baseline.

To demonstrate the merits of our proposed method in time cost, we counted the total numbers of visiting for domain-specific SLU modules. While preselection baseline executes domain-specific SLU modules once, post-selection baseline visits domain-specific SLU modules for the same number of times as the number of domains. In our proposed method, the specific SLU was visited by a maximum number of times as the domain number and a minimum of one time.

For 354 test utterances, the specific SLU analysis was run 693 times in our method (6.31 s) and 1,416 times in post-selection approach (13.48 s). Combining domain filtering with the cascade structure decreased the number of visiting classifiers by 51% compared to post-selection approach. Because most of the answer domains are arranged for high ranking in the domain-ordering phase, the domain filtering of the correct answer domain was considered relatively early.

12.4 Conclusions

This paper presented a two-step approach for efficient domain selection in multi-domain dialog system. Our approach listed the domain candidates in descending order of scores computed by a preselection model and then performed the content-based domain filtering to select the final domain. The feasibility of the method was demonstrated by the experimental results that our proposed approach achieved better performance and higher efficiency than baseline approaches.

In this work, we only considered domain-specific features extracted from SLU modules. For future work, we plan to incorporate richer features from whole dialog manager for each target domain.

Acknowledgments This research was supported by the MKE (Ministry of Knowledge Economy, Korea), under the ITRC (Information Technology Research Center) support programs supervised by the NIPA (National IT Industry Promotion Agency) NIPA-2012-(H0301-12-3002) and NIPA-2012-(H0301-12-3001).

References

1. Lin, B., Wang, H., Lee, L.: A distributed architecture for cooperative spoken dialogue agents with coherent dialogue state and history. In: ASRU-99, Keystone (1999)
2. Pakucs, B.: Towards dynamic multi-domain dialogue processing. In: Proceedings of Eurospeech, Geneva (2003)
3. Komatani, K., Kanda, N., Nakano, M., Nakadai, K., Tsujino, H., Ogata, T., Okuno, H.G.: Multi-domain spoken dialogue system with extensibility and robustness against speech recognition errors. In: Proceedings of the SIGDial, Sydney (2006)
4. Lee, C., Jung, S., Kim, S., Lee, G.G.: Example-based dialog modeling for practical multi-domain dialog system. Speech Commun. **51**(5), 466–484 (2009)
5. Nakano, M., Sato, S., Komatani, K., Matsutama, K., Funakoshi, K., Okuno, H.G.: A two-stage domain selection framework for extensible multi-domain spoken dialogue systems. In: SIGDial, Portland (2011)

Part IV
Human-Robot Interaction

Chapter 13
From Informative Cooperative Dialogues to Long-Term Social Relation with a Robot

Axel Buendia and Laurence Devillers

Abstract A lot of progress have been made in the domain of human-machine dialogue, but it is still a real challenge and, most often, only informative cooperative kind of dialogues are explored. This paper tries to explore the ability of a robot to create and maintain a long-term social relationship through more advanced dialogue techniques. We expose the social (Goffman), psychological (Scherer) and neural (Mountcastle) theories used to accomplish such kind of complex social interactions. From these theories, we build a consistent model, computationally efficient to create a robot that can understand the concept of lying and have compassion: a robotic social companion.

13.1 Introduction

Dialogue cooperativity is crucial to task-oriented spoken human-machine dialogue [4]. Spoken language dialogue systems have always relied on cooperative users, so it is mandatory that the system's dialogue must be cooperative as well. According to Grice [22], any interactive exchange between a speaker and an addressee supposes a minimum of agreement and cooperative effort. Grice's rules are important; people follow them when they want to be explicitly informative and direct. But they also often break the rules in social interaction. Social interactions require certain abilities like social understanding and the mind theory. Social understanding enables

---

A. Buendia (✉)
Spir.Ops, 8 passage de la bonne graine, CNAM CEDRIC, 292 rue St Martin, 75003 Paris, France
e-mail: axel.buendia@spirops.com

L. Devillers
Department of Human-Machine Interaction, LIMSI-CNRS, University Paris-Sorbonne 4, Orsay cedex, France
e-mail: devil@limsi.fr; laurence.devillers@paris-sorbonne.fr

J. Mariani et al. (eds.), *Natural Interaction with Robots, Knowbots and Smartphones: Putting Spoken Dialog Systems into Practice*, DOI 10.1007/978-1-4614-8280-2_13,

planning ahead and dealing with new circumstances. While the mind theory makes it possible to anticipate the mental state of another person. The feeling given to people in daily interaction is exposed to ruptures. On a sociological point of view, this idea is really important. The disagreement among communicative norms between people and the violation of expectation are two important concepts: most people expect friends, even strangers with truthful relations without deceits.

The development of an artificial creature (a robot) that interacts with human beings on a daily basis poses several important intrinsically connected challenges to the perception, dialogue modelling and artificial intelligence. This artificial creature should have the cognitive abilities, the sensing components and the dialoguing capabilities to enable it to develop a social behaviour and communicate with human beings, with the appropriate level of abstraction based on the context. Such cognitive abilities imply the development of new representations and new AI architecture. The development of abilities such as anticipation, expectation, memorization and continuous training process are also necessary to create relationship with the robot. The necessity to develop a long-term human-robot interaction theory is essential. It should be inspired by theories and models in sociology, in psychology and also neurosciences. This theory should also be inspired by empirical groundings for really improving the domain.

Using a large variety of qualitative methods, the sociologist Erving Goffman developed in his research several theories about the elements of social interaction:

- The "rites of interaction" [20] show a Durkheim perspective that considers everyday life interactions as a miniature ceremony. In this ceremony the sacred nature of the society takes refuge within actors with mutual respect. In this perspective, the society is nested in each actor in the form of collective representations. The notion of "face" as the self-image corresponding to approved social attributes. Our face is emotionally invested and can be lost, maintained or enhanced during interactions.
- Our experiences are recorded in a laminated universe and made multiple realities socially centred: each one of us has his own perspectives and has his own "frames" [21] which allow us to interpret the situations and its context.
- The same individual can have different simultaneous "social roles" during an interaction. Goffman empirical investigations give evidence that it is possible to pass from one facet to another during the same interaction. He shows that the only condition is that this passage does not question the central role of the individual in the interaction. Goffman also defined the notion of "role distance" [19] as a relationship between real life role experiencing and the real role that has important implications.
- In most interactions, a behaviour of tolerance (called by Goffman "working acceptance") between the partners exists allowing the interaction to be maintained even when standards are broken.

Concepts such as the face, the frames, the social roles and the working acceptance seem really important for a better model of the social interactions. But it is not straightforward to derive computational models from these theories.

Social behaviour and relationship are connected to the expression and regulation of emotions, that are a bridge between the individual and their connections to others. Emotions have not only a physiological, cognitive and psychological basis, but also are influenced and constrained by cultural norms and beliefs. In literature, a lot of different models have appeared [28], from complex ones like in psychology the component process model (CPM) based on appraisal theories [33] to more simple ones like [6]. Emotions for social interactions have to fulfil two roles:

- Influencing decision processes: emotions change our decision process, favouring certain types of behaviours (anger tends to encourage violent behaviours). Emotions can also modify the decision process in a more subtle way, just by altering the usual way an action is performed (facial micro expressions, quicker moves, etc.).
- Showing a desired or non-completely desired state of mind: according to Goffman, people tend to play a role during social interactions. This role must be maintained to avoid ruptures. It is then really important to try to show coherent emotional state.

Most of the time those models are separated. In [6], they use a unique model to describe both aspects, but this model lacks the diversity needed to create various social interactions, like during a game (bluff) or a questioning (losing control under the pressure). The question of control remains: how the internal emotional state influences the external perceived state and what is the influence of the intentions? Emotional state is by nature influenced by the environment, the perceived/anticipated events, etc. It is very volatile. Designing a dynamic model is as important as listing the dimensions of the emotional state. Real-life emotions are often a mixture of different affective states [16]. In most publications, the dynamic model is the same across different emotional dimensions and it is parametrized by more static attributes, sometimes referred to as the personality [10]. In [14] the emotional and interactional profile of the user is dynamically built during the dialogue to drive the answering strategy of a robot. The question of mutual influence between the different axes is an interesting one [12]. Most of the models try to create an independent axis to avoid this problem. But when we mix an internal model with an external one, we have to specify those interdependent relations.

In order for a human-robot interaction to proceed in a natural manner, robots must adhere to some human social norms. Giving these skills to robots is going to depend on many purposes. It is necessary to understand the main roles and uses of the robot in the future. Creating adaptive robots interacting with human beings through language and understanding of the social and emotional dimensions will allow to imagine robot uses that vary from entertainment to therapeutic applications (for an elderly living alone at home, more or less autonomous, even depressives or with the Alzheimer's disease). What can we say on this long-term social and emotional relationships with a robot? In short, the relationships that we think we might have with robots tend to be deceptive and illusory. Without the ability to understand complex social interaction with human beings and without the capacity to detect complex behaviour such as deception, robots would be unable to deal with long-term relationship. Several projects aim at conceiving a robot with emotions and language in order to support dependent people at home. The aim of those projects is also to explore the acceptability of a robot in ecological situations for long-term relationship. The French ROMEO project [14, 17] aims to design a social humanoid robot able to assist elderly and disabled persons at home in everyday life activities and able also to play games, for example, with the grandchildren of the user. The ANR ROBADOM project [9] also aims at designing a robot in order to support the elderly with mild cognitive impairment, living alone at home and suffering or not from depression. First experiments are carried out for exploring the acceptability of a robot such as Robovie [24] that can mimic many human abilities like compassion, lying, exerting authority, claiming responsibility and even making jokes.

The goal seems clear and the inspiration from various social interaction theorists sounds valid and promising. But the gap is huge between social theories and designing a control model. To overcome this gap, we try to get inspiration from theories in other domains. We are also inspired by psychological models such as the CPM of affective states proposed by Scherer [34] for the appraisal of emotional events. Neuroscience brings new approaches to how the neocortical brain is architectured. Mountcastle [29] describes a two-way approach of the neocortical columns already discovered: a bottom up to extract semantic and symbols from signals and a top down to create anticipation and expectation. This architecture opens interesting ways to create new architectures that could be able to cop up with the complexity of social interactions.

All these developments suggest that the evolution of the human-robot interactions is moving in a positive direction, but until now none of these systems are autonomous nor able to work in a real context. In this article, we propose some research perspectives based on ideas and theories from psychology, sociology and neuroscience to propose a long-term human-robot interaction system. Section 13.2 discusses about robots and perceptions, while Sect. 13.3 focuses on the robot understanding. Section 13.4 presents an overview of the human abilities (compassion, joking, lying) applied to a robot. Finally, article conclusion with some discussions regarding the next stages in the evolution of the human-robot interaction systems.

13.2 Robots and Perceptions

Sensors gather raw information to feed the decision process. Their role is fundamental. A lot of research is focusing on the exploitation of those raw data (object/face tracking/recognition, voice recognition/understanding, etc.). Those researches produce a lot of independent modules. Each module produces an output, sometimes coupled with a reliability measure.

13.2.1 Towards a Bio-inspired Scheme to Build an Event Representation

Each sensor produces raw data. To analyse these data, the system uses processes that will produce new outputs. These processes represent the main components (atoms) of the perceptive architecture. A new atomic process can then use the previous computed outputs. Those dependencies create a hierarchy of the processes. This hierarchy is representative of the refined quality of information. Each process may have one or several roles from:

- Information refiner: by refining the used information, the process is creating semantically richer information. For example, from camera buffer, the process outputs the identification of the individual in front of the robot.
- Information aggregator: by aggregating several information, the process is building a new more complex information. For example, processes responsible for the creation of events.
- Information consolidator: by comparing several information and their respective reliabilities, the process improves the reliability and veracity of existing information. For example, by comparing the voice recognition outputs and the face recognition outputs, the process guesses the correct identification of the individual talking to the robot. It is important to note that the consolidator can also decrease the veracity of an information.

Those roles imply that each process, when outputting a new information, has to propose all its hypothesis with the corresponding global reliability/veracity and not just the best one. By providing those meta information (reliability, veracity), the architecture is able to optimize all the information. These processes are the atoms of the perception structure. The lattices of hypothesis formed by all the perception atoms construct a kind of column that could be compared to the bio-inspired neocortical columns described by [29]. The described functions are interesting and the prediction functionalities described in this paper are really important to implement in this architecture and are described in Sect. 13.3.2. This column metaphor presents several advantages:

- Modularity: this structure allows the assembly of several sources of data without any knowledge of the processes that generated them. Only the interface (format of the outputs) should remain.
- Distributiveness: each module could be implemented by different companies/research facilities.
- Bio-inspired: studies tend to converge towards the notion of semantic columns created through the neocortical parts of the brain. These columns are interconnected.

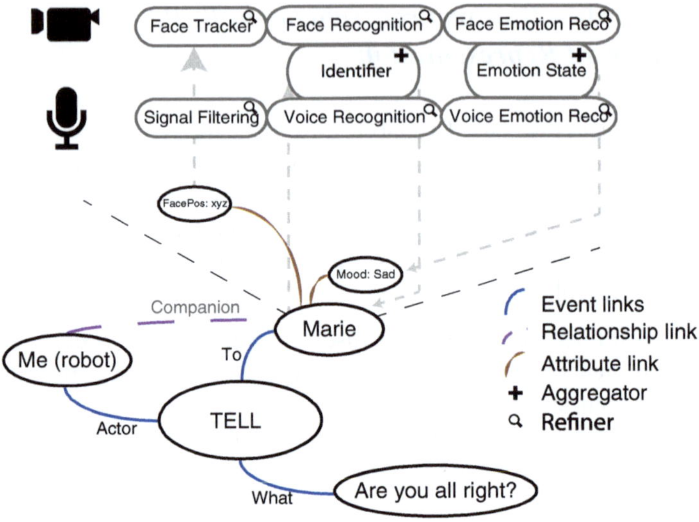

Fig. 13.1 An event representation with the bio-inspired perceptive architecture

In this architecture, there is an important raw-data provider: the memories manager. Its responsibility is to keep track of all the information. It could be used by each atom to create continuous time optimizations or reinforce reliability. How this module works is more detailed in Sect. 13.3.4.

It becomes possible to build complex data like event description (The robot is talking to Marie saying, "Are you all right?") (see Fig. 13.1). To build such information, aggregators are used to assemble the different pieces of the event. The information contained in events is very refined and corresponds to rich semantic information. Most of the time, the kernel of an event is the action that originated it (including acts of language). All the elements that describe the event are like electron orbiting around this action kernel. It is important to notice that events are mainly links between already existing memorized nodes. Highly probable electrons are:

- Actor: the entity that has performed the action.
- Target: the object of the action. It could even be the same entity that performed the action.

The kernel and each electron refer to semantic concepts, and as with a zoom, they can be specified with specific attributes. For example, the action of talk can be specified with mood (modulating the way of the talk like tone, etc.), familiarity (evaluating the used familiarity for the talk), etc. Those attributes are most likely application dependent, just as are the list of possible actions and electrons. These events are highly semantic and can be referred as the atomic pieces of the semantic description. They hold usually the action-oriented facts, beliefs and reasoning. Those semantically rich events play an important role in the creation of a social bind.

13.2.2 Social and Affective Signals Perception

One human convention is the use of expressive social and affective signals as an integral part of social interactions such as "You looks sad" or "I need to be comforted". People's long-term relationships depend heavily on shared emotional experiences [18]. There are many cases where the voice is not the only clue available to identify emotion and social dimensions. The study of how the paralinguistic and multi-modal clues can be perceived and how they can impact the human-robot interaction, through developing an underlying long-term theory of vocal and multi-modal interaction, is a real challenge.

Real-life emotions are often mixtures of simultaneous affective states: emotion, mood and affective disposition. There is no clear typology of the different emotion mixtures. In a previous study on human-human dialogues collected in a medical emergency call centre [15], the most frequent conflictual mixtures (i.e. with positive/negative valence) involved relief/anxiety and positive/stress. Such conflicting affective states are often observed near the end of the dialogue, when the person knows that help is coming, but still remains fearful about his condition. Two types of emotions are also described in [11]: the speaker emotion based on her/his internal emotional state named cause type and the effect that she/he would be likely to have on the listener, named effect type. When we perceive a mixture of emotions, are they linked to cause type or effect type?

Most researchers only focus on finding solutions for specific traits (emotion, personality, etc.) detection rather than considering several simultaneous levels of features: from long-term features to short-term features. Schuller et al. in [35] focus on the distinction of these different features:

- Long-term features are, for example, affective dispositions towards the others or personality features [10] such as in [8].
- Medium-term features are more or less temporary states like the health state and also structural social and interactional signals such as role in interaction or positive/negative attitude [38].
- Finally, short-term features correspond to the mode of interaction: emotion or mixture of emotion-related states [12, 15] such as stress, interest, confidence, uncertainty, deception, politeness, frustration, and sarcasm.

An emotional episode can be brief, sometimes lasting only a few seconds. If it lasts for some hours then it is a medium-term feature such as mood and not an emotion. Actually, very few systems use in the same time several levels of these features such as emotion (short-term feature), mood (medium-term feature) and personality dimensions (long-term feature) in their detection system [14] or even try to mix several affective states that are commonly present in real-life emotional experience [15]. In order to build a vector of social and affective signals (we called this vector the emotional and interactional profile of the user), some of the first experiments dynamically extract paralinguistic clues during a dialogue to drive the answering strategy of the robot [14, 38].

These multi-level detection and associated strategies developed by LIMSI have been tested with the NAO emotional system first with a wizard-of-oz, then with a real system. The AI system developed by SpirOps is also used. The retroaction loop will be also used in our future researches for extracting features with knowledge of the context described in the emotional event representation. It is obvious that the profile (i.e. the social and affective vector) must also aggregate information from multi-modal clues and the reliability measures linked to these different information in order to improve the performance of the detection system.

13.3 Understanding

13.3.1 Emotional Evaluation of Events

Events are a symbolic translation of what happens in the environment of the robot. This translation is often basic and a more complex interpretation is required to fully understand the events meaning like for Sect. 13.3.2. Those events are compatible with the dynamics of emotions: they include the actor, the target and the action which are all essential to compute the emotional potential. They still have to be interpreted for complex events. For example, the action "shrug" is usually viewed as a passive action, without much emotional potential. But if a participant shrugs for a decisive question, the emotional potential of this question is somewhat transferred to the "shrug" action, leading to an important emotional potential. Once again, it is a matter of interpretation. And once again to simplify the problem, designers can trade in generality for hard coded behaviours. The event structure is useful to value its emotional potential; understand commands; generate emotional response; detect intents of the others; and match with our goals.

The CPM proposed by Scherer [34] defines an emotion as a sequence of state change occurring in five organic systems in an interdependent and synchronized way to answer to the evaluation of an external or internal stimulus. The five systems are the cognitive (activity of the central nervous system), the psycho-physiological (peripheral answers), the motivational (trend/tendency), the engine (movement, facial expression and vocalization) and the subjective feeling system. The model suggests four major appraisal objectives that an organism needs to reach to adaptively react to a salient event:

- How relevant is this event for me? Does it directly affect me or my social reference group? (relevance)
- What are the implications or consequences of this event and how do they affect my well-being and my immediate or long-term goals? (implications)
- How well can I cope with or adjust to these consequences? (coping potential)
- What is the significance of this event for my self-concept and for social norms and values? (normative significance)

This way of influence is quite obvious: the CPM uses a sequential process for describing how the different modules of the organism react to an emotional event. How to use the CPM for real-life emotion in real context with cause and effect type [11]? It is not straightforward to adapt the CPM for mixtures of emotions or for mixtures of affective states. Furthermore, an individual is rarely without any affective states or feeling when a new event appears. How the CPM mixes/synchronizes the appraisal of the new event with an old feeling is not described by the model. A social robot sensitive to emotions should not take only punctual affective and social traits into account but also should have a memory of the emotional and interactional profile of the user along the interactions in order to have a chance of being more relevant in its behavioural responses [14].

13.3.2 Intents, Goals and Models of Others

During a social interaction several systems are used to conduct this interaction. Each participant is driven by its own goals. Those goals are inherent to the decision process and will play a role in the selection of the right action (including talk). Those goals might be conflicting. Some represent the primary goals such as survive and feed. Others are more socially oriented. According to Goffman [19], saving face and performing the expected role are important goals for social interactions. Those social goals may even be in conflict with primary goals (to be accepted in a group people may do foolish actions). According to Goffman [21], the notion of context also called the frames implies the description of the roles of the different participants. But those roles are not only dependent on the social interaction nature but also on the role that each participant grants himself and the role that each participant grants to other participants for that kind of interaction. Those roles are going to change all along the interaction.

Whatever the goals are, they lead to an action which is performed. The other participants will then try to interpret this action to find its meaning. This part is quite complex, it may imply a voice recognition, semantic translation (which are not in the scope of this article). Even if it does not, for a shrug, for example, the meaning of this shrug for the robot depends on the context, which is an open concept quite complex to describe. The first clue is the implicit meaning of the action. This is the first level reading. For example, a reception robot announcing to a newcomer an event for the evening. As this newcomer shrugs, the robot understands that he is not interested in this event and so it decides not to continue to describe this event.

To be even more efficient, the robot has to evaluate the consequences of each of its actions. The decision process should then rely on anticipation processes that try to explore the plausible futures. Implementing those kinds of anticipation is not in the scope of this article. They can be generic and very complex to implement or hard coded and very simple (but less flexible). Nevertheless, it is important to note that those kinds of effect anticipations are interesting for emotions (cf. Sect. 13.3.1), for planning (often coupled with goal-oriented architectures) and for complex reflexivity. They can be seen as the top-down information flow as described by [29]. This flow tries to anticipate the future information issued by one of the perceptive atoms. This top-down flow is often coupled with specific sequences of actions (some kind of plans) that are executed as long as the guessing is correct. The perceptive architecture is then completed with two new processes:

- Information seer: its role is to anticipate the future information provided by one or several of the other processes. It is often associated with a plan process, but can also be used by the decision system to evaluate consequences of actions.
- Plan: its role is to apply a plan and verify all along that the context remains favourable. To check the context, the plan process is using one or several seers.

For example, the reception robot is asked where its boss is. The robot is about to answer that it does not know, but it anticipates the disappointment of its interlocutor and then decides to show embarrassment during its answer.

The third level reading implies the notion of intents. The goal is to understand the intents of the action performer, what are his underlying goals. This reading is quite useful for collaborating interactions. By guessing the goals of the participants, the robot might anticipate what it has to do, making the whole interaction more fluent and the robot performance more human [3]. This level can be achieved with a reflexive decision process which could be reversed: observing actions to guess the goals that triggered them. To be more effective, the robot tries to map what it thinks to be the state of the action performer, including its role (according to Goffman) in the current social interaction. For example, a reception robot, when asked by a newcomer where the boss is, has to anticipate the intents of this newcomer, which in this case should be a localization request. The consequences of this request imply the execution of Go-to-goal action. The effect of this action is the newcomer in the Boss Office. If the boss has told the reception robot not to be disturbed for the next hour, the robot finds a conflict. It then can choose to avoid answering.

13.3.3 Emotions and Decision

Decisions are influenced by emotions. Everyday life provides us with a lot of examples, from the angry driver that yells at another driver to the audience screaming during a horror film. As exposed in [13], this influence may be of one or two kinds:

- Decision influence: the influence triggers a new decision (screaming during a horror film). To achieve that, the decision process should directly integrate the emotions into its computations.
- Performance influence: emotions partially influence the main decision by altering the performance (raising your voice while you are angry, a light smile when you get a royal flush at the river during a poker play while you are trying to remain stony face). This part raises the problem of procedurally generated moves, according to various continuous parameters.

In the other way, emotions are influenced by the decision process in two ways [13]:

- Control of external emotions: for example, during poker people try to control their external emotion. This control is more or less effective and tries to maintain the role you play during a social interaction and to avoid ruptures. The decision process should then address the external emotional state, combined with some kind of innate and uncontrollable influence from the internal emotional state. The nature of this combination is a focus of interest for future work.
- Cognitive abilities included in the decision process: emotions are sometimes instinctive and sometimes are the product of our complex mind [25]. In the last case, the kind of emotions the robot can feel heavily depends on its ability to use advanced cognitive abilities. For example, the fear of death requires a hard coded fear or the capacity to anticipate death and to reason about this anticipation. Hope is an anticipation of a "reward". Disappointment is often built on hope followed by a non-fulfilment.

The importance of showing emotions is demonstrated in [6] where the authors show the increase of the behaviour readability in the case of visible and coherent emotions. This readability is essential for social acceptance and greatly eases social interactions [7].

13.3.4 Memories Management

The memories manager is responsible for information storage. It is used to keep a history of these information throughout time. Its first role is to time-stamp each information by providing a time descriptor (start date, duration is often used, frequency may be important, etc.). Storing all these data is sometimes problematic. Several solutions have been proposed to solve this problem. The first is to filter what the robot will memorize. The action to memorize is filtered by the decision process. The decision is based on some kind of evaluation function. Those functions usually include:

- Emotional potential of the information: [32] links the emotional potential to the probability of memorizing it. This value is relative to the CPM (Scherer).
- Link with the robot goals: information linked to the robot goals are more likely to be memorized. The problem is then how to evaluate this link. Are the current active goals more important than the others? How to measure the distance of an information to a goal? Even if this connexion seems redundant with CPM, this one strictly represents the links from the information top the goals, on a purely "used by" basis.
- Any application-specific dependencies: information about people close to the robot (remember when John has to take his pills).

The second way is to erase some of the memories. This process is quite definitive and should be avoided if possible. The complexity is to create an evaluation function that will select the memories to delete (forget). This function should include:

- The initial potential used to filter the memory during memorizing: this score is good to keep as it is an important clue of the original print of the memory.
- The use of the memory: this notion addresses several aspects. The most evident is how often the robot used the memory since its creation. This ratio (number of uses/age) denotes the use of the memory by the robot. Another aspect is the number of other memories to which the memory is linked to. Most of the time, links exist between memories that tries to represent semantic relations. These relations could be logical relations (e.g. causality), spatial relations (isInside, isNextTo, etc.) or even conceptual relations (isTypeOf). If the memory is an important node inside the memories network, deleting it suppresses a lot of information (all its links).

The third is more interesting because it minimizes the loss of information. It is based on the principle of generalization. By using generalization operators, the memories manager is able to merge two or more memories together. The principle is based on the events graph. Each part of the graph could be generalized:

- Time merging: when an event lasts in time, the graph that represents it is susceptible to change. Several slightly different versions of the graph would be in memory. If the versions are not different, the merge is simple; the system just changes the duration of the event (end date). If they are different:
  - Different attributes: the divergent attributes are deleted.
  - Different nodes: if two nodes are different, the system will try to merge them by keeping everything that they have in common. For example, the robot puts a first glass into the dishwasher and memorizes this event. Then the robot puts a plate into the dishwasher. All those events concern different targets, they will be merged keeping only common attributes (type: dishes).
  - Links are kept and reconnected to the new graph.
- Cycle merging: quite similar to time merging, this merge occurs when the same memory occurs at fixed repeated intervals. Those memories are merged and new attributes are used to describe the properties of the cycle.

To choose which generalization to perform, a function evaluates the loss of information of this generalization. Losing an attribute may cost less information than a node. It is important to create different weights for the different attributes/nodes. Some are more important than others. For example, the name (or identification) of a person is more important than its hair colour (in most contexts). When merging attributes, different ones are deleted. But for two identical attributes (like age) with different values (like 4 and 6 years old), it is possible to keep this attribute and generalize the values (like young). This needs generalization trees or operators. These operators are also important to generate conversation and to tune the granularity of the information the robot uses. For example, when the robot tells the location of the grocery, if its location is nearby, the robot uses navigation indications. If the robot is in another district, it tells the street address. If it is in another city, it specifies the city.

13.4 What About Lies, Compassion, Jokes

Lies are used to hide information or to achieve our goals. Deception is a major relational transgression that often leads to distrust between relational partners. Deception violates relational rules and is considered to be a negative violation of expectations. It is somehow non-ethical to give the capacity of deception to a robot companion during all contexts. But lying appears as a normal component of human social interaction [36]. By examining the social practice of deception in managing communication, researchers in [23] made several observations that have key design implications for interpersonal communication technologies. For example, the concept of butler lying has been introduced to describe these kinds of deceptions that help manage our interactions. Susan Stuart in [37] claims that the capacity for deception is necessary for a theory of mind and a theory of mind is necessary for complex social interaction: "Without this ability we would be unable to deceive and detect deception in the actions of others, and our ability to interact within our social group would be greatly impaired".

Empathy is the capacity to recognize emotions that are being experienced by another. Compassion is also of a great importance for social interactions and relationship. Compassion is useful to read emotions properly and to mirror them [1,30]. A robot that makes "jokes" is a matter of context and of correlation with the current subject. It depends also on the mood of the dialogue.

First experiments are carried out to explore the acceptability of Robovie [24] mimicking many human abilities like compassion, lying, exerting authority, claiming responsibility and even making jokes. Human being behaviours such as lies, compassion and jokes imply that the robot has the ability to represent and understand some complex human being behaviours. It is not straightforward to design a robot with such abilities. For example, creating a "good" lie requires several mechanisms. The simplest implies avoidance strategies to allow the robot to refuse to answer. These strategies only work partially and only for hiding information. They are triggered by the necessity to hide an information from a specific interlocutor. This trigger can be simulated by a confidentiality/trust mechanism. Building a credible lie requires more advanced mechanisms. The robot needs a representation of the information known by its interlocutor. Having a representation of its goals may also be useful, as it is easier to believe something that tend to satisfy our goals. If we consider a more pragmatic approach driven by the wished effect, there are less generic technical short cuts or tricks which only work on the short term.

The future of robots in our society is certainly not to be seen as a replacement of human beings but as a new tool to simulate memory, educational assistance and mediation processes, and as an accompanying tool in the society for elderly dependent persons or others like autistic children. As a matter of fact, we can consider that a robot can lie to better stimulate someone like for educational purposes. In edutainment software and tutoring systems, user states such as uncertainty [27] or even deception can be employed to adapt the system-teaching pace [26]; generally, paralinguistic clues are essential for tutors and students to make learning successful [31].

13.5 Discussion

Being able to create advanced social interactions requires a lot of elements. According to Goffman, the robot has to keep several active goals on top of its usual primary goals. These goals represent its role in the different social interactions. These extra active goals are also useful in preventing the robot from the ruptures that could break its role and make it lose face. The robotic nature adds new constraints, like the necessity to cope up with unreliable information. To address all those elements and constraints, we imagined a bio-inspired architecture that mixes a bottom-up approach to refine semantically rich information from sensors' raw data with a top-down anticipatory planning system. We couple this perceptive architecture with a goal-oriented decision system able to cope up with uncertainty and manages several hypothesis at the same time. Goffman's roles are then integrated as new social goals.

To ease the social interaction and implement the "face" notion (according to Goffman), we rely on the anticipatory mechanisms of the perceptive architecture and the reflexive property of the goal-oriented decision system. Combined, these two mechanisms provide interesting detection of intents. The reflexive system is structured by the Goffman roles. The robot tries to find the role(s) that its interlocutors are playing. These roles aggregated with other information (such as emotions, personality traits) form the profile of the interlocutor. This profile coupled with the current context is used to guess the interlocutor dispositions. In this way, the context is built from information and structured according to Goffman interactions descriptions "frames". This structure enables us to limit the system and to make it computationally acceptable.

Emotions play a central role in this architecture. They are used to ease social interactions and to manage the short-, mid- and long-term state of the robot, giving interlocutors the illusion of life, not only for a brief demo, but also for a long-term relationship. Coupled with memories, the robot is able to create a bond with its interlocutors, reinforcing its social acceptance. Its ability to show compassion and to lie helps him understanding its interlocutors' behaviours and motives and to avoid the ruptures. The question of what emotions, moods and personality traits should be in the model is a complex question. As a first step, we have proposed to build an emotional and interactional profile of the user using emotions, moods and personality traits [14]. We can imagine giving a unified model of emotion (static and dynamic models) and let designers create their own according to their needs.

Each application should focus on what is strictly necessary. The models described here are really complex and limit the possibilities making them more manageable, easier to debug and constitutionally less expensive. The CPM (Scherer) used answers the question of emotion evaluation for an instant event. How to merge short-term emotions with long- or mid-term emotions is still an open question.

The whole architecture is issued from several projects. The first one is DEEP [5], started in 2005 and implemented the event storage and generalization strategies. Since 2005, we worked on several robotic and/or virtual reality projects. Along those projects we try to refine the different decisional technologies to achieve a character that is socially richer (robot or virtual). The perception module and the dynamic user profile are also issued from several projects such as the ANR Affective Avatars project [2] which aim at detecting emotions during speech to drive the selection during non-verbal behaviours of user's avatar and the ROMEO project [17] which aimed to develop a humanoid robot that can act as a comprehensive assistant for persons suffering autonomy loss.

This paper describes the different elements needed to build a system able to cope up with complex social interactions, perception and dialogues. We try to follow the different fields of cognitive theories: neural, sociological and psychological. The goal is to find a computational model that could be created from these theories. This goal is often achieved by detailing these theories as deeply as possible to obtain a coherent system. During the French ROMEO project and now with his continuation ROMEO 2, we have the opportunity to work on this architecture that will mix interdependent perception decision processes in a unique and effective way.

References

1. Andre, E., Klesen, P., Gebhard, P., Allen, S., Rist, T.: Integrating models of personality and emotions into lifelike characters. In: Paiva, A. (ed.) Affective Interactions: Towards a New Generation of Computer Interfaces, pp. 150–165. Springer, New York (2001)
2. ANR affective Avatars project: http://ddata.over-blog.com/xxxyyy/0/06/46/58/Cahier-ANR-1-Nomadisme.pdf
3. Bartneck C., Croft E., Kulic D.: Measuring the anthropomorphism, animacy, likeability, perceived intelligence and perceived safety of robots. In: Metrics for Human-Robot Interaction Workshop in Affiliation with the 3rd ACM/IEEE HRI 2008, vol. 471, pp. 37–44 (2008)
4. Bernsen, N.O., Dybkjær H., Dybkjær E.L.: Cooperativity in human-machine and human-human spoken dialogue. Discourse Process. **21**(2), 213–236 (1996)
5. Bosser, A.-G., Levieux, G., Sehaba, K., Buendia, A., Corruble, V., De Fondaumière, G., Gachet, A., Gal, V., Natkin, S., Sabouret, N.: Dialogs taking into account experience, emotions and personality. In: Ma, L., Rauterberg, M., Nakatsu, R. (eds.) International Conference on Entertainment Computing. Lecture Notes in Computer Science, vol. 4740, pp. 356–362 (2007)
6. Breazeal, C.: Emotion and sociable humanoid robots. Int. J. Human-Comput. Stud. **59** 119–155 (2003)
7. Breazeal, C.: Function meets style: insights from emotion theory applied to HRI. IEEE Trans. Syst. Man Cybern. C: Appl. Rev. **34**(2), 187–194 (2004)
8. Chastagnol, C. Devillers, L.: Personality traits detection using a parallelized modified SFFS algorithm. In: Proceedings of Interspeech 2012 (2012)

9. Chetouani, M., Wu, Y.H., Jost C., Le Pevedic B., Fassert, C., Cristancho-Lacroix, V., Lassiaille, S., Granata, C., Tapus, A., Duhaut, D., Rigaud, A-S.: Cognitive services for elderly people: The ROBADOM project. In: ECCE Workshop Robots That Care 2010, The European Conference on Cognitive Ergonomics (2010)
10. Costa, P.T., McCrae, R.R.: Revised NEO personality inventory (NEO-PI-R) and NEO five-factor inventory (NEO-FFI) manual. In: Psychological Assessment Resources. Odessa, FL (1992)
11. Cowie, R.: Emotional states expressed in speech. In describing the emotional states expressed in speech. In: Proceedings of ISCA ITRW on Speech and Emotion: Developing a Conceptual Framework for Research, pp. 224–231 (2000)
12. Cowie, R., Douglas-Cowie, E., Martin, J.-C., Devillers, L.: The essential role of human databases for learning in and validation of affectively competent agents. In: Scherer, K., Bänziger, T., Roach, E. (eds.) A Blueprint for an Affectively Competent Agent Cross-Fertilization Between Emotion Psychology, Affective Neuroscience, and Affective Computing, pp. 151–165. Oxford University Press, Oxford (2011)
13. Damasio, A.: The Feeling of What Happens: Body and Emotion in the Making of Consciousness, was named as one of the ten best books of 2001 by New York Times Book Review. Harcourt Brace, New York (2001)
14. Delaborde, A., Devillers, L.: Use of Nonverbal Speech Cues in Social Interaction Between Human and Robot: Emotional and Interactional Markers, in Affine 2010, ACM (2010)
15. Devillers, L., Vidrascu, L., Lamel, L. : Challenges in real-life emotion annotation and machine learning based detection. Neural Networks **18**(4), 407–422 (2005)
16. Devillers, L., Vidrascu, L., Layachi, O.: Automatic detection of emotion from vocal expression. In: Scherer, K.R., Banziger, T., Roesch, E. (eds.) A Blueprint for an Affectively Competent Agent: Cross-Fertilization Between Emotion Psychology, Affective Neuroscience, and Affective Computing, pp. 232–244. Oxford University Press, Oxford (2010)
17. FUI national Romeo project: http://projetromeo.com
18. Gockley, R., Simmons, R., Forlizzi, J.: Modeling affect in socially interactive robots. In: The 15th IEEE International Symposium on Robot and Human Interactive Communication (ROMAN06) (2006)
19. Goffman, E.: Role Distance, dans Encounters. Bobbs Merrill, Indianapolis (1961)
20. Goffman, E.:Interaction Ritual: Essays on Face-to-Face Behavior. Anchor Books, New York (1967). ISBN 0-394-70631-5
21. Goffman, E.: Frame Analysis: An Essay on the Organization Of Experience. Harper and Row, London (1974). ISBN 978-0-06-090372-5
22. Grice, H.-P.: Logic and conversation. In: Cole, P. (ed.) Syntax and Semantics: Speech Acts, vol. 3, pp. 41–58. Academic, New York (1975)
23. Hancock,J., Birnholtz, J., Bazarova, N., Guillory, J., Perlin, J., Amos, B.: Butler lies: awareness, deception and design. In: Proceedings of the ACM (2009)
24. Kahn, P.H., Jr., Kanda, T., Ishiguro, H., Gill, B.T., Ruckert, J.H., Shen, S., Gary, H.E., Reichert, A.L., Freier, N., Severson, R.L.: Do people hold a humanoid robot morally accountable for the harm it causes? In: Proceedings of the 7th ACM/IEEE International Conference on Human-Robot Interaction, pp. 33–40. New York (2012)
25. Ledoux J.: The Emotional Brain. Weidenfeld and Nicolson, London (1998)
26. Litman, D., Forbes, K.: Recognizing emotion from student speech in tutoring dialogues. In: Proceedings of ASRU (2003)
27. Litman, D., Rotaru, M., Nicholas, G.: Classifying turn-level uncertainty using word-level prosody. In: Interspeech09, pp. 2003–2006 (2009)
28. Marsella, S. , Gratch J., Petta, P.: Computational models of emotion. In: Scherer, K.R., Banziger, T., Roesch, E. (eds.) A Blueprint for an Affectively Competent Agent: Cross-Fertilization Between Emotion Psychology, Affective Neuroscience, and Affective Computing. Oxford University Press, Oxford (2010)

29. Mountcastle V.: An organizing principle for cerebral function: the unit model and the distributed system. In: Edelman, G., Mountcastle, V. (eds.) The Mindful Brain, pp. 7–50. MIT Press, Cambridge (1978)
30. Ochs, M., Pelachaud, C., Sadek, D.: An Empathic Virtual Dialog Agent to Improve Human-Machine Interaction. AAMAS (2008)
31. Price, L., Richardson, J.T.E., Jelfs, A.: Face-to-face versus online tutoring support in distance education. Studies in Higher Education **32**(1), 1–20 (2007)
32. Richards, J.M., Gross, J.J.: Emotion regulation and memory: The cognitive costs of keeping one's cool. J. Personality Soc. Psychol. **79**(3), 410–424 (2000)
33. Scherer, K.R.: Appraisal theories. In: Dalgleish, T., Power, M. (eds.) Handbook of Cognition and Emotion, pp. 637–663. Wiley, Chichester (1999)
34. Scherer, K.R., Schorr, A., Johnstone, T.: Appraisal Considered as a Process of Multilevel Sequential Checking. Oxford University Press, USA (2001)
35. Schuller, B., Steidl, S., Batliner, A., Burkhardt, F., Devillers, L., Müller, C., Narayanan, S.: Paralinguistics in speech and language - state-of-the-art and the challenge. In: Computer Speech and Language (CSL). Special Issue on Paralinguistics in Naturalistic Speech and Language, p. 39. Elsevier (2012) [IF: 1.353, 5-year IF: 1.489 (2010)]
36. Spence, S.A., Farrow, T.F., Herford, A.E., et al.: Behavioral and functional anatomical correlates of deception in humans. Neuroreport **12**, 2349–2353 (2001)
37. Stuart, S.: The role of deception in complex social interaction. Cogito **12**(1), 25–32 (1998)
38. Tahon, M., Delaborde, A., Devillers, L.: Real-life emotion detection from speech in human-robot interaction: experiments across diverse corpora with child and adult voices. In: Proceedings of the Interspeech 2011, pp. 3121–3124 (2011)

Chapter 14
Integration of Multiple Sound Source Localization Results for Speaker Identification in Multiparty Dialogue System

Taichi Nakashima, Kazunori Komatani, and Satoshi Sato

Abstract Humanoid robots need to head toward human participants when answering to their questions in multiparty dialogues. Some positions of participants are difficult to localize from robots in multiparty situations, especially when the robots can only use their own sensors. We present a method for identifying the speaker more accurately by integrating the multiple sound source localization results obtained from two robots: one talking mainly with participants and the other also joining the conversation when necessary. We place them so that they can compensate for each other's localization capabilities and then integrate their two results. Our experimental evaluation revealed that using two robots improved speaker identification compared with using only one robot. We furthermore implemented our method into humanoid robots and constructed a demo system.

14.1 Introduction

We are currently developing a multiparty dialogue system with humanoid robots. A multiparty dialogue system enables speech interaction with more than two participants [13].

There are two main problems in the multiparty dialogue systems that have been developed to date. The first problem is that most systems require expensive devices. This imposes that many applications be developed only in specific settings. Multiparty dialogue systems should not only understand what participants are saying but also identify where they are and who they are addressing. Several studies have been done to identify and keep track of such participant behavior with high-definition cameras [4], ultrasound proximity sensors [7], wide-angle

T. Nakashima (✉) • K. Komatani • S. Sato
Graduate School of Engineering, Nagoya University, Nagoya, Japan
e-mail: taichi_n@nuee.nagoya-u.ac.jp; komatani@nuee.nagoya-u.ac.jp; sslab.nuee.nagoya-u.ac.jp

J. Mariani et al. (eds.), *Natural Interaction with Robots, Knowbots and Smartphones: Putting Spoken Dialog Systems into Practice*, DOI 10.1007/978-1-4614-8280-2_14,

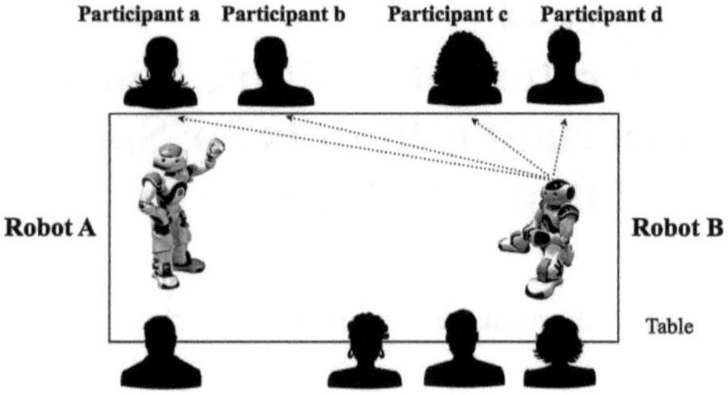

Fig. 14.1 Our conversation setting. We place two robots on a table and multiple participants sit around it; not all seats are occupied

cameras [2], or in a room equipped with many video cameras and microphones [5]. The second is that the interaction is sometimes suspended because participants do not often understand what they can ask to the system. Since the problem occurs in spoken dialogue systems with single participants [3], this must happen in multiparty dialogue systems too.

Our approaches to the problems are the following:

- We use only microphones and cameras equipped on the robot itself, i.e., we do not presuppose a special environment that has rich sensors.
- We use two robots. They join the conversation and keep it going by talking to each other when the human participants do not speak. Furthermore, since the sensors equipped on the robots are poor, we use two robots so that they can compensate for each other's poor capability.

We set up a meeting situation as a test bed. Multiple participants sit around a table and talk with the robots, as shown in Fig. 14.1. This situation simplifies several difficulties in multiparty dialogues, e.g., the positions of the participants are naturally narrowed.

This paper focuses on *speaker identification*, identifying where a speaker is, which is the first step toward accomplishing a multiparty dialogue. Speaker identification enables robots to head toward the participants when answering their questions. Heading toward the participants is essential for robots that participate in a multiparty dialogue because it enables the participants to understand the role of the addressees [8] and to feel involved in the interaction [1]. In addition to such robot behavior, identifying each participant individually is required when, for example, the robot encourages a participant who has not yet spoken to speak.

We use sound source localization for speaker identification. Sound source localization is a technique of estimating where a sound is coming from. There are two main problems in the setting shown in Fig. 14.1:

1. Some positions of the participants are difficult to localize for the robots when they are using only their own sensors.
2. Noise may cause incorrect localization. As a result, the localization results do not always indicate the direction where a speaker is.

In the conversation setting depicted in Fig. 14.1, the angular difference between two participants becomes smaller when their positions are more distant from a robot. For example, in Fig. 14.1, the angular difference between Participants a and b is small from Robot B's position. Thus, it is difficult to identify these two participants as two individuals. In addition to that, incorrect localization due to the robot's own motor noise is unavoidable when the microphones equipped on it are used.

We solve these two problems by placing two robots on the table, as shown in Fig. 14.1 and integrating sound source localization results from them. As a result, the two robots are able to compensate for each other's localization capabilities because the small angular difference between the two participants from one robot's position is large from the other. For example, the small angular difference between Participants a and b from Robot B's position is large from Robot A's position and it is easy for Robot A to separately localize the two participants. We then integrate the localization results by representing them with the weight of their power indicating sound pressure. We also define a confidence measure based on the power that indicates the accuracy of the localization results.

This paper is organized as follows. Section 14.2 provides an overview of related work and places our research within this field. Section 14.3 presents our method of integrating multiple sound source localization results for speaker identification. Section 14.4 explains evaluation experiments and shows that using two robots improves speaker identification compared with using only one robot. Section 14.5 presents an overview of the demo system which we implemented our method. Section 14.6 gives discussion about future work.

14.2 Related Work

The simplest approach to identifying speakers in multiparty situations is to provide microphones to all participants [6]. The microphones and their positions enable the system to accurately identify where speakers are and when they speak. This approach needs the microphones to be prepared for each participant before the conversation starts. Another approach is to perform conversation in smart room environments where rich sensors are distributed [11]. Many sensors in the room keep track of users' behavior and identify speakers. This approach is valid only in specific environments and is not for general purpose. Therefore, a lot of research have been carried out to identify speakers with a robot equipped with sensors [12].

A lot of research generally used image processing for speaker identification when using multimodal information such as audio or visual inputs from the robot's sensors. Haider and Al Moubayed [4] proposed a method of identifying speakers by detecting lip movements. The system developed by Bohus and Horvitz [2] identifies speakers by tracking participants and recognizing their gestures on the basis of image processing. These methods are valid only when the participants are always in the field of view of the system's camera. If the participants are outside the field of view or in a dark environment, it is difficult to identify speakers. Bennewitz et al. [1]

proposed a method to maintaining a probabilistic belief about participants in the surroundings on the basis of face detection even if they are not in the field of view of the robot's camera. This method also needs to detect the participants once in the field of view of the camera and it is hard to identify speakers who have never been in it.

We use mainly sound source localization results to identify speakers. It is difficult to keep track of participants in the field of view of the robot's camera at all times in our conversation setting where the robots are on a table around which multiple participants sit. Moreover, the field of view of the robot's camera is narrow. Our approach using localization results enables us to identify speakers even if they have never been in the field of view.

14.3 Integration of Multiple Sound Source Localization Results

We integrate sound source localization results obtained from the two robots for speaker identification.

Both robots obtain sound source localization results that represent the azimuth of the sound source and its power, that is, the sound pressure. In the localization results, the counterclockwise direction is assumed to be positive and the front of the robot is set to $0°$ (e.g., the direction of the robot's left hand is $90°$). We integrate the localization results obtained from the two robots in the following three steps:

1. We unify the coordinate systems for both robots.
2. We express the localization results with the weight of their power.
3. We sum up converted localization results obtained from the two robots.

These steps enable us to integrate the localization results. We also define a confidence measure based on power, which indicates the accuracy of the localization results.

First, we unify the coordinate systems of both robots shown in Fig. 14.2 because they have different ones. The variables used in Fig. 14.2 are defined below:

- (x_R, y_R): a coordinate system that both robots have. Here, $R \in \{A, B\}$ and A and B correspond to Robots A and B in Fig. 14.2. The origin of the coordinate system is the robot's head, i.e., the center of microphones equipped on the robot. The positive direction of the x_R axis is in the front of the robot and the positive y_R axis indicates its left side. We assume that both robots are at the same height and we do not consider the vertical difference between them.
- W (cm): the horizontal distance between the two robots.
- L (cm): the horizontal distance between the origin of the coordinate system of the robot and possible positions of the participants.

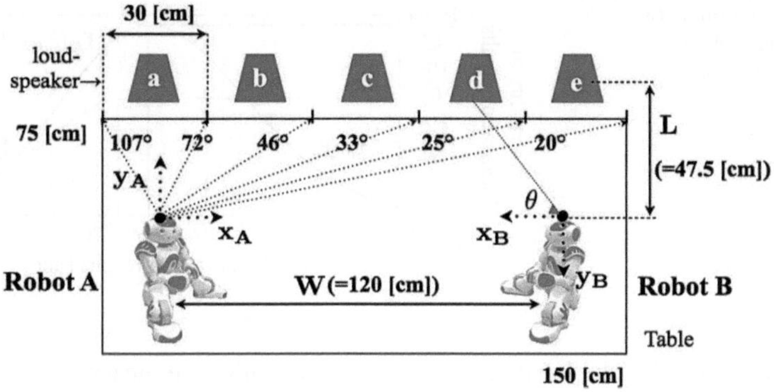

Fig. 14.2 Position of robots and loudspeakers. Loudspeakers are placed where speakers can sit

We unify the coordinate system of Robot B with that of Robot A. Where Robot B obtains a sound source localization result θ (degree) as input, the position of the sound in Robot B's coordinate system (x_B, y_B) is expressed in Eq. (14.1):

$$(x_B, y_B) = \left(\frac{l}{\tan\theta}, l\right)$$

$$l = \begin{cases} L & , if\ \theta > 0 \\ -L & , if\ \theta < 0 \end{cases} \tag{14.1}$$

The position of the sound (x_B, y_B) in the coordinate system of Robot A is expressed in Eq. (14.2), where α is the angular difference between the coordinate systems of both robots and (u, v) is the origin of the coordinate system of Robot B in that of Robot A:

$$\begin{pmatrix} x_A \\ y_A \end{pmatrix} = \begin{pmatrix} \cos\alpha & -\sin\alpha \\ \sin\alpha & \cos\alpha \end{pmatrix} \begin{pmatrix} x_B \\ y_B \end{pmatrix} + \begin{pmatrix} u \\ v \end{pmatrix} \tag{14.2}$$

When robots are placed as shown in Fig. 14.2, $L = 48.5$, $\alpha = 180$, $(u, v) = (W, 0)$, and $W = 120$. Thus, we unify sound source localization results from both robots. Note that now we manually measure L, α, and W.

Second, we express the localization results with the weight of the power. It allows us to consider how reliable each result is. We define probability density function $f_r(\theta)$ [Eq. (14.3)] from a localization result θ_r (degree) and its power p_r (dB), assuming that the ambiguity of a sound source localization result follows a normal distribution. Here, r indicates ID, e.g., Robot A's localization result is θ_A. In addition, σ_r^2 is the variance and indicates how much variation exists in the localization results. Its value will be determined below according to the power p_r:

$$f_r(\theta) = \frac{1}{\sqrt{2\pi\sigma_r^2}} \exp\left(-\frac{(\theta - \theta_r)^2}{2\sigma_r^2}\right) \tag{14.3}$$

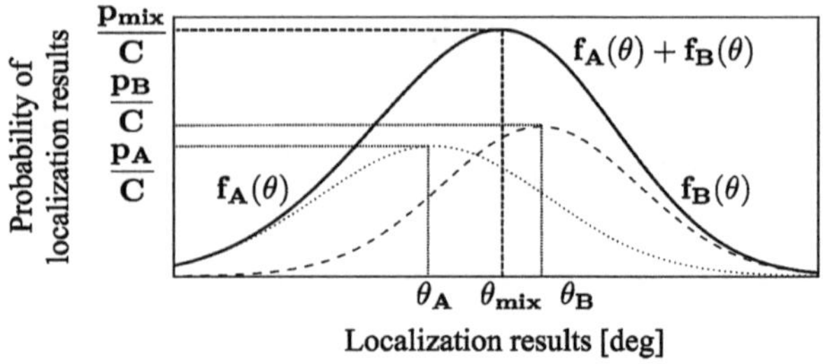

Fig. 14.3 Example of integrating probability density functions

We use the power of the localization results as a confidence measure that indicates its accuracy, assuming that the power of the localization results caused by noise is low. We define that the maximum probability $f_r(\theta_r)$ is proportionate to the power p_r of the localization result θ_r [Eq. (14.4)]. This definition means that the larger power p_r indicates the higher probability that the localization result is θ_r. Here, C in Eq. (14.4) is a constant value and its value is determined empirically:

$$f_r(\theta_r) = \frac{1}{\sqrt{2\pi\sigma_r^2}} = \frac{1}{C}p_r \tag{14.4}$$

We define σ_r from Eq. (14.4) so as to follow the normal distribution, as shown in Eq. (14.5). Here, σ_r is inversely proportionate to the power p_r, that is, the larger the power p_r is, the less likely the localization results are spread out:

$$\sigma_r = \frac{C}{\sqrt{2\pi}}\frac{1}{p_r} \tag{14.5}$$

An example of $f_r(\theta)$ is depicted in Fig. 14.3. The horizontal axis plots the localization results θ (degree), and the vertical axis plots the probability of localization results.

Finally, we obtain an integrated localization result and its power by summing up localization results obtained from the two robots after the above steps. When we obtain the localization results θ_A (degree) and θ_B (degree) and their powers p_A (dB) and p_B (dB) from Robots A and B, we define $f_A(\theta)$ and $f_B(\theta)$, respectively. By applying $f_A(\theta)$ and $f_B(\theta)$ to Eqs. (14.6) and (14.7), we obtain the integrated localization result θ_{mix} (degree) and its power p_{mix} (dB):

$$\theta_{mix} = \arg\max_{\theta}(f_A(\theta) + f_B(\theta)) \tag{14.6}$$

$$p_{mix} = C(f_A(\theta_{mix}) + f_B(\theta_{mix})) \tag{14.7}$$

We set a threshold for the power after the integration. We delete the result if the integrated power is smaller than it. Thus, we expect to reduce the number of incorrect localization results due to noise. Note that we determine the threshold through experiments.

14.4 Evaluation Experiments

We conducted experiments to evaluate whether using two robots improved speaker identification compared with using only one robot. In the experiments, we played speech sounds from loudspeakers and identified them.

14.4.1 Settings of Experiments

We prepared a table (150 × 75 cm) and placed five loudspeakers in positions where participants could sit, as shown in Fig. 14.2. The loudspeakers were arranged at intervals of 30 cm, and the area of the loudspeaker was ±15 cm from its center. We regard a localization result as correct when it is in the range of the loudspeaker actually playing the sound. The ranges are depicted in Fig. 14.2 from Robot A's position.

We used the robot audition software HARK [9], which outputs localization results (degree) on the basis of the MUltiple SIgnal Classification (MUSIC) method [10] and its power (dB) for every frame (= 0.01 s). The MUSIC method localizes sound sources by using impulse responses (transfer function) between a sound source position and each microphone. We used four microphones equipped on humanoid robot NAO's[1] head. The impulse responses for calculating the transfer function had been recorded at 36 points, at intervals of 10°, 1 m away from the microphones. Consequently, the angular resolution of the localization results was 10°.

14.4.2 Data and Evaluation Measure

We played audio files from loudspeakers and collected evaluation data. Each audio file was recorded by one participant and its duration was about 10 s. It included five utterances whose durations were 1.0 s on average. We recorded such files from four participants and played them from the five positions (a–e, in Fig. 14.2). We evaluated the sound source localization result per utterance, i.e., for 100 utterances.

We used *Precision* (P), *Recall* (R), and F*-measure* (F) as evaluation measures. We define them as follows:

$$P = \frac{\text{Number of frames when localization result was correct}}{\text{Number of all detected frames}} \quad (14.8)$$

[1]http://www.aldebaran-robotics.com/en/.

Table 14.1 Results of identifying loudspeakers (SPK)

	Robot A			Robot B			Integration		
SPK	*P*	*R*	*F*	*P*	*R*	*F*	*P*	*R*	*F*
a	0.56	0.89	0.69	0.00	0.00	–	0.57	0.85	0.68
b	0.49	0.65	0.56	0.00	0.00	–	0.40	0.50	0.45
c	0.00	0.00	–	0.13	0.13	**0.13**	0.38	0.49	**0.43**
d	0.06	0.03	**0.04**	0.63	0.83	0.72	0.48	0.67	0.56
e	0.09	0.03	**0.05**	0.50	0.69	0.58	0.39	0.61	0.48
ALL	0.33	0.32	0.33	0.39	0.33	0.36	0.45	0.62	0.52

P, *R* and *F* denote *Precision*, *Recall* and *F-measure*, respectively

$$R = \frac{\text{Number of frames when localization result was correct}}{\text{Number of all speech frames}} \tag{14.9}$$

$$F = 2\left(\frac{1}{P} + \frac{1}{R}\right)^{-1} \tag{14.10}$$

Here, the number of all speech frames was 2,591 (frames).

14.4.3 Results

We evaluated whether using the two robots improved speaker identification compared with using only one robot. The results of identifying loudspeakers at the five positions (a to e) are summarized in Table 14.1. The three cases are listed in the table: when only Robot A was used, when only Robot B was used, and when the results from the two robots were integrated. The *F-measures* of ALL were the best when the thresholds of the power were 24, 25.5, and 25 dB when only using Robot A, only using Robot B, and integrating with $C = 800$, respectively.

As indicated by the results when only Robot A or B was used, it was difficult to identify loudspeakers that were far from the robots. For example, Robot A was not able to identify loudspeakers at c, d, or e which were far from it. This can be seen as low F-measures, which are printed in bold (or dashes when no correct results were obtained), in the column of Robot A in the table. As well, F-measures for positions a, b, and c from Robot B were low, which are dashes or printed in bold. The reason for this is that the range of the loudspeaker, in which the results are regarded as correct, becomes narrower from Robot A's position when the robot's position is farther from it; thus, it is difficult to localize the area. Moreover, the identification accuracy differed between the two robots. This is because the performance of the robot's microphones differed.

The integration results indicate that the system is able to identify the areas that one robot alone cannot identify. In particular, the loudspeaker at position c which neither robot was able to identify was identified with similar *F-measure*

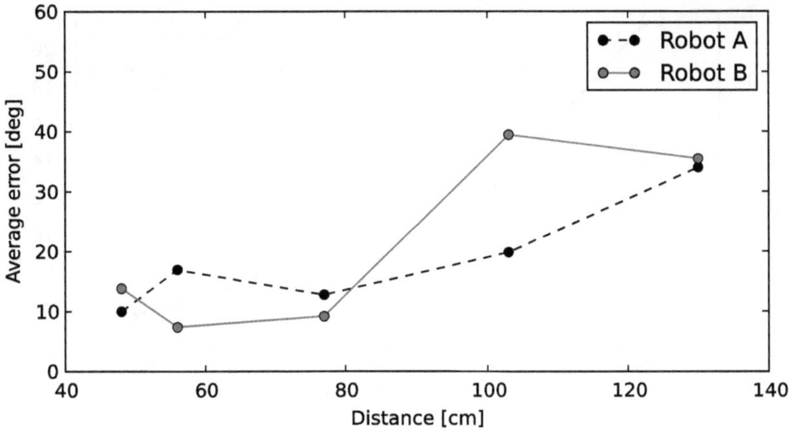

Fig. 14.4 Average errors for robots A and B by distance

as the loudspeakers in other positions, which is printed in bold in the column of integration in Table 14.1. Apart from the loudspeaker at position c, the *F-measure* of the loudspeakers at the other positions was slightly decreased. The reason is that both localization results were not always correct and incorrect ones gave the wrong effects in some cases.

Here, we investigate the causes of the poor performances for the distant loudspeakers. Figure 14.4 plots the average errors in degrees of each robot by distance, where the horizontal axis is distance in centimeters and the vertical axis is the average error in degrees. The average error was calculated by the angular difference between the localization results and the center of the loudspeaker. The graph indicates that the average errors were large when the loudspeakers were distant from the robots. Namely, the localization results were spread out if the sound sources were far from the microphones and were not able to localize them accurately. This result implies that such performances of localization are attributed to the difficulty of localizing the loudspeakers which are distant from the robot.

14.4.4 Sound Source Localization Results by Power

We evaluated whether the integrated power (p_{mix}) is valid as a confidence measure that indicates the accuracy of localization results. Figure 14.5 plots the error rates and the average errors in degrees by the power (8 dB). The horizontal axis plots the power in units of 8 dB, the left vertical axis plots the error rate, and the right vertical axis plots the average error (degree). Here, the error rate was calculated by using the number of incorrect localization results, which were not in the area of the loudspeakers while the speech sounds were played. The average error was calculated

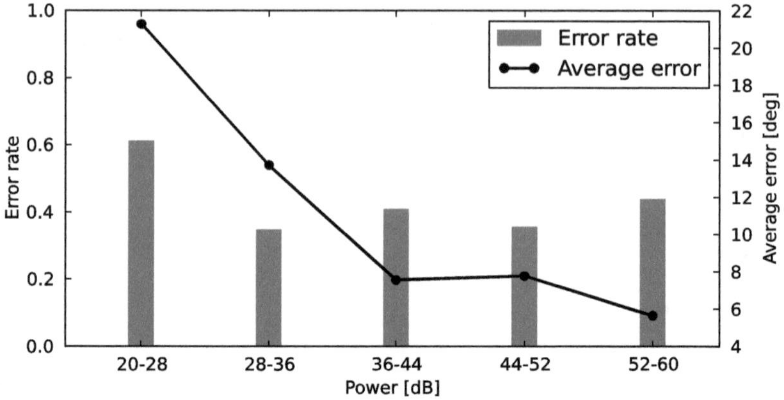

Fig. 14.5 Error rates (*left* axis) and average errors (*right* axis) by power

by using the difference between incorrect localization results and the center of the loudspeaker.

As we can see from Fig. 14.5, the error rate was high when the power was low. The average error indicated that the localization results with high power had small average error. This indicates that when the power of localization results was high, there were less incorrect localization results. Even if sound source localization results were incorrect, they indicated that the loudspeakers were near. As a result, the power of the sound source localization results can be used to distinguish whether the localization results are accurate or not. When the power is low, the system should check whether a participant exists there or not, e.g., by executing face detection.

14.5 Demo System

We implemented our method into humanoid robots and constructed a demo system. Figure 14.6 shows multiple participants interacting with our robots. The task of our system is to introduce our laboratory, and the participants can ask questions about it (e.g., its research field). Our system has four main characteristics:

- The system can identify a speaker and the robot heads toward the participant to answer his/her question in a multiparty setting (upper photograph of Fig. 14.6).
- The system uses the integrated power (p_{mix}) as the confidence measure of the localization result. When the integrated power is low, the robot checks whether a speaker exists in the direction or not by executing face detection with the camera equipped on the robot instead of answering immediately (lower photograph of Fig. 14.6). Conversely, when the localization result has more power, the robot heads toward the participant to answer his/her question immediately. The use of

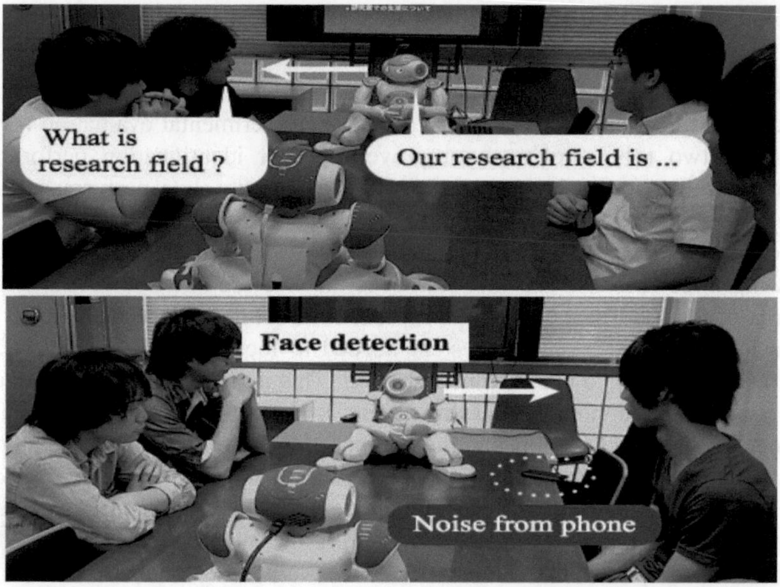

Fig. 14.6 Multiple participants interacting with our system. *Upper* photograph shows that the system identifies speaker and heads toward participant to answer his question. *Lower* photograph shows that system carries out face detection in the direction that the localization result with low power indicates

power from the localization results is expected to reduce incorrect identification caused by noise and to respond to high confidence utterances immediately without unnecessary checking.

- The two robots execute speech recognition and select automatic speech recognition (ASR) results obtained from the one which is nearer to the speaker on the basis of the localization results. This mechanism is expected to obtain more accurate ASR results. Furthermore, when the system is not able to obtain precise ASR results, this mechanism enables the system to generate a prompt to the participant so that they can speak to the nearest robot of the two.
- The two robots talk with each other when they detect no utterances from participants for a certain period of time to keep the conversation going.

A video recording of multiple participants interacting with our robots is available online.[2]

[2]http://sslab.nuee.nagoya-u.ac.jp/en/?page_id=112.

14.6 Discussion and Conclusion

We presented a method to integrate multiple sound source localization results for speaker identification in a multiparty dialogue. Experimental evaluation revealed that using two robots relatively improved speaker identification performances compared with using only one robot. However, the absolute performance needs to be improved for conducting multiparty dialogues; the *F-measure* was 0.52 (Table 14.1). The difficulty of the sound source localization is attributed to the following three reasons. The first reason is that we use strict correct answers of the area of the loudspeakers. In our experimental settings, there were small areas of the loudspeakers where the system needed to localize (e.g., in Fig. 14.2, the loudspeaker at points d and e from Robot A). This is because we need to identify each participant individually and this ability is important for multiparty dialogue systems. The second reason is that the number of the microphones equipped on the robot's head is only four and the distance between them is small. Such restrictions make it difficult for the system to obtain accurate localization results. The third reason is that the loudspeakers are distant from the microphones. Compared with the headset microphone, the influence of room reverberation or environmental noise is unavoidable.

There are two main areas in our future work to improve speaker identification. The first is to use other types of evidence of a speaker's existence, such as image processing results from the robot's camera or previous localization results as well as the localization results at a point in time. The previous localization results are accumulated during the conversation. Integrating other kinds of evidences with the current method is expected to improve speaker identification. The second is to collect interaction data between our system and human participants and evaluate speaker identification performance. In our experimental evaluation, the presence of participants is simulated by using loudspeakers positioned where participants may be.

In addition to improving speaker identification, we plan to implement automatic calibration of position parameters. The parameters consist of the distance between robots, the distance between robot and participants, and so on. We currently have to measure such information manually. We should design a calibration technique of obtaining the information automatically by robots themselves.

References

1. Bennewitz, M., Faber, F., Joho, D., Schreiber, M., Behnke, S.: Integrating vision and speech for conversations with multiple persons. In: Proceedings of IEEE/RSJ the International Conference on Intelligent Robots and Systems (IROS), pp. 2523–2528 (2005). doi: 10.1109/IROS.2005.1545158
2. Bohus, D., Horvitz, E.: Models for multiparty engagement in open-world dialog. In: Proceedings of the SIGDIAL 2009 Conference, pp. 225–234 (2009)

3. Gruenstein, A., Seneff, S.: Releasing a multimodal dialogue system into the wild: User support mechanisms. In: Proceedings of the 8th SIGdial Workshop on Discourse and Dialogue, pp. 111–119 (2007)
4. Haider, F., Moubayed, S.A.: Towards speaker detection using lips movements for human-machine multiparty dialogue. In: FONETIK 2012 (2012)
5. Jovanovic, N., op den Akker, R., Nijholt, A.: Addressee identification in face-to-face meetings. In: Proceedings of the 11th Conference of the EACL (2006)
6. Matsuyama, Y., Taniyama, H., Fujie, S., Kobayashi, T.: Framework of communication activation robot participating in multiparty conversation. In: Proceedings of AAAI Fall Symposium, Dialog with Robots, pp. 68–73 (2010)
7. Moubayed, S.A., Beskow, J., Blomberg, M., Granström, B., Gustafson, J., Mirnig, N., Skantze, G.: Talking with furhat - multi-party interaction with a back-projected robot head. In: FONETIK 2012 (2012)
8. Mutlu, B., Shiwa, T., Kanda, T., Ishiguro, H., Hagita, N.: Footing in human-robot conversations: how robots might shape participant roles using gaze cues. In: Proceedings of the 4th ACM/IEEE International Conference on Human Robot Interaction, pp. 61–68 (2009)
9. Nakadai, K., Takahashi, T., Okuno, H.G., Nakajima, H., Hasegawa, Y., Tsujino, H.: Design and implementation of robot audition system 'HARK' - open source software for listening to three simultaneous speakers. Adv. Robotics **5**, 739–761 (2010)
10. Schmidt, R.O.: Multiple emitter location and signal parameter estimation. IEEE Trans. Antenn. Propagat. **34**, 276–280 (1986). doi: 10.1109/TAP.1986.1143830
11. Stiefelhagen, R., Yang, J., Waibel, A.: Modeling focus of attention for meeting indexing based on multiple cues. IEEE Trans. Neural Networ. **13**, 928–938 (2002)
12. Traum, D.: Issues in multi-party dialogues. In: Advances in Agent Communication. Lecture Notes in Artificial Intelligence, vol. 2922, pp. 201–211. Springer, Berlin (2004)
13. Traum, D., Rickel, J.: Embodied agents for multi-party dialogue in immersive virtual worlds. In: Proceedings of the First International Joint Conference on Autonomous Agents and Multiagent Systems, pp. 766–773 (2002)

Chapter 15
Investigating the Social Facilitation Effect in Human–Robot Interaction

Ina Wechsung, Patrick Ehrenbrink, Robert Schleicher, and Sebastian Möller

Abstract The social facilitation effect is a well-known social-psychological phenomenon. It describes how performance changes depending on the presence or absence of others. The current study investigates if the social facilitation effect can also be observed while interacting with anthropomorphic robots.

15.1 Introduction

While most robots are not developed to socially interact with humans but rather to accomplish a given work task (e.g. assembly line robots), the use of so-called social robots is slowly increasing, in various areas, e.g. as museum assistants [1] or as toys [2]. Hence, if social robots will be part in future daily life as social actors, the investigation of human–robot interaction (HRI) is getting increasingly important. According to Duffy [3], such social robots need to have anthropomorphic, i.e. human-like qualities in order to be capable of meaningful social interactions. Still this does not mean they should exactly look like humans. Moreover, robots looking very human but "behaving" non-human may be perceived as strange or eerie [4], an effect described as the *uncanny valley* [5]. This term refers to the fact that with increasing resemblance to human appearance, the perceived familiarity increases until a certain level of human-likeness is reached at which small differences in appearance lead to an uneasy feeling. Familiarity then drops and increases again if the human-likeness is perfect. Just recently, neural correlates of the uncanny valley effect have been observed [6]. The authors conclude that the uncanny valley effect may be based on perceptual mismatch: For very human-like robots, the

I. Wechsung (✉) • P. Ehrenbrink • R. Schleicher • S. Möller
Quality and Usability Lab, Telekom Innovation Laboratories, TU Berlin, Ernst-Reuter-Platz 7, 10587 Berlin, Germany
e-mail: ina.wechsung@telekom.de; patrick.ehrenbrink@gmail.com; robert.schleicher@telekom.de; sebastian.moeller@telekom.de

J. Mariani et al. (eds.), *Natural Interaction with Robots, Knowbots and Smartphones: Putting Spoken Dialog Systems into Practice*, DOI 10.1007/978-1-4614-8280-2_15,

brain "expects" equally human-like movements, this is often not case. If those expectations are not met, the uncanny valley effect may occur [6]. The current study investigates if the presence of robots differing in the level of anthropomorphism influences human performance. Theoretical background of the study is the *social facilitation effect*. The social facilitation effect is a well-studied social-psychological phenomenon describing how performance changes depending on the presence or absence of members of the same species. In the following section, previous work regarding the social facilitation effect in human–robot interaction is reviewed. Moreover, theories that help to explain and to predict this effect are described.

Based on that, we describe our own study that varied the level of anthropomorphisms and task complexity. After discussing the results, their implication for HRI are discussed.

15.2 The Social Facilitation Effect

The social facilitation effect was first reported by Triplett in 1898 [7]. He observed that bicyclers are faster when they are cycling together with others compared to cycling against the clock, and that children are faster in reeling in fishing lines if other children are present [7]. Although Triplett focused on the effects of competition rather than on the effects of presence of other humans, this study is often seen as the starting point of research on social facilitation [8].

Allport, who also coined the term social facilitation, showed that Triplett's observation could also be shown for non-competitive situations [9]. Moreover, Allport was the first who stated the assumption that the social facilitation effect is dependent on the task type [9]. While performance is enhanced in the presence of other humans for tasks of low complexity, performance decreases for difficult, complex tasks. The latter is also known as *social inhibition*. Zajonc and colleagues [10] found similar results even for cockroaches and explained those findings as follows: The presence of other members of the same species induces higher arousal. Higher arousal leads to more dominant reactions, which tend to be the reactions well known by the subjects. In easy, well-learned tasks the dominant, well-learned reaction is often correct, while for complex tasks a solution determined for the individual situation would be more appropriate and the dominant or default reaction is often false. The theory can be illustrated with the following example: If something is getting burnt in the oven, the dominant and right reaction is to take it out of the oven. If the food is already on fire, taking it out of the oven may be wrong: Leaving it in the oven would be the better option as in the oven the oxygen necessary for a fire is rapidly decreasing and the fire will quickly go off. Taking it out may result not only in burnt food but also in burnt oven knobs.

Besides Zajonc's so-called *drive theory* other researchers suggested social comparison and attentional resources to explain the social facilitation effect (cf. [11] or [12] for an overview). Social comparison theories consider the anticipation of evaluation by others, the desire to present oneself in a specific way (impression management) or the intention to perform in compliance with social standards as explanations for the facilitation effect [12]. Attention theories assume that the presence of others leads to a shift of cognitive resources [12]. Sanders et al. [13], for example, claim that the presence of others is distracting and that this distraction results in an attentional conflict, which in turn increases the arousal. Later, Baron added the concept of *cognitive overload*, which means that the presence of others does not only increases arousal but also the cognitive load.

According to drive and attention theories, the level of arousal is crucial for a social facilitation effect to occur. Hence, a robot leading to higher arousal should also lead to a stronger social facilitation effect. Previous research could show that robots can trigger a social facilitation effect (e.g. [14]). However, to the authors knowledge, the level of human-likeness as not been systematically investigated so far. According to Mori's uncanny valley, a very human-like robot is likely to be perceived as strange and eerie and should thus lead to higher arousal compared to a less human-like robot.

Hence, in the current study we used three robots, differing in their level of human-likeness, to investigate their influence on the social facilitation effect.

15.3 Method

The following section describes the materials, set-up and procedure used in our study.

15.3.1 Robotic heads

Three different robot heads were built. Two of them were based on a plaster mask of a real human face. In the following, we will use their informal names for their description. For $head_{human}$ the original mask form was used. For $head_{cartoon}$ the human features were altered in accordance with the design recommendation suggested by DiSalvo et al. [14]. Their results imply that in order to let robots appear less human, their head should be wider than it is tall. Moreover, the distance between the eyes should be slightly wider than the diameter of the eye.

The third "head", referred to as $head_{box}$ in the following, showed no anthropomorphic features. It was a box, covered with skin-coloured textile fabric, and was approximately of the size of a human head.

$Head_{human}$ and $head_{cartoon}$ were placed on an artificial body which also contained a loudspeaker. The mechanics and electronics of those both heads were built using *LEGO Mindstorms NXT* kit. The heads could raise the eyebrows and move their eyes as well as their mouth. The body could not move. In the condition $head_{box}$, the loudspeaker was placed inside the box itself.

In all conditions the robots served as the artificial experimenter as soon as the actual experiment started. They were operated from an adjoining room. For the

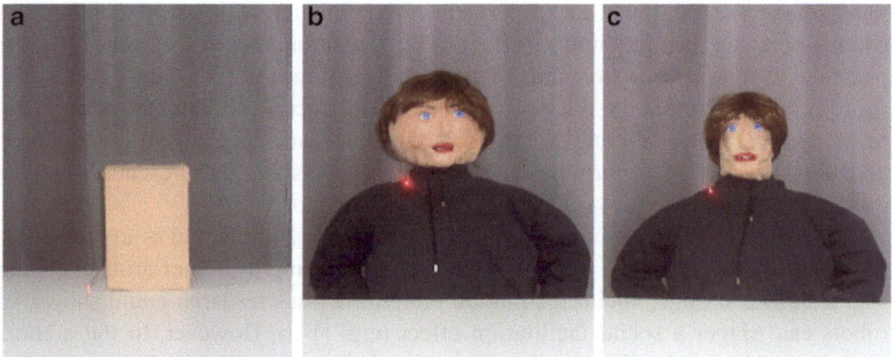

Fig. 15.1 Robot heads used in the study: (**a**) $head_{box}$, (**b**) $head_{cartoon}$ and (**c**) $head_{human}$

instructions the robots gave during the experiment, the—*bits3 de male unit selection general*—voice available from the text-to-speech system *Mary*[1] was used. It was the most gender-neutral voice of the *Mary* system and furthermore optimized for German, the language in which the experiment was conducted.

All robots are depicted in Fig. 15.1. The robots were changed for each participant but not within a participant.

15.3.2 Participants

Initially, 41 German-speaking subjects were invited to the study. Five participants were excluded from the analyses due to technical problems with the set-up, e.g. system crashes, a person disturbing the experiment or the wrong task order. Moreover, one person was 45 years old, although we stated in the call for participation that only persons aged between 18 and 35 years were permitted for the study. Hence, this participant was left out of further analysis. The data of the 35 remaining subjects was examined with respect to outliers. This led to the exclusion of five further subjects based on the criteria of $X_{individual} > 3rd\ Quartile\ of\ X_{total} + 1.5{*}InterQuartileRange\ of\ X_{total}$.

Another subject rated the condition $head_{box}$ as the most human-like condition and was also excluded from further analysis.

15.3.3 Tasks

Arithmetic tasks including subtraction and addition were chosen as such tasks are easy to vary in complexity. The *easy tasks* consisted of one two-digit number and

[1] http://mary.dfki.de/.

one single-digit number. No task included a carry operation in the units. An example for a task without a carry operation is "13 + 3", an example for a task with a carry operation is "13 + 9". The *medium tasks* were pairs of two-digit numbers and again they did not involve a carry operation, e.g. "11 + 43". The *complex tasks* were pairs of three-digit numbers and did require a carry operation in the units, e.g. "374 + 597". For all complexity levels, three subsets were prepared. For each subset the duration of the tasks was set to two minutes. Each subset contained more tasks than one could possibly complete in five minutes. This was checked in a pretest. For each complexity level the error rate was calculated.

Furthermore, the participants had to carry out a monitoring task to ensure that they were constantly aware of the robot. Therefore we installed a red LED on the robots (cf. Fig. 15.1) and asked the participants to immediately contact the experimenter in the other room if the LED starts blinking. They were told that the blinking indicates a very low battery status and that the battery is not fully charged due to the multiple previous experiments. However, this was not true: The LED would never start blinking. Participants were informed about this deception, immediately after they finished the experiment.

15.3.4 Measures

As a manipulation check for human-likeness we adapted the procedure from [15] and asked participants to sort 14 different robots, presented on 4 × 4 cm cards, with respect to their human-likeness. In addition to the three robots in the experiment, 11 further robots were used, e.g. *Kismet*,[2] *Asimo*[3] and *Nao.*[4]

To assess arousal we used the respective dimension of the *Self-Assessment-Manikin* (SAM) questionnaire. The SAM is a non-verbal measurement instrument initially developed to rate the affective quality of pictures or similar stimuli on three dimensions: valence, arousal and (social) dominance [16]. In the version we used, each dimension comprises nine human-like cartoon characters along a nine-point scale. For arousal, the manikins range from a sleepy figure to a highly aroused figure. Respondents were asked to choose the manikin most corresponding to their emotional state; it was also possible to mark between two figures.

Mental effort was measured using the *SEA* scale (*Subjektiv Erlebte Anstrengung*), a unipolar instrument ranging between 0 and 220 with higher values indicating higher effort [17]. The SEA scale is the German version of the *SMEQ* also known as *RSME* [18, 19]. We chose the SEA as it is a lightweight instrument shown to have excellent psychometric properties [19] even in comparison to more elaborate measures [20].

---

[2] http://www.ai.mit.edu/projects/humanoid-robotics-group/kismet/kismet.html.

[3] http://world.honda.com/ASIMO/.

[4] http://www.aldebaran-robotics.com/en/.

We also used two additional questionnaires, assessing the interaction with the robots, but the results will not be reported in the current paper.

15.3.5 Procedure

To avoid a social facilitation effect triggered by the human experimenter, the interaction with the participants was kept at a minimum. They were first asked to fill in the demographic questionnaire and the consent form while the experimenter was pretending to work on a computer. Next they were brought into the room where the experiment took place. They were placed at the table where the robot was sitting. Furthermore a camera was standing in the room; it was not hidden but placed behind the participants, so they could only see it when turning their head. The camera was necessary for the human experimenter to observe the progress, so that he could remotely start the different programs of the robot. Furthermore he was checking if the participants were complying with the instruction the robot gave. However, this was not told to the participants.

The robot was introduced as the artificial experimenter, who will guide them through the experiment. At this point, the robot was already switched on. Next, the experimenter switched on the LED but told the participant that he switched on the robot. He pretended to wonder why the LED was red, indicating a low battery level. He then asked the participant to monitor the robot, in order to prevent the robot to run out of power. He then left the room and started the first program from the adjoining room. The robot explained the task to the participant. The robot asked the participant to take the first sheet from the paper stack (*instructions questionnaires*). The paper stack contained the task sheets and the questionnaire. The first sheet was the training trial.

After the training trial, the robot instructed the participant to take the next sheet. The participant was asked to not remove the cover sheet and to not start calculating until the robot says "Start!". They further were told that they should stop calculating when the robot says "Stop" (*instructions test*). The stop signal was given after two minutes. This was repeated for the next two sets, which were from the same difficulty levels. Then the robot asked the participants to fill in the SEA and the SAM questionnaires for the respective task complexity condition (*instructions questionnaire*). This was repeated for all task complexity conditions. After all conditions were finished, the experimenter entered the room and switched out the LED. He asked if everything went fine and presented the human-likeness card-sorting task. Finally, the participants were paid and debriefed. An example procedure is presented in Fig. 15.2.

Fig. 15.2 Example of experimental procedure. *Dark blue boxes* indicate that only the artificial experimenter was present. *Light blue boxes* indicate that the human experimenter was present

15.4 Results

15.4.1 Manipulation Check

A Friedman test showed that the human-likeness ratings differed between the three heads, $\chi^2(2, \mathrm{N}=29)=46.41, p<0.01$. The highest rank, which indicates the highest human-likeness, was given to $head_{human}$ and the lowest rank to $head_{box.}$ The results of the post hoc tests (Wilcoxon signed-rank tests with adjusted alpha-level) showed that each head significantly differed from the others ($p<0.05$).

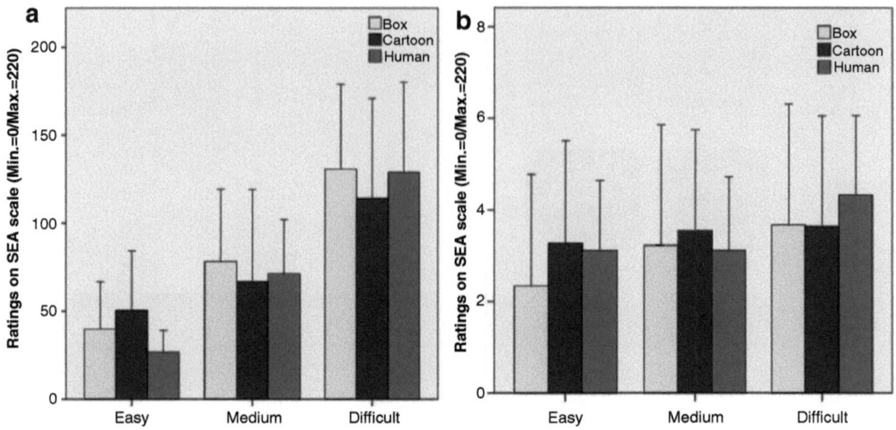

Fig. 15.3 Mean ratings on (**a**) the *SEA* scale and (**b**) the *SAM* arousal scale by task complexity and robot head. *Error bars* display one standard deviation

15.4.2 Arousal

A repeated measure ANOVA with task complexity as within factor and robot head as between-subject variable did only show a main effect for task complexity, $F(2{,}52) = 8.83$, $p < 0.01$, $eta^2 = 0.23$. Differences between the robot heads were not observed, $F(2{,}26) = 0.130$, $p < 0.879$, $eta^2 = 0.01$. Also an interaction effect between robot head and task complexity was not found, $F(4{,}52) = 1.33$, $p = 0.162$, $eta^2 = 0.09$ (cf. Fig. 15.3). However, here the effect size (eta^2) was indicating a medium effect. Thus, the non-significant result might be caused by a too small sample.

15.4.3 Mental Workload

Also for the ratings on the SEA scale, only a main effect for task complexity was observed, $F(2{,}52) = 73.54$, $p < 0.01$, $eta^2 = 0.717$. Again, neither a main effect for robot head, $F(2{,}26) = 0.103$, $p = 0.803$, $eta^2 < 0.01$, nor an interaction effect between robot head and task complexity was found, $F(4{,}52) = 1.47$, $p = 0.225$, $eta^2 = 0.031$(cf. Fig. 15.3).

15.4.4 Task Performance

Regarding error rate a significant main effect for task complexity, $F(2{,}52) = 35.68$, $p < 0.01$, $eta^2 = 0.541$, as well as for robot head was observed, $F(2{,}26) = 4.20$,

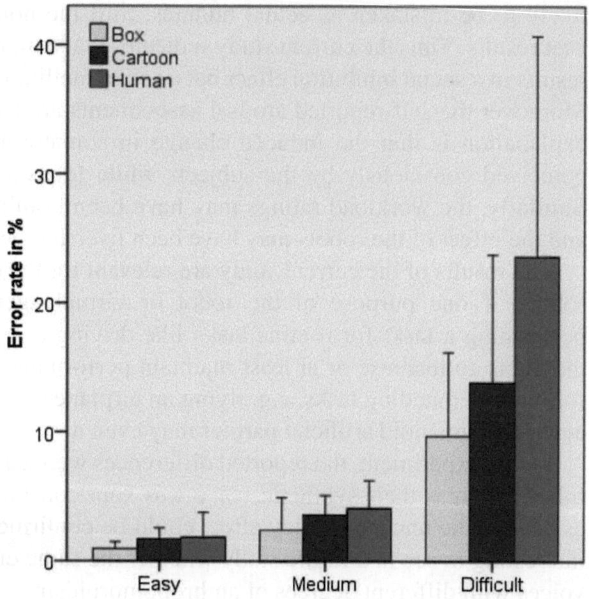

Fig. 15.4 Mean error rate in percentage by task complexity and robot head. *Error bars* display one standard deviation

$p = 0.026$, $eta^2 = 0.082$. Sidak-corrected post hoc tests showed that $head_{human}$ significantly differed from $head_{box}$. $Head_{box}$ and $head_{cartoon}$ did not differ from each other and $head_{cartoon}$ did also not differ significantly from $head_{human}$.

Furthermore, an interaction effect between task complexity and robot head could be shown, $F(4,52) = 2.70$, $p = 0.040$, $eta^2 = 0.323$. However, performance was always best in the box condition; thus the anthropomorphic heads did not facilitate performance in the easy condition, but as expected $head_{human}$ leads to the highest error rate in the difficult condition (cf. Fig. 15.4).

15.5 Discussion and Conclusion

The current study investigated the influence of three robots, differing in their level of human-likeness, on the social facilitation effect or social inhibition effect, respectively. While the robots did not lead to differences in self-reported arousal and mental workload, an influence on performance could be shown. As expected, the most anthropomorphic robot led to the worst results for complex tasks. Furthermore, only the most anthropomorphic robot differed from the non-anthropomorphic robot. The medium human-like robot did not differ from any of the other conditions. This indicates that, at least for performance measures, a higher degree of human-likeness is more likely to trigger a social inhibition effect. This may either be due to such robots being deeper in the uncanny valley or because they are more likely to be mistaken as actual humans. Still the non-human-robot always led to best results. Thus the current study indicates that a higher degree of human-likeness results in a social inhibition effect but a social facilitation effect could not be shown. Moreover the self-reported arousal assessments are rather inconsistent. A possible explanation is that the induced change in somatic arousal was too subtle to be perceived consciously by the subjects while focusing on the mental arithmetics. Similarly, the workload ratings may have been dominated by the task complexity and the effect of the robots may have been overshadowed by this.

The results of the current study are relevant for the design of socially interacting robots, if one purpose of the robot or virtual agent is to support humans in performing a task: for routine tasks like driving a car, an anthropomorphic robot may help to increase or at least maintain performance through its presence, while for more demanding tasks, e.g. flying an airplane or monitoring a complex system, an all-to-humanoid artificial partner may even decrease the actual performance.

In this experiment, the reported differences were caused by the appearance of the robot, whereas their synthetic voice was kept constant. After the social inhibition as well as the uncanny valley effect could be confirmed for this set-up, it would be interesting to see in a future study whether the same effect can also be observed for voices with different degrees of anthropomorphism.

References

1. Nourbakhsh, I.R., Kunz, C., Willeke, T.: The mobot museum robot installations: a five year experiment. In: Proceedings of the IEEE/RSJ International Conference on Intelligent Robots and Systems IROS 2003, Las-Vegas (2003)
2. Billard, A.: Robota: clever toy and educational tool. Robot. Auton. Syst. **42**(3–4), 259–269 (2003)
3. Duffy, B.R.: Anthropomorphism and the social robot. Robot. Auton. Syst. **42**(3–4), 177–190 (2003)
4. MacDorman, K.F.: Androids as an experimental apparatus: Why is there an uncanny valley and can we exploit it? In: CogSci-2005 Workshop: Toward Social Mechanisms of Android Science, Stresa, pp. 106–118 (2005)
5. Mori, M.: The uncanny valley. Energy **7**(4), 33–35 (2005) (MacDormand, K.F., Minato, T. Transl.) (Original work published 1970)
6. Saygin, A.P., Chaminade, T., Ishiguro, H., Driver, J., Frith, C.: The thing that should not be: Predictive coding and the uncanny valley in perceiving human and humanoid robot actions. Soc. Cogn. Affect. Neurosci. **7**(4), 413–422 (2011)
7. Triplett, N.: The dynamogenic factors in pacemaking and competition. Am. J. Psychol. **9**, 507–533 (1898)
8. Strubbe, M.J.: What did Triplett really find? A contemporary analysis of the first experiment in social psychology. Am. J. Psychol. **118**, 271–286 (2005)

9. Herfordt, J.E.: Soziale Erleichterung: Die Erleichterung kognitiver Prozesse durch die bloße Anwesenheit einer weiteren Person. Digitale Dissertation (Social facilitation: The facilitation of cognitive processes by the mere presence of another person. digital Ph.D. thesis), University Freiburg (2008). http://www.freidok.uni-freiburg.de/volltexte/5711/pdf/Herfordt_Diss_komplett.pdf
10. Zajonc, R.B., Heingartner, A., Herman, E.M.: Social enhancement and impairment of performance in the cockroach. J. Pers. Soc. Psychol. **13**, 83–92 (1969)
11. Guerin, B.: Social Facilitation. Cambridge University, Cambridge (1993)
12. Aiello, J.R., Douthitt, E.A.: Social facilitation from Triplett to electronic performance monitoring. Group Dyn. Theory Res. Prac. **5**(3), 163–180 (2001)
13. Sanders, G.S., Baron, R.S., Moore, D.L.: Distraction and social comparison as mediators of social facilitation effects. J. Exp. Soc. Psychol. **14**, 291–303 (1978)
14. Riether, N., Hegel, F., Wrede, B., Horstmann, G.: Social facilitation with social robots? In: Proceedings of the Seventh Annual ACM/IEEE International Conference on Human–Robot Interaction (HRI'12), pp. 41–48. ACM, New York (2012)
15. DiSalvo, C.F., Gemperle F., Forlizzi, J., Kiesler, S.: All robots are not created equal: the design and perception of humanoid robot heads. In: Proceedings of the 4th Conference on Designing Interactive Systems: Processes, Practices, Methods, and Techniques (DIS'02), pp. 321–326. ACM, New York (2002)
16. Bradley, M.M., Lang, P.J.: Measuring emotion: The self-assessment manikin and the semantic differential. J. Behav. Ther. Exp. Psy. **25**, 49–59 (1994)
17. Eilers, K., Nachreiner, F., Hänecke, K.: Entwicklung und Überprüfung einer Skala zur Erfassung subjektiv erlebter Antrengung (Development and evaluation of a scale to assess subjectively perceived effort). Zeitschrift für Arbeitswissenschaft **40**, 215–224 (1986)
18. Zijlstra, F.R.H.: The construction of a scale to measure perceived effort. Ph.D. thesis, Delft University of Technology, Delft (1985)
19. Sauro, J., Dumas, J.: Comparison of three one-question, post-task usability questionnaires. In: Proceedings of CHI 2009, pp. 1599–1608 (2009)
20. De Waard, D.: The measurement of drivers' mental workload. Ph.D. thesis, University of Groningen, Traffic Research Centre, Haren (1996)

[illegible] Person [illegible] (digital [illegible]) [illegible]

[illegible]

[illegible] (2007)

[illegible] (2012)

[illegible]

[illegible] (2004)

[illegible]

Chapter 16
More Than Just Words: Building a Chatty Robot

Emer Gilmartin and Nick Campbell

Abstract This paper presents the motivation for, design and current implementation of a robot spoken dialogue platform, created to aid exploration of multimodal human-machine dialogue. It also describes the design of a dialogue used to collect a database of interactive speech while the robot was exhibited a over a three-month period at the Science Gallery in Dublin. The system was wizard-controlled and collected samples of informal, chatty dialogue—normally difficult to capture under laboratory conditions for human-human dialogue and particularly so for human-machine interaction. The system is being further developed to facilitate further data collection and experimentation.

16.1 Conversation Is Not Only an Exchange of Words

Speech meditates human interactions in all areas of life. Some conversations have a clear purpose such as communicating important information (warning) or creating change (giving an order), while, in others, the goal of the exchange is not so much to transfer linguistic information as to cement social bonds.

Earlier accounts of human spoken interaction considered speech as the (often imperfect) verbal transmission of thought or text [3, 11], with syntactic and semantic content forming the core message while other components (e.g. prosody, timing, gesture) were seen as wrapping which added expressive value. More recent work has considered human spoken dialogue as a dynamic joint activity where participants collaborate to accomplish a common goal or resolve a coordination problem [4, 9]. Dialogue is viewed as a series of exchanges designed to carry out a goal, and much experimental work has relied on corpora created from task-based dialogues such as

E. Gilmartin (✉) • N. Campbell
Speech Communications Lab, Trinity College Dublin, Dublin, Ireland
e-mail: gilmare@tcd.ie; nick@tcd.ie

J. Mariani et al. (eds.), *Natural Interaction with Robots, Knowbots and Smartphones: Putting Spoken Dialog Systems into Practice*, DOI 10.1007/978-1-4614-8280-2_16,

tangrams or map tasks [7]. This view does not cover ongoing ebbing and flowing "chat" where content may be trivial or phatic. The function of this chat is to cement bonds or form the "soundtrack" to co-presence rather than to transmit information or collaborate on a task [10]. Most conversations contain elements of chat and task-based dialogue and successful dialogue modelling should account for both.

Natural dialogue design must also consider the multimodality of natural conversation. Unimodal (speech channel only) is a very restricted form of dialogue, overrepresented in dialogue systems as a result of the preponderance of telephone-based systems. Natural dialogue comprises several modalities or streams of information, including facial expression, gaze and gesture recognition (visual channel), in addition to audio information. Dialogue systems intended to interact naturally with humans must take on an active listening and watching role.

Historically, speech technology has concentrated on means of production (TTS) and recognition (ASR) of linguistic or propositional content encoded in speech, allowing users to perform tasks that previously required keyed input and read output. Early systems were essentially text querying of databases with speech bolted on. More recently focus has shifted towards making interaction more human-like, using more expressive speech, adding backchannels, adaptation and more realistic turntaking. While modern interactive systems can process some of the linguistic aspects of human communication, they are not yet capable of processing the complex dynamics involved in social interaction and fail to capture the coordination and adaptation on the part of interlocutors [5]. A multimodal system that uses chat as part of its interaction would more closely model human-human interaction and could aid better and more efficient human-machine interaction.

16.2 Chat Versus Task-Based Dialogue

We consider spoken interaction an amalgam of two activities—task-based dialogue and chat, sharing the same speech apparatus. We can outline a continuum from completely phatic "chat" to content-critical spoken communication—for example, pleasantries between two people at a bus stop would fall near the chat end of the spectrum while air traffic control communication would fall at the content critical task-based end (see Fig. 16.1).

This view is consistent with theories of language evolution which see speech as information transfer, and those based on language as a social activity [6]. Chat and task-based dialogue are contrasted below.

Goal. In chat, the goal of conversation is to establish co-presence, build or maintain interpersonal bonds and keep channels of communication open. This goal may extend far beyond the life of an individual session or dialogue, and participants may be unaware of why they are chatting. In task-based dialogue, goals by definition are the accomplishment of the task and are more discrete, short term and clearly defined. Contrast (1) a customer booking a flight in a travel agent with (2) a resident

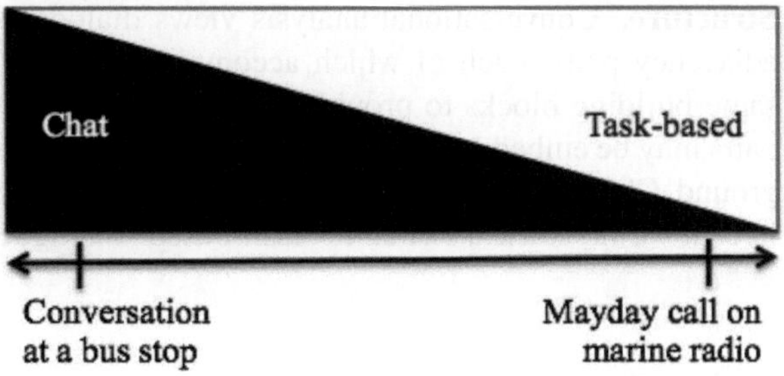

Fig. 16.1 A continuum for chat and task-based dialogues

having a short daily chat with the concierge in the lobby of a block of flats. In the travel agents, the goal is a successful business transaction. Both participants are aware of this goal, and the task will typically succeed or fail in the course of one interaction. In the lobby, the goal is to maintain a positive social relationship, participants probably do not have this goal explicitly in mind, and the relationship continues over an extended period of time. Real-life dialogues do not fall completely into chat or task-based categories, but move up and down the spectrum as necessary, throughout the duration of the interaction. For instance, in a business meeting, important exchanges where content is critical are interspersed with lighter intervals where good relations are maintained or reaffirmed.

Cognitive Processing. The "dual process" theory of cognition can be applied to spoken dialogue. In this model a phenomenon can result both from an implicit unconscious process (commonly known as System 1) and from an explicit conscious one (System 2) [8]. The action of creating a response to content-critical information in a task-based dialogue is analogous to System 2, while the processing of chat can be shallow and fast, as per System 1. The advantage to the organism of System 1 processing is economy of cognitive load.

Context. Much conversation analysis is based on a left-context reactive view of dialogue where interlocutors consciously use linguistic information received to formulate their next contribution. This may hold true for task-based dialogue to some extent. However, dialogue toward the chat end of the spectrum relies more on feed forward control or prediction, where the speaker does not need to consciously analyse the content of the previous utterance, and may produce utterances "unconsciously" at appropriate intervals based on past experience.

Content. The importance of information content varies with dialogue type. Small talk or chat is light, without controversial content, and often phatic. Where information transfer is the object of conversation, as in task-based exchanges, content may be positive, negative or controversial. In this kind of interaction, the content may be unwelcome to the receiver, but the importance of accomplishing the task overrules social considerations. In an emergency or in an operating theatre (for example), orders are given tersely and may appear impolite to an onlooker.

Structure. Conversational analysis views dialogue as consisting of often nested adjacency pairs, each of which accomplishes a task. Chat, in our view, uses the same building blocks to provide co-presence. In a task-based dialogue, adjacency pairs may be embedded to several levels as participants collaborate to build common ground. Chat, on the other hand, is more shallow. The occurrence of deep embedding in chat would cause a switch to "task-based" dialogue.

16.3 Human-Machine Chat

To implement multimodal chat and task-based human-machine dialogues, we designed a robot platform to engage strangers in friendly conversation and tested it over three months with visitors to the Science Gallery, Dublin.

The robot, Herme, was built using LEGO Mindstorms NXT technology, programmed in Python (see Fig. 16.2). Below we further describe the design of the dialogue and system hardware and implementation in a Wizard of Oz scenario.

Dialogue. The type of dialogue we were interested in collecting was informal chat, as described above. Here, content is secondary to social interaction. We designed a script with chat and task-based elements with reference to previous work on making human-machine dialogue more realistic [1]. This script was a non-branching sequence of predetermined utterances. The strategy used short "volleys" where the robot asked a question or made a statement followed by a related question,

Fig. 16.2 Herme and Him, two robots (one female, one male) that observe people while they are talking to them. Built from LEGO, they support cameras and microphones for interactive speech synthesis

succeeded by a wait for the interlocutor to respond, with the robot then providing interjections of "really", "oh" or "why" to establish and maintain the illusion of attention and backchannelling.

By thus maintaining the initiative throughout the interaction, we were able to substitute "polite listening" for any form of "understanding" of the visitor's reply. Several participants commented on this, likening the experience to "talking to someone at a party". The dialogue was designed to contain an initial phase of friendly chat followed by a task—id number collection—and returning to chat and joke telling. We were lucky to hit upon an effective dialogue sequence very early on in the research and gained much insight into the successful maintenance of a conversation through critical timing after monitoring people's responses to the utterances.

System Hardware. The capture environment consisted of a prominent corner booth with a large display screen over a waist-high platform containing the robots, supervised by gallery attendants who made sure that participants understood that their interaction with the robot was recorded. There were two Sennheiser MKH60 P-48 shotgun microphones mounted at the top of the main screen, alongside a Logitech C-910 HD webcam that provided a top-down overview of the interaction. During the latter half of the exhibition we added a movement sensor to trigger onset and offset of the conversations as an additional control sensor. On the platform were two robots, one male and one female (see Fig. 16.2), with the female (on the left) interacting with visitors while the male guided another Logitech HD webcam to ensure that the interactions were recorded from a more inclusive angle. The main (female) robot camera stayed zoomed in to observe the face of the main interlocutor. Microphones on the webcams provided a close-up source of sound to be used in conjunction with that of the shotgun microphones. An iSight camera mounted at the corner of the display gave the remote operator a wide overall view.

Inside the booth, two Mac-Minis controlled the robot and a Unix workstation collected and stored the data and provided the Skype interface for the wizard.

Synthesis was made by the default Apple synthesiser using Princess voice, shifted upwards acoustically by a Roland Sound Canvas UA-100. The machines ran continuously, streaming all data to disk while the Gallery was open.

Wizard of Oz. The final dialogue was used as the basis for the Wizard of Oz study. During these sessions the operator waited until the face recognition software had identified a person in Herme's range and then stepped through the pre-prepared script. The operator pressed a key to produce the first utterance and could then press to produce each successive utterance or stop the dialogue. The operator was free to wait as long as he/she liked between utterances but instructed to press to respond when it felt natural. The Wizard paradigm was used to capture natural inter-speaker gap lengths for the dialogue—these gaps are currently being used as models for improved automated gap timing. Participants were encouraged to stay and continue the conversation at several stages throughout the dialogue. Interjections such as "I like your hair!" surprised people, but kept them interested. "Do you know any good jokes?" usually elicited a negative response, to which the robot laughed, but the subsequent "tell me a knock-knock joke" was in almost all cases dutifully complied with, as was the polite (and often genuinely amused) listening to the robot's joke in turn. The robot's laugh was "captivating" [2].

Visitors collaborated with the robot to keep talking and we collected 483 signed consent forms, despite the robot's total lack of speech recognition and inability to respond reactively to propositional content. The participants' willingness to maintain conversations is consistent with a System 1 imperative to build and maintain social bonds. Conversations ran into difficulty when information transfer became important (obtaining signed consent for use of the recordings) and the inability of the robot to negotiate meaning led to interlocutors coming "out of step". As participants were not paid subjects but random visitors who walked in off the street, their only incentive to continue talking to a piece of plastic was the result of deeply ingrained social norms.

16.4 Summary and Future Work

This work describes the design and implementation of a robot platform for the extraction of data and acquisition of knowledge related to spoken interaction, by capturing natural language and multimodal/multisensorial interactions using voice-activated and movement-sensitive sensors in conjunction with a speech synthesiser. The platform proved a low-cost way to collect massive amounts of real-world data.

The system is being improved and a single-machine portable system is under construction for fieldwork. The data collected are currently being analysed to gain insight into aspects of human-machine interaction in domains including prosody, gesture, and dialogue structure. The corpus is now available to interested researchers via the Internet.

Acknowledgements This work is being carried out at TCD with thanks to Science Foundation Ireland (SFI Stokes Professorship Award 07/SK/I1218) and PI (FastNet project) grant 09/IN.1/I2631.

References

1. Bickmore, T., Cassell, J.: How about this weather?-social dialogue with embodied conversational agents. In: Proceedings of the AAAI Fall Symposium on Socially Intelligent Agents. North Falmouth, MA (2000)
2. Campbell, N.: On the use of nonverbal speech sounds in human communication. In: Proceedings of the 2007 COST Action 2102 International Conference on Verbal and Nonverbal Communication Behaviours (2007)
3. Chomsky, N.: Aspects of the Theory of Syntax. MIT Press, Cambridge (1965)
4. Clark, H.: Using Language. Cambridge University Press, Cambridge (1996)

5. Douxchamps, D., Campbell, N.: Robust real time face tracking for the analysis of human behaviour. In: Proceedings of the 4th International Conference on Machine Learning for Multimodal Interaction (2007)
6. Dunbar, R.: Grooming, Gossip, and the Evolution of Language. Harvard University Press, Cambridge (1998)
7. Garrod, S., Anderson, A.: Saying what you mean in dialogue: A study in conceptual and semantic co-ordination. Cognition **27**(2), 181–218 (1987)
8. Kahneman, D.: Thinking, Fast and Slow. Farrar Straus and Giroux, New York (2011)
9. Lewis, D.: Convention: A Philosophical Study. Wiley-Blackwell, New York (2002)
10. Malinowski, B.: The problem of meaning in primitive languages. Supplementary in the Meaning of Meaning (1923)
11. Saussure, F.D.: Course in General Linguistics. McGraw Hill, New York (1916)

Chapter 17
Predicting When People Will Speak to a Humanoid Robot

Takaaki Sugiyama, Kazunori Komatani, and Satoshi Sato

Abstract We tackle the novel problem of predicting when a user is likely to begin speaking to a humanoid robot.

Human speakers usually take the state of their addressee into consideration and choose when to begin speaking to the addressee, and our idea is to use this convention with a system that interprets audio input.

The proposed method predicts when a user is likely to begin speaking to a humanoid robot by machine learning that uses the robot's behaviors—such as its posture, motion, and utterance—as input features.

We create a data set manually annotated by three human participants indicating in real time whether or not they would be likely to begin speaking to the robot. We collect the parts to which the three commonly give the same labels and use these parts as the training and evaluation data for machine learning.

Results of an experimental evaluation showed that our model correctly predicted 88.5% of the common parts in an open test. This result is similar to the results of a cross-validation, demonstrating that our model is not dependent on a specific training data set. A possible application of the model is the elimination of environmental noises that occur at timing when a cooperative user is not likely to begin speaking to a robot.

17.1 Introduction

Incorrect or unnecessary responses are a critical problem in speech interaction with humanoid robots. Such responses are caused by surrounding noises when a humanoid robot is in real environments, particularly those in which a robot uses

T. Sugiyama (✉) • K. Komatani • S. Sato
Graduate School of Engineering, Nagoya University, Nagoya, Japan
e-mail: takaak_s@nuee.nagoya-u.ac.jp; komatani@nuee.nagoya-u.ac.jp; ssato@nuee.nagoya-u.ac.jp

J. Mariani et al. (eds.), *Natural Interaction with Robots, Knowbots and Smartphones: Putting Spoken Dialog Systems into Practice*, DOI 10.1007/978-1-4614-8280-2_17,

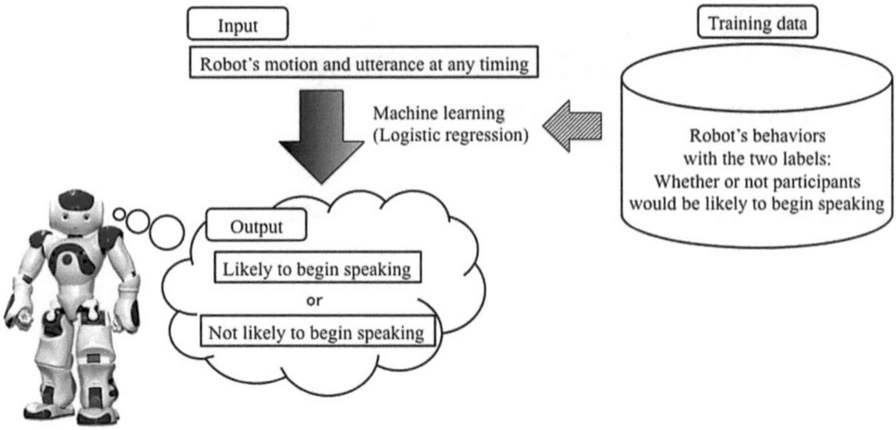

Fig. 17.1 System overview

its own microphone for automatic speech recognition. A robot needs to ignore unnecessary sounds, such as environmental noises, user's unintentional mutters, and laughter. It is natural to focus on input sounds to ignore such sounds, and there have been several studies investigating this [6, 14]. For example, Lee et al. [7] proposed a method to distinguish human voices from noises by using a Gaussian mixture model (GMM) based on acoustic features [5]. Several other studies have also used visual information to detect a user's vocal activity [1].

We adopt a novel approach to ignoring noises: we shift our focus from input sounds to the recipient state, i.e., the robot's state. This idea is inspired by the fact that human speakers usually take the state of their addressee into consideration and chooses when to begin speaking to the addressee. We assume that this convention can be applied to human-robot interaction and have built a model that uses the robot's state to predict when the user is likely to begin speaking. If the robot can predict the user's timing, it can know the timing when a cooperative user is likely to begin speaking. Conversely, audio received when the user is unlikely to begin speaking can be discarded as non-speech with high probability. The proposed model can be used in conjunction with existing approaches focusing on input sounds to help distinguish user utterances.

An overview of the proposed method is shown in Fig. 17.1. Our approach is twofold:

1. We formulate the problem by a machine learning framework.
2. We collect training data for the machine learning by using multiple participants.

First, we define the robot states to be used as input features. These states are defined at any timing by using the robot's behaviors, such as its posture, motion, and utterance. We then predict whether the user is likely to begin speaking. That is, we cast the problem as a binary machine learning task, where the input consists of features describing the robot's state and the output indicates whether the user is "likely to begin speaking" or "not likely to begin speaking." We use a logistic regression function as a classifier.

Second, we collect annotated data for machine learning. Different people may have different ideas about when to begin speaking, and so we asked several participants to label some real-time behaviors. While watching a sequence of behaviors of a humanoid robot, they indicated in real time whether or not they would be likely to begin speaking. We use the parts to which participants give the same labels as the training and evaluation data for machine learning. In essence, this enable us to extract the parts for which several participants are likely to begin speaking, thereby eliminating the effect of individuality.

Our prediction target, i.e., when people will begin speaking to a humanoid robot, corresponds to the transition relevance place (TRP) in human-human conversation. TRP was first advocated by Sack et al. [11] and is now a famous notion in the human conversation analysis field. It indicates places where the addressee is likely to begin speaking, i.e., to take a turn. Several studies analyzing human conversation have also revealed that nonverbal behaviors are useful cues for turn-taking [2, 5]. These findings have been introduced into human-machine interaction. Skantze and Gustafson [12] monitored users attention by tracking their head movements. Vertegaal et al. [13] constructed a conversational system that uses eye movements to estimate to which agent the user is listening or speaking. In the proposed method, we also use multimodal information provided not by a human user but by a humanoid robot. Because the robot behaviors are controllable by the system developer, the proposed model has the potential to control user behaviors as well as to ignore noises.

The rest of this paper is organized as follows. We define the prediction target—that is, whether or not a user is likely to begin speaking—in Sect. 17.2. In Sect. 17.3, we describe how to collect the annotated data used for machine learning. Input features and experimental evaluation based on the collected data are described in Sect. 17.4. In Sect. 17.5, we conclude the paper.

17.2 Definition of Whether a User Is Likely to Begin Speaking or Not

Our model predicts whether a user is likely to begin speaking to the robot or not. This is the output of our model and it is used as the teaching signal for machine learning. An overview of our model is shown in Fig. 17.1.

We assume the following three conditions in this paper:

1. The content that the user is trying to convey to the robot is not urgent.
2. The user regards the robot as a social being.
3. Only one user is participating in the conversation.

First, the contents of an utterance are crucial to determine whether or not a user is actually likely to begin speaking to the robot. For example, a very urgent content, e.g., asking the robot to call an ambulance, would be communicated regardless of the robot's state. We here assume that the content a user tries to convey must not be urgent. Something like asking the robot to turn on the air conditioner would qualify.

Second, we assume that a user treats the robot not completely as a machine but as a social being; that is, we assume that the user feels a kind of anthropomorphism. It is well known that robots such as Geminoid [3] and Repliee R1 [8], which closely resemble humans, make users feel anthropomorphism [4, 9]. If a robot moves only when instructed, people begin speaking to it without really considering its state. Here, we use a humanoid robot that speaks and moves like a human, thereby satisfying this assumption. This tendency has also been demonstrated in the famous psychological experiment by Reeves and Nass [10].

Finally, we assume that only one user participates in a given conversation. This is just for the sake of simplicity. If several users are participating in the same conversation, whether or not the user is likely to begin speaking varies depending on each user's position. For example, suppose a robot turns towards user A and stops silently, and user B is located to the left of the robot. In such a case, we expect that user A will begin speaking, but user B does not, because the robot is heading towards user A. To simplify the problem, we only consider cases with a single user. Handling multiple users will be one of our future works.

17.3 Preparing Target Data

17.3.1 Building Robot's Actions

As factors affecting whether a user is likely to begin speaking to the robot, we take into consideration the robot's posture, motion, and utterance, specifically whether or not the robot utters, moves, or turns towards the user. We only use factors that the robot can automatically obtain because these will be used as input features in online interactions.

Whether a user is likely to begin speaking cannot be determined by just one factor because an actual robot's behaviors contain these factors in a *continuous* and *compounded* manner. By *continuous* we mean that the robot exhibits another behavior directly after making one behavior. If the robot exhibits two or more behaviors continuously, a user may consider the relationship between them when deciding whether or not to speak. For example, a user may be likely to begin speaking when the robot turns towards him/her after speaking towards another direction. By *compounded* we mean that the robot exhibits behaviors containing many factors at a time. In such behaviors, whether the user is likely to begin speaking differs in accordance with the combinations of factors. For example, a user

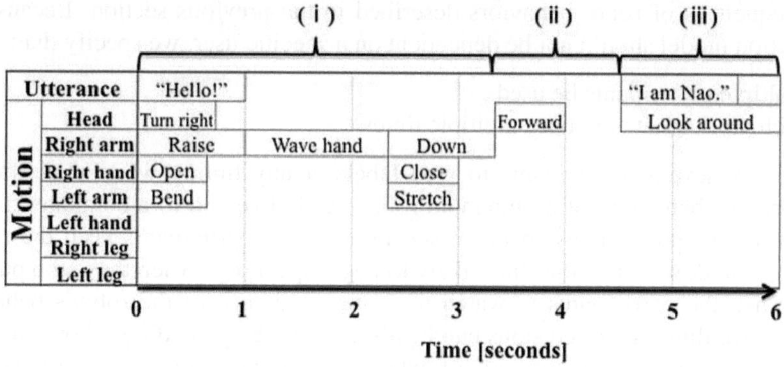

Fig. 17.2 Part of robot behaviors

is usually likely to begin speaking when the robot says nothing. However, people may not be likely to begin speaking when the robot bows and makes no utterance.

We made two sequences of robot behaviors that contain the various factors in a continuous and compounded manner. We use a humanoid robot called Nao made by Aldebaran Robotics[1] and use VoiceText as the text-to-speech (TTS) engine.[2] The content of the two sequences is a self introduction by Nao. Hereafter, we call these two sequences α and β. Sequence α was 150.0 s long and sequence β was 259.3 s long. We made sequence α first and then sequence β because sequence α is simpler and speech information is dominant in it. Sequence β is longer and contains more varied combinations of factors than sequence α. We use these sequences as training and evaluation data sets after human annotation, as described in the next section.

A part of the robot behaviors is depicted in Fig. 17.2 as an example. In (i), the robot waves its hand, turns right, and says "Hello." Specifically, the robot moves its head, then right arm, and then right and left hands. Since we assume that a user is in front of the robot, the robot is not facing the user in this part. In the next part (ii), the robot faces forward—that is, the direction of the user—and does not move or speak. In (iii), the robot speaks for about 1 s and looks around. Here, (i) and (iii) exhibit compounded factors and (i) to (ii) exhibit continuous factors.

17.3.2 Data Collection Using Participants

We collect a data set to which human participants give labels, indicating whether or not they would be likely to begin speaking to the robot. The target data is the

[1] http://www.aldebaran-robotics.com/.

[2] http://voicetext.jp/.

two sequences of robot behaviors described in the previous section. Because the prediction model should not be dependent on a specific user, we specify that:

- Multiple participants be used
- Each participant annotates multiple times

We ask several participants to give labels at any timing while they watch a sequence of behaviors of a humanoid robot. A GUI rigged to a computer display shows and records "likely to begin speaking" when participants keep the mouse button push down and show "not likely to begin speaking" when they stop pushing it. We ask the participants to watch the entire sequence of the robot's behaviors because we think their decisions can be affected by the preceding robot behaviors. There are three participants, all of whom are students in our laboratory. The procedure for the data collection was as follows:

1. Participants were instructed on the experiment procedure and the GUI usage and then allowed to practice with the GUI for a while.
2. Participants sat down in front of the robot.
3. Participants watched the robot behaviors several times. This was to prevent a participant from being surprised when watching the robot's motion for the first time.
4. Participants annotated whether or not they would begin speaking by using the GUI. They watched the same sequence of the robot behaviors three times.

Written instructions were provided before the experiment to avoid any inaccuracies stemming from potentially unreliable oral instructions. Participants could ask questions as needed. Participants were also given the following instruction to imagine a concrete situation:

> Please indicate when you can say "Hey" to the robot to ask it to speak a little bit more loudly.

17.3.3 Analysis of Collected Data

Here, we analyze the differences among participants on the basis of these data sets. More specifically, we investigate whether there are common parts at which the three participants are likely to begin or not begin speaking. This is to check whether the collected data could be used as training data for the machine learning.

We obtained three data sets from each participant after the data collection described in Sect. 17.3.2 (each participant watched the same sequences three times). We use the second data set for each participant because we feel the participants may not be skilled enough to use the GUI during the first trial. We also think the participants might forget to give labels for the first one, although we tried to prevent them from being surprised. As for the third set, some participants seemed tired by

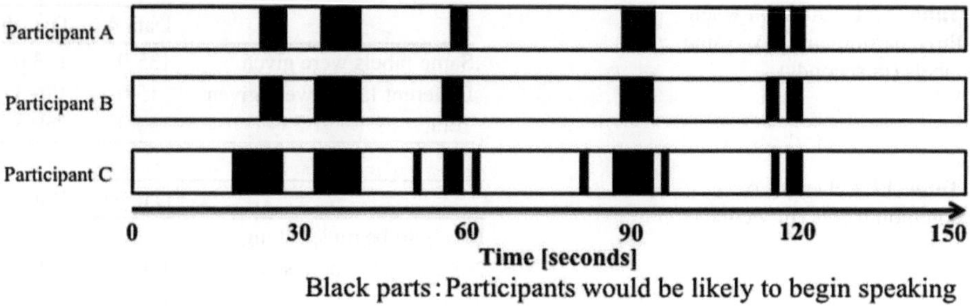

Fig. 17.3 When three participants would be likely to begin speaking for data α

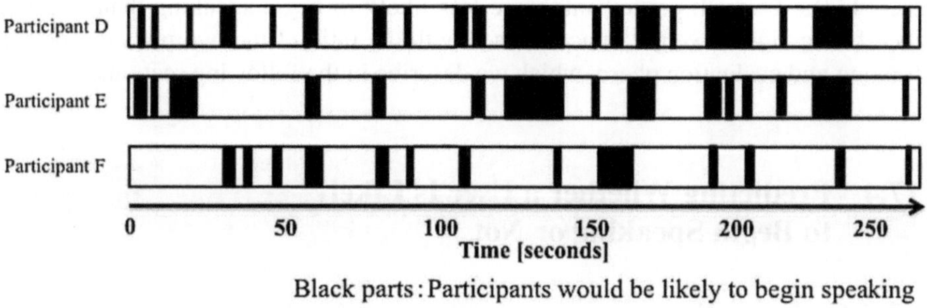

Fig. 17.4 When three participants would be likely to begin speaking for data β

this time. Consequently, we use the two sequences from the second trial and denote them as data α and β. Each data set contains the annotations of all three participants.

The labels given by the three participants are shown in Figs. 17.3 (data α) and 17.4 (data β). These figures show the time at which each participant pushed the mouse button; that is, when they would be likely to begin speaking. The black parts denote when the participants would be likely to begin speaking and the white parts denote when they would not be. The horizontal axis is the time of the robot behaviors.

We investigate whether the three participants give the same labels to the same behaviors. For example, in data α (Fig. 17.3), all three participants would be likely to begin speaking at 25, 40, 60, 90, 115, and 120 s during the robot behaviors. There were also several common parts where they were not likely to begin speaking. The results show that, in general, there are several common parts mixed in with a few that the participants do not agree on. We use the common parts as the training and evaluation data.

Here, we discuss the details of the collected data. Table 17.1 shows that the three participants gave the same labels for 135.0 of 150.0 s in data α and 143.0 of 259.3 s in data β. The details of parts given the same labels are shown in Table 17.2. In data α, they would be likely to begin speaking for 14.2 s and not likely to begin speaking for 120.8 s during the 135.0 s. In data β, they were likely to begin speaking for 16.1 s and not likely to begin speaking for 126.9 s during

Table 17.1 Duration when three participants gave same labels (in seconds)

	Data α	Data β
Same labels were given	135.0	143.0
Different labels were given	15.0	116.3
Total	150.0	259.3

Table 17.2 Details of common labels (in seconds)

	Data α	Data β
Likely to begin speaking	14.2	16.1
Not likely to begin speaking	120.8	126.9
Total	135.0	143.0

the 143.0 s. Overall, participants were less likely to begin speaking than not. We therefore assigned weight in accordance with the ratio of the two parts during the training and evaluation phase, which we describe in the following sections.

17.4 Predicting Whether a User Is Likely to Begin Speaking or Not

17.4.1 Input Features

Here, we explain the input features used in the logistic regression to predict whether the user is likely to begin speaking or not. The nine features are listed in Table 17.3. These features are obtained every 0.1 s and the prediction is thus also performed every 0.1 s.

Feature x_1 represents the elapsed time from the end of the robot's previous utterance. This feature, which we call the *speech interval*, is calculated for every timing t, i.e., when it is possible for a user to start speaking. Feature x_1 is defined as

$$x_1 = \begin{cases} t - (t_i + t_0) & (x_1 > 0) \\ 0 & (otherwise) \end{cases} \tag{17.1}$$

Here, t_i is the end time of the previous robot utterance i and t is the current time. t_0 is a constant offset reflecting the time until the user perceives the end of a robot utterance. We set t_0 to 1.1 after a preliminary experiment. An outline of the speech interval is depicted in Fig. 17.5, where the horizontal axis is time and the robot utterances are indicated by the black bars.

Features x_2 and x_3 represent whether a robot utterance is a question or not: $x_2 = 1$ when the previous robot utterance is interrogative and $x_3 = 1$ when the utterance ends with a rising intonation. These features remain "1" until the next robot utterance starts.

Features x_4 to x_7 represent the robot's motion and are defined by changes in the joint angles of the robot. The robot we used, Nao, has 26 joint angles and their

Table 17.3 Input features obtained from robot behaviors

x_1	Speech interval	Elapsed timing from the end of robot utterance
x_2	Utterance pattern	Whether robot utterance is interrogative
x_3	Prosody	Whether robot utterance ends with rising intonation
x_4	Motion (head)	Angle difference with previous frame (0.1 s before)
x_5	Motion (left arm)	Angle difference with previous frame (0.1 s before)
x_6	Motion (right arm)	Angle difference with previous frame (0.1 s before)
x_7	Motion (legs)	Sum of angle differences of legs with previous frame (0.1 s before)
x_8	Head/eye direction (horizontal)	Angle position from the front
x_9	Head/eye direction (vertical)	Angle position from the front

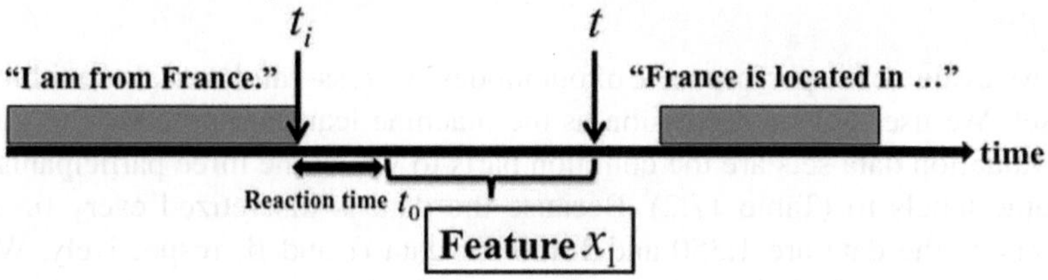

Fig. 17.5 Feature x_1: speech interval

angle positions can be obtained via its API. Here, "change" is defined as the absolute difference of the angle position from the previous frame (i.e., 0.1 s before). We sum up the difference for each part (head, left arm, right arm, and legs) to roughly represent the robot's motion and eliminate small noises from the position sensors.

Features x_8 and x_9 represent the robot's head/eye direction and are defined by using the angle positions of the robot's neck, which can also be obtained via its API. These features are defined as the absolute angle difference (in radians) between the user's position and the robot's head/eye direction, since both represent whether the robot turns towards the user or not. We assume that the user sits down in front of the robot, so these features are simply angle positions from the front. The outline of the robot's head/eye direction is shown in Fig. 17.6. The left and right figures are for features x_8 and x_9, respectively.

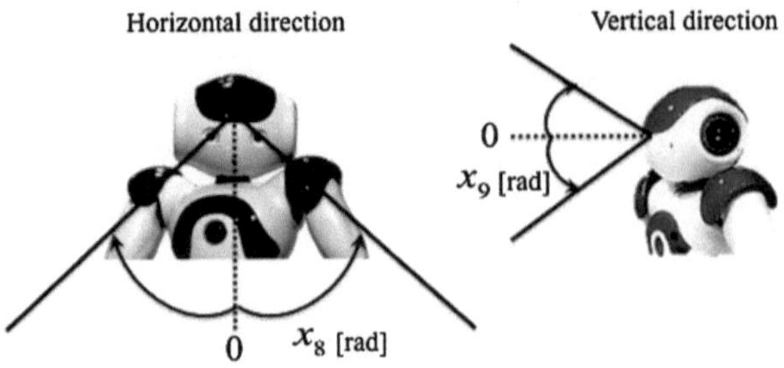

Fig. 17.6 Features x_8 and x_9: robot's head/eye direction

Table 17.4 Prediction accuracy by our proposed method and subsets of the features (%)

	Data α	Data β
All features (proposed)	87.4	92.1
Utterance only	85.9	91.2
Motion only	63.2	76.0
Head/eye direction only	63.4	71.5

17.4.2 Experimental Evaluation by Cross-Validation

First, we evaluate the performance of our model by cross-validation, i.e., within each data set. We use logistic regression as the machine learning method. The training and evaluation data sets are the common parts to which the three participants gave the same labels to (Table 17.2). Because the data is discretized every 0.1 s, the numbers of the data are 1,350 and 1,430 for data α and β, respectively. We use the common agreed labels as the teaching signals of the target variable; that is, we assign "1" when the agreed label is "likely to begin speaking" and "0" otherwise. The explanatory variables are the nine input features described in the previous section. We gave weight to the cases of "likely to begin speaking" in accordance with the numbers of the two labels (8.5 and 7.9) for data α and β, respectively. The performance was measured in terms of prediction accuracy, i.e., the ratio of the number of correctly predicted labels to the number of all agreed labels. A stratified tenfold cross-validation was performed.

The prediction accuracies of the proposed method were 87.4% and 92.1% for data α and β, respectively (Table 17.4). We also show how much the individual features contributed to the performance. We classify the nine features into three groups: "Utterance," "Motion," and "Head/eye direction." "Utterance" includes x_1 (speech interval), x_2 (utterance pattern), and x_3 (prosody). "Motion" includes x_4 (head motion), x_5 (left arm motion), x_6 (right arm motion), and x_7 (legs motion). "Head/eye direction" includes x_8 (horizontal head/eye direction) and x_9 (vertical head/eye direction). Their prediction accuracies are also listed in Table 17.4. The prediction accuracies of "Utterance" were 85.9% and 91.2% for data α and β, respectively. "Utterance" was the most effective for the prediction. "Motion" and "Head/eye direction" were less effective when used by themselves but were helpful when used together with other features.

Table 17.5 Comparison of models with different training data (open vs. cross-validation)

Training data	Relationship with test data	Prediction accuracy for data α
Data β	Open	88.5% (1,195/1,350)
Data α	Cross-validation	87.4% (1,180/1,350)

17.4.3 Experimental Evaluation for Open Data

We performed an additional experiment in which we completely separate the evaluation data set from the training data set. This shows that our model is effective for data sets other than the original training data set. We train the model using data sets α and β and then evaluate them on data α. This evaluation within data α is performed also by tenfold cross-validation. The case using data β is an open test. The results are summarized in Table 17.5. The prediction accuracies of the open test and the cross-validation were 88.5% and 87.4%, respectively.

The accuracy for the open test, 88.5%, was only 1.1% higher than that for the cross-validation. Since it was almost equivalent to that of the cross-validation, we think the model trained with a specific data set can also be effective for another data set. This result demonstrates that our proposed model does not depend on a specific training data set.

17.5 Discussion and Conclusion

Human speakers usually take the state of their addressee into consideration; that is, they do not begin speaking at a random timing. After assuming a user begins speaking to a humanoid robot in a similar manner, we constructed a model for predicting the timing a user is more likely to begin speaking to the robot. More specifically, we predict it by machine learning that uses humanoid robot's behaviors (its posture, motion, and utterance) as the input features. We first evaluated our model by tenfold cross-validation and confirmed the prediction accuracies in each test set: 92.1% and 87.4%. We then further evaluated our model using completely different data sets for the training and test, i.e., an open test. The prediction accuracy was 88.5%, which was almost equivalent to the case of cross-validation (87.4%). This result demonstrates that our model does not depend on a specific training data set.

The results of this study show that the robot's motion, posture, and utterance are useful features for predicting whether a user is likely to begin speaking. A possible application of the proposed model is to eliminate environmental noises that occur at times when a cooperative user is not likely to begin speaking to the robot. We can now consider the recipient state, as well as information obtained from audio when making predictions.

Our work in this study is the first step of larger work. The following issues still remain:

1. Variety of robot behaviors. We used only two behavioral sequences. More data will make the results more comprehensive.
2. Variety of participants. We used common parts annotated by just three participants; using more participants is desirable. In addition, any cultural differences among the participants would be interesting.
3. Interactive data collection. All the experiments in this paper were performed offline. It will be interesting to determine what happens when participants actually utter something and the robot replies, instead of just clicking a mouse.

We plan to develop an interaction robot that ignores unnecessary environmental noises. In other words, we will integrate our model with the existing statistical model that classifies input sounds into human voices and noises [7]. We think this integration will enhance the performance of both models and that this kind of models will prove indispensable for robust human-robot interaction.

References

1. Bregler, C.: Eigenlips for robust speech recognition. Int. Comput. Sci. Inst. **2**, 669–672 (1994)
2. Duncan, S.: Some signals and rules for taking speaking turns in conversations. J. Pers. Soc. Psychol. **23**, 283–292 (1972)
3. Ishiguro, H., Nishio, S.: Building artificial humans to understand humans. The JPN Soc. Artif. Organs **10**(3), 133–142 (2007)
4. Kanda, T., Ishiguro, H., Imai, M., Ono, T.: Development and evaluation of interactive humanoid robots. Int. Conf. Robot. Autom. **92**(11), 1839–1850 (2004)
5. Kendon, A.: Some functions of gaze direction in social interaction. Acta Psychol. **26**, 22–63 (1967)
6. Kim, W., Ko, H.: Noise variance estimation for Kalman filtering of noisy speech. IEICE Trans. Inf. Syst. **E84-D**(1), 155–160 (2001)
7. Lee, A., Nakamura, K., Nisimura, R., Saruwatari, H., Shikano, K.: Noise robust real world spoken dialogue system using GMM based rejection of unintended inputs. In: Proceedings of INTERSPEECH, pp. 173–176 (2004)
8. Minato, T., Shimada, M., Ishiguro, H., Itakura, S.: Development of an android robot for studying human-robot interaction. In: Proceedings of IEA/AIE Conference, pp. 424–434 (2004)
9. Mori, M., Macdorman, F., Kageki, N.: The uncanny valley. The Robot. Autom. Mag. **19**(2), 98–100 (2012)
10. Reeves, B., Nass, C.: The Media Equation: How People Treat Computers, Televisions, and New Media as Real People and Places. Cambridge University Press, Cambridge (1996)
11. Sacks, H., Schegloff, A., Jefferson, G.: A simplest systematics for the organization of turn-taking for conversation. Language **50**(4), 696–735 (1974)
12. Skantze, G., Gustafson, J.: Attention and interaction control in a human-human-computer dialogue setting. In: Proceedings of the SIGDIAL Conference, pp. 310–313 (2009)
13. Vertegaal, R., Slagter, R., Veer, G., Nijholt, A.: Eye gaze patterns in conversations: there is more to conversational agents than meets the eyes. In: Proceedings of the SIGCHI conference on Human factors in computing systems, pp. 301–308 (2001)
14. Yoon, S., Chang., D.: Speech enhancement based on speech/noise-dominant decision. IEICE Trans. Inf. Syst. **E85-D**(4), 744–750 (2002)

Chapter 18
Designing an Emotion Detection System for a Socially Intelligent Human-Robot Interaction

Clément Chastagnol, Céline Clavel, Matthieu Courgeon, and Laurence Devillers

Abstract The long-term goal of this work is to build an assistive robot for elderly and disabled people. It is part of the French ANR ARMEN project. The subjects will interact with a mobile robot controlled by a virtual character. In order to build this system, we collected interactions between patients from different medical centers and a Wizard-of-Oz operated virtual character in the frame of scenarii written with physicians and functional therapists. The human-robot spoken interaction consisted mainly of small talking with patients, with no real task to perform. For precise tasks such as "Finding a remote control," keyword recognition is performed. The main focus of the article is to build an emotion detection system that will be used to control the dialog and the answer strategy of the virtual character. This article presents the Wizard-of-Oz system for the audio corpus collection which is used for training the emotion detection module. We analyze the audio data at the segmental level on annotated measures of acoustically perceived emotion but also at the interaction level with global objective measures such as amount of speech and emotion. We also report on the results of a questionnaire qualifying the interaction and the agent and compare between objective and subjective measures.

C. Chastagnol (✉) • C. Clavel • M. Courgeon
Department of Human-Machine Interaction, LIMSI-CNRS, University of Orsay 11, Orsay, France
e-mail: cchastag@limsi.fr; clavel@limsi.fr; courgeon@limsi.fr

L. Devillers
Department of Human-Machine Interaction, LIMSI-CNRS, University Paris-Sorbonne 4, Orsay cedex, France
e-mail: devil@limsi.fr; laurence.devillers@paris-sorbonne.fr

J. Mariani et al. (eds.), *Natural Interaction with Robots, Knowbots and Smartphones: Putting Spoken Dialog Systems into Practice*, DOI 10.1007/978-1-4614-8280-2_18,

18.1 Introduction

This work is in the field of Intelligent User Interface with the aim of building more natural dialogs with robots and machines. Robots and machines have to be sensitive to human emotions and social signals to be socially intelligent. Natural human-machine interaction must use verbal and nonverbal communication. Verbal communication encodes the semantics of the message, whereas nonverbal communication can modify the semantics of the message and give information about emotions and personality but also social and interactional information. It is conveyed along various intertwined channels such as facial expressions, prosody, and gestures. We focus in this article on the steps necessary to build the emotion detection module of a spoken dialog system (SDS) for a social robotic companion.

The field of robotic companions has been very active lastly. Such robots have been designed for various applications, including at-home medical care and assistance [8] and autism therapy for children [13]. Social interactions with machines have also been investigated using virtual characters, e.g., for learning [11] and as an exercise coach for older adults [2]. Wizard-of-Oz setups can be used to early test parts of a system by replacing the missing parts by human operators [1]. They are now often used when building SDS or collecting emotional data [6, 9]. Emotion recognition is particularly interesting for some applications such as call-centers [7] and previous works have already tried to integrate it in SDS [10], along with virtual characters [14]. Recently some works explored the detection of emotional states such as interest [15] or involvement in a conversation [12], which are closely related to emotions but also to the problem of evaluating the user satisfaction and engagement for a given system. There is currently no standard way to evaluate SDSs, although some theoretical frameworks such as PARADISE [17] have been developed.

This work is part of the French ANR ARMEN project that tries to design and build a prototype for a robotic companion (RC) for elderly and disabled people. The RC will ultimately be able to look for lost objects and pick them up, small-talk with the users in the frame of specific topics, and call for help if the user is not feeling well. It is made of a mobile platform with a robotic arm. Users will control it by interacting in a natural, spoken way with a virtual character (VC) displayed on a separate screen that stays with the user (it can be fixed on a bed or a wheelchair) so that the interaction with the RC is maintained even if the mobile platform is away in another room. We work on the SDS, more precisely on the emotion recognition part, and the VC to some extent. To build the system, we first had to collect data as close to reality as possible. Patients from several medical centers taking part in the project interacted with the VC, controlled in a Wizard-of-Oz fashion, around several short scenarii designed to induce emotions. We present the experimental protocol, the collected data, and first results. Section 18.2 describes the specifications of the Wizard-of-Oz system. In Sect. 18.3, details on the experimental protocol needed to acquire spontaneous emotional data are given. The collected corpus is presented in Sect. 18.4. First results for the emotion recognition module are given in Sect. 18.5. A questionnaire was administered to the patients at the end of their interaction with the VC. We use this information and compare it to objective measures made on the annotated data such as dialog quality measures or global perceived emotions in Sect. 18.6. The conclusion and some perspectives on future work are presented in Sect. 18.7.

18.2 Data Collection System

We describe here the data collection system that will help us in building the emotion detection module and the complete SDS. We first present the architecture of the SDS. Since the interaction will happen as small talk in the frame of specific topics, we therefore chose to design a SDS around the structure of a dialog tree with closed questions and the virtual character driving most of the interaction. The SDS relies first on an automatic segmenter to get the user's vocal statements. The output audio segments are processed by an emotion recognition system and a word-spotter. The spoken recognition system is performed by another partner of the project. The keyword recognition system will be used in correlation with the emotion detected to control the decision in the dialog tree. Detecting emotions in the user's voice will also help having a more natural interaction with the VC by triggering facial expressions in answer. The best follow-up answers to a user statement will be decided by a decision module by combining the outputs of the emotion recognition system and of the word-spotting system. We decided not to use a complete ASR system and go for a word-spotter with a more limited vocabulary instead to enhance robustness because the final users have very varied voices, with sometimes vocal pathologies, "aged" or unvoiced voices, etc. In addition to the dialog, the dialog trees also encode beforehand the most probable emotions as well as the target-words for a given dialog node. A schematic view of the different components of the final SDS is given in Fig. 18.1.

In order to collect data, we built a system in a Wizard-of-Oz fashion, with every module from the final SDS (except the memory, storing the fixed scenarii, and virtual character) simulated by a human operator.

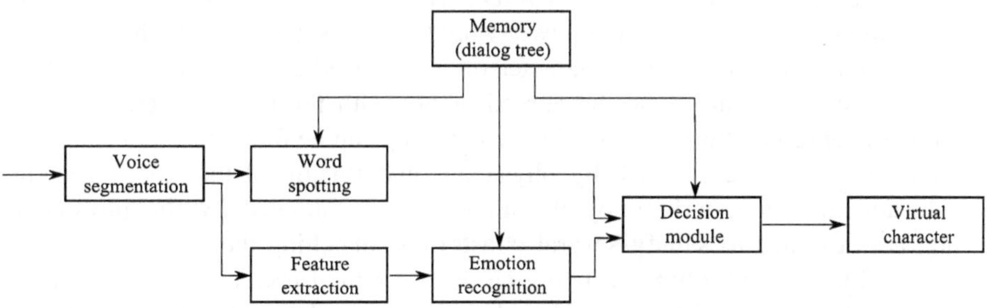

Fig. 18.1 Architecture of the final interaction system. Every module except the memory and the virtual character were simulated by a human operator in the Wizard-of-Oz experiment

The VC is based on the MARC (multimodal affective reactive characters) platform, developed at LIMSI-CNRS [5]. The female model Mary was used after a poll following a first data collection in June 2010; the experimental protocol was very similar but used only a synthetic voice instead of a VC. In this data collection, the VC was used as a digital muppet in a Wizard-of-Oz way: the operator was triggering answers that animated the VC. An interface was implemented to control the VC: it would display the dialog tree for the current scenario and the operator would simply select the most appropriate answer in a closed choice after a user statement, thus acting as a perfect audio processing and decision module for the VC. The interface would then send commands to animate the VC and make it play prerecorded sounds with on-the-fly lips synchronization; the commands described facial expressions using the BML language. The different facial expressions, also chosen by the operator, were designed at LIMSI-CNRS according to previous perception studies.

18.3 Experimental Protocol

The data collection happened on-site in medical centers. The subjects were brought by nurses to a room where the equipment was set up. The equipment consisted in a laptop displaying the virtual character with loudspeakers, a second laptop for the remote and wireless operation of the virtual character, a camera on a tripod, and a wireless AKG lapel microphone with an M-Audio external sound card, used to record the audio with the software Audacity on the second laptop.

Two people were conducting the data collection: an interviewer and an operator, not speaking with the subjects but setting up the equipment and operating the virtual character without the subject knowing. The experiment would take place as follows: first the interviewer would greet the subject and explain to him/her the purpose of the experiment (collecting data for a future assistive robot) while the operator was attaching the lapel microphone onto the subject's clothes, adjusting the camera's height, checking that the audio and video were working, and starting the recordings. Then the subject would interact with the virtual character in the frame of four different scenarii, detailed below (each scenario being explained beforehand by the interviewer). At the end, the interviewer would ask the subject to rate the quality of the interaction and the virtual character on a list of adjectives, using a five-level Likert scale to indicate if he/she agreed or not with the proposed adjectives. A schematic representation of the experimental setup can be found in Fig. 18.2.

The scenarii were outlined by physicians and functional therapists from the reeducational center, written by the authors, and validated by the physicians. They are designed to satisfy several constraints: matching the test cases for the functionalities of the robot and being close to the final user experience, eliciting emotions to collect useful training data, being easy to relate to for the subjects, and offering variability but staying in a limited dialog to ensure robustness. There were four scenarii. In the first one ("introduction" scenario), designed to put the subjects

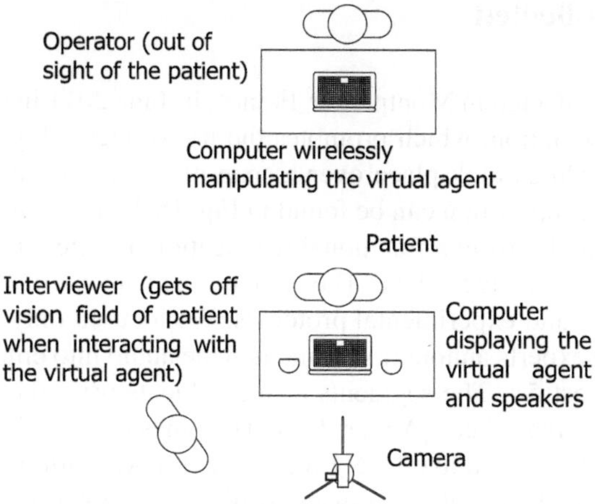

Fig. 18.2 Diagram of the experimental setup

at ease and build a beginning of proximity, the virtual character and the subject would introduce themselves. Then in a short training phase, the VC would ask the subject to try to pronounce the phrase "My voice conveys emotions" with different emotional intonations (anger, happiness, sadness). Then pictures of staff and other patients of the center known by the subject, taken at various positive occasions (birthdays, games, activities, etc.), were presented and the virtual character would ask the subject to name the people, if they were friends, explain the situations, and if he/she had good memories of them. In the second scenario ("pillbox" scenario), the subject was reminded by the virtual character that he/she should take his/her medication and acted as if the robot was going to look after his/her pillbox. In the meantime, it would small-talk, asking about how he/she felt, elaborating a little and then asking about what kind of events and activities were organized at the medical center and how he/she liked it. In the third scenario ("alert" scenario), the subject had to imagine that he/she was not feeling alright and needed to call for help. The virtual character had to call for help after having understood the subject and then tried to reassure him/her. In the fourth and last scenario ("remote" scenario), the subject would ask the VC to find the remote control because he/she wanted to watch television. After acting as if the robotic platform went to look for the remote control, the VC would ask the subject what he/she wanted to watch and elaborate on this.

At the end of the interaction, a questionnaire was administered to the subjects. A list of adjectives was presented to them and they had to tell, on a five-level Likert scale, if the adjective was rightly describing the VC; the whole interaction was rated similarly over four adjectives. Additionally, the subjects were asked if they would like to interact again with the VC and if they would like to own such a system. The results of the questionnaire are presented and analyzed in Sect. 18.6.

18.4 Data Collected

This corpus was collected in Montpellier, France, in June 2011 in collaboration with the Approche association, which promotes the use of technology to help disabled people. The recordings took place over a span of 3 days and involved 25 people from 25 to 91 (age repartition can be found in Fig. 18.3), 16 of them coming from a retirement home and 9 from a functional reeducational center. The corpus consists of 8.7 h of audio and video data. The patients had no particular knowledge in computer sciences and experimental protocols. We focused only on the audio data in this work. Two experts annotators segmented the audio into emotionally coherent segments of at most 5 s. The segments of the subjects were then annotated using a simple scheme: five labels (Anger, Fear, Happiness, Neutral, Sadness, plus an additional Junk label) and a five-level Likert scale for Activation. The segmentation was done with the Transcriber software while the annotation was done with a home-designed software. For the segmentation, the data was split in two parts; each annotator would annotate one part and then swap for cross-validation. Each segment was then annotated once by each annotator. In the final corpus, only the consensual segments (approximately 63%) were kept.

Some objective measures of involvement in the interaction were derived from the segmentation Transcriber files such as statistics on the number of subject turns, number of answers to a question from the VC, speech duration, and response time. In particular, the response times were not directly annotated, but rather estimated from the segmentation files. After sampling the segmentation Transcriber files, we considered that the response was less than a second if there was no silent segment between a VC segment and a subject segment, with an average of 500 ms. Otherwise, we used the duration of the silent segment and added 500 ms for the imprecision of the segmentation. Other measures such as statistics on the interventions of the interviewer during a scenario or the overlapping segments were extracted. The values of Activation and the emotional labels were also collected from the annotation files.

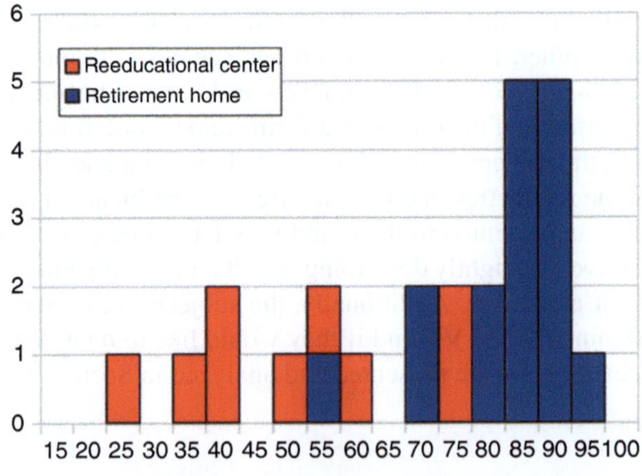

Fig. 18.3 Speakers repartition by age and medical center

Table 18.1 Details on the training sets

Set name	Armen	JEMO
Number of segments	545	1,491
Number of speakers	25	59
UAR score on four classes	45.1%	68.0%
Number of Anger segments	92 (17%)	291 (19%)
Number of Happiness segments	236 (43%)	359 (24%)
Number of Neutral segments	136 (25%)	534 (36%)
Number of Sadness segments	81 (15%)	307 (21%)

18.5 Results for the Emotion Recognition Module

We used widely accepted techniques to benchmark our corpus of emotions. We first deleted the Fear class because it had too few instances and subsampled the Neutral class by randomly choosing N segments, where N is the average of the number of instances in the Anger, Happiness, and Sadness classes. We then extracted acoustic parameters from the instances using the openEAR toolkit with the Interspeech 2009 feature set, containing 384 features [16]. We used the resulting set to train a SVM classifier with a RBF kernel [4]. A grid search optimization was done on the C and gamma parameters, with fivefold cross-validation. The recognition results are expressed in terms of unweighted average recall (UAR).

The results (45.1% UAR on four classes) show that working on spontaneous emotional data is hard, especially with subjects who have difficult voices. To give a comparison, we applied the same procedure to the JEMO corpus, previously presented in Brendel et al. [3] and extended since. It is a corpus of spontaneous and induced emotions collected in the frame of a game in lab conditions. It contains 1,491 instances from 59 speakers, aged from 18 to 60 (additional details are given in Table 18.1). We obtained a score of 68.0% UAR on the same four classes.

18.6 Results of the Questionnaire and Objectives Measures at the Dialog Level

As presented above, the interaction consisted mainly of small talking with patients, with no real task to perform. There is no standard way to evaluate such interactions, but we are mainly interested in how the users perceived the interaction and in their satisfaction. So we administered them a questionnaire at the end of their interview on these subjects. We present the results of the questionnaire in Sect. 18.6.1. This kind of information is also useful to drive the interaction or evaluate it automatically. We think that the satisfaction or engagement in the interaction, from a more global point of view than segment-level recognition of emotion, could be estimated from objective measures, extractible from the audio signal, and information on the user

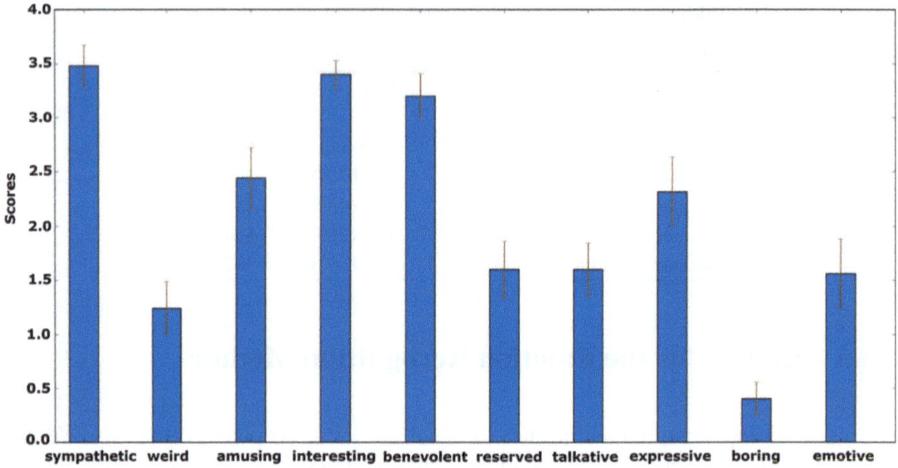

Fig. 18.4 Virtual character traits attribution levels

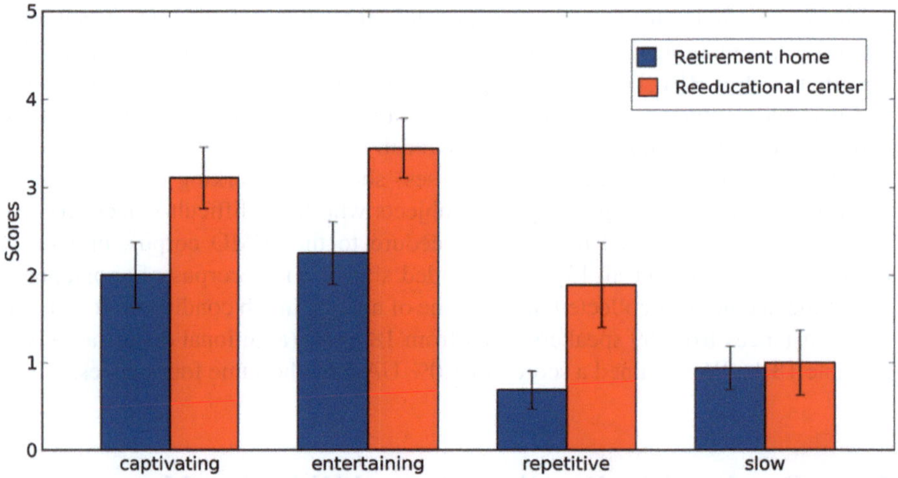

Fig. 18.5 Perceived quality of the interaction

such as age and gender. We extracted these measures and, as a first step, computed their correlations with the items of the questionnaire to see if some could be shown to verify our intuition. The results of this small study are presented in Sect. 18.6.2.

18.6.1 Questionnaire Results

The questionnaire helped us to understand two dimensions: how the users perceived the VC in terms of personality and how they perceived the interaction (cf. Figs. 18.4 and 18.5). The former was perceived positively by the subjects, with a high attribution of positive qualifiers and a low attribution of negative ones. The latter was also deemed positive, but some differences can be noted between the subjects coming from the functional reeducational center, younger on average than the subjects coming from the retirement home (cf. Fig. 18.3). The subjects from the reeducational center found the interaction more captivating and entertaining, though a bit repetitive, and they all declared to be willing to interact again with the VC. The subjects from the latter were more reserved on this subject, but they found the interaction less repetitive and slow.

18.6.2 Correlations Between the Questionnaire and Objectives Measures

We selected the cross-correlations that had an absolute value of over 0.4 (the critical value for two-tailed tests at alpha = 0.05 being 0.396 in our case). Some of them are presented in Table 18.2. The correspondence between the ID and the extracted measure can be found in Table 18.3. Several points can be highlighted from these results. First we confirm expected points, e.g., the number of turns for the subject is positively correlated with the number of quick answers (short response time), with a value of 0.58. Also subjects that talk more and more often generate a higher number of overlapping segments (correlation values around 0.60). Age seems to be linked to the number and total duration of overlapping speech.

Some results reveal interesting links between the objective measures of involvement in the interaction and the perceptively annotated levels of Activation: subjects who express higher average and maximum Activation values have a higher number of speaker turns, a higher total speech duration; they also generate a higher number of overlapping segments and answer more frequently with a short response time. If we compare the Activation values with the questionnaire answers, we see that subjects with these characteristics also find the VC more communicative, talkative, and less boring. This seems to have little link to age (only the minimum Activation value is negatively correlated to age with a value of −0.48) and none to gender. Besides, subjects with a higher proportion of segments annotated as Anger show a higher average and maximum Activation value; they also perceived the VC to be more talkative, less interesting, and the interaction to be slower. Subjects with a higher proportion of Happiness segments answered more quickly and found the VC to be more communicative and amusing. On the contrary, subjects with a higher proportion of Neutral segments have a higher average response time.

The questionnaire answers seem to be quite correlated to all of our extracted measures. For instance, subjects with a higher average response time found the VC to be less sympathetic and more bizarre. The age of subjects is negatively correlated with the VC being perceived as amusing and emotive, and the interaction being perceived as captivating and entertaining. Somehow paradoxically to this, age is negatively correlated with the interaction being perceived as repetitive. This is clearly visible in Fig. 18.5 where the evaluation of the interaction is plotted for the two different centers, where the age repartition was very different (cf. Fig. 18.3).

Table 18.2 Cross-correlations between the objective measures of involvement, emotion, and the questionnaire answers

	1	2	3	4	5	6	7	8	9	10	11	12	13	14
1	1.000	**0.693**	**0.575**	−0.157	**0.436**	**0.701**	−0.224	0.279	0.194	−0.184	**0.639**	**0.599**	0.273	0.161
2	**0.693**	1.000	0.356	−0.152	0.363	**0.507**	−0.105	0.056	0.239	−0.044	**0.589**	**0.572**	0.193	0.300
3	**0.575**	0.356	1.000	**−0.709**	**0.421**	**0.406**	−0.080	0.375	**0.460**	**−0.583**	0.206	0.204	0.116	0.155
4	−0.157	−0.152	**−0.709**	1.000	−0.207	−0.012	−0.091	−0.217	−0.331	**0.506**	−0.129	−0.140	0.008	−0.077
5	**0.436**	0.363	**0.421**	−0.207	1.000	**0.644**	0.291	**0.514**	0.241	−0.361	**0.535**	**0.427**	0.062	−0.130
6	**0.701**	**0.507**	**0.406**	−0.012	**0.644**	1.000	−0.332	**0.461**	0.204	−0.286	**0.471**	0.376	0.310	−0.017
7	−0.224	−0.105	−0.080	−0.091	0.291	−0.332	1.000	0.035	−0.152	−0.053	−0.120	−0.089	**−0.478**	−0.145
8	0.279	0.056	0.375	−0.217	**0.514**	**0.461**	0.035	1.000	0.076	−0.328	0.149	0.099	−0.005	−0.266
9	0.194	0.239	**0.460**	−0.331	0.241	0.204	−0.152	0.076	1.000	**−0.400**	0.192	0.165	−0.172	0.198
10	−0.184	−0.044	**−0.583**	**0.506**	−0.361	−0.286	−0.053	−0.328	**−0.400**	1.000	−0.210	−0.259	−0.073	−0.115
11	**0.639**	**0.589**	0.206	−0.129	**0.535**	**0.471**	−0.120	0.149	0.192	−0.210	1.000	**0.956**	**0.408**	0.212
12	**0.599**	**0.572**	0.204	−0.140	**0.427**	0.376	−0.089	0.099	0.165	−0.259	**0.956**	1.000	**0.414**	0.289
13	0.273	0.193	0.116	0.008	0.062	0.310	**−0.478**	−0.005	−0.172	−0.073	**0.408**	**0.414**	1.000	0.072
14	0.161	0.300	0.155	−0.077	−0.130	−0.017	−0.145	−0.266	0.198	−0.115	0.212	0.289	0.072	1.000
15	0.188	0.392	0.137	−0.080	**0.437**	0.157	0.163	0.254	**0.413**	0.068	0.317	0.163	−0.010	0.039
16	0.144	0.152	0.117	−0.183	**0.507**	0.190	0.131	**0.445**	0.301	−0.122	0.357	0.201	−0.133	−0.389
17	0.258	0.065	0.334	**−0.471**	0.275	0.009	0.340	0.168	0.256	−0.367	0.311	0.298	−0.202	−0.105
18	−0.201	−0.059	−0.245	**0.552**	−0.085	0.092	−0.190	−0.149	−0.007	0.274	−0.379	−0.380	−0.119	−0.074
19	−0.153	−0.176	0.180	−0.268	0.182	−0.166	0.131	−0.028	**0.426**	−0.208	−0.047	−0.091	**−0.464**	−0.042
20	−0.139	−0.321	−0.121	−0.028	−0.096	−0.202	0.179	**−0.410**	0.269	−0.262	0.109	0.181	−0.059	−0.025
21	−0.033	−0.186	0.080	0.072	0.106	−0.016	0.222	0.019	0.229	0.060	−0.271	−0.275	**−0.494**	**−0.420**
22	0.378	0.241	0.333	−0.229	0.369	0.090	0.306	0.274	0.211	−0.263	0.247	0.285	**−0.400**	−0.181
23	−0.076	−0.178	0.073	0.050	0.297	0.068	0.094	0.288	0.336	−0.115	−0.052	−0.159	**−0.433**	**−0.514**
24	−0.223	0.023	−0.203	0.326	−0.341	−0.113	−0.153	−0.111	0.139	0.214	−0.354	−0.328	**−0.507**	0.161
25	−0.076	0.098	0.098	0.063	0.243	0.060	0.287	**0.441**	−0.033	−0.155	−0.052	0.073	−0.156	0.122

Please refer to Table 18.3 for the IDs

Table 18.3 Details on some of the extracted measures

ID	Measure extracted
1	Number of turns
2	Total speech duration
3	Number of answers with a quick response time (<1s)
4	Average response time
5	Average activation value
6	Maximum activation value
7	Minimum activation value
8	Proportion of segments annotated as Anger
9	Proportion of segments annotated as Happiness
10	Proportion of segments annotated as Neutral
11	Number of turns marked as overlap
12	Total overlap time
13	Age
14	Gender (1: male, 2: female)
15	Virtual character being perceived as communicative
16	Virtual character being perceived as talkative
17	Virtual character being perceived as sympathetic
18	Virtual character being perceived as weird
19	Virtual character being perceived as amusing
20	Virtual character being perceived as interesting
21	Virtual character being perceived as emotive
22	Interaction being perceived as captivating
23	Interaction being perceived as entertaining
24	Interaction being perceived as repetitive
25	Interaction being perceived as slow

Overall there are almost no differences between genders, with exceptions for some perceived traits of the VC, which is judged less emotive, and the interaction less entertaining, by women than by men. Obviously a control condition with a male VC would be very interesting to balance this result.

18.7 Conclusion and Perspectives

We presented our work on designing an emotion detection module for a robot companion dialog system. A corpus has been collected with real end users in medical centers. The patients were interacting with a Wizard-of-Oz version of the final system. The recorded audio data was manually segmented and annotated for perceived emotions and used to train a first emotion recognition system. We presented first results on the recognition of emotions, showing that the data we collected with real users, sometimes with pathological voices, is hard compared to more classical databases. Other interactional clues were extracted and analyzed with respect to a questionnaire on the perception of the virtual character by the users.

Further experiments will test the automatic decision module which will use knowledge from the keyword speech recognition module and the emotion detection module. Also, we intend to design a protocol to evaluate the VC expressivity and the strategy of the dialog. Then, the evaluation of the complete system will require special attention. There is no widely accepted evaluation method for SDSs at this point and the existing methods usually focus on task-oriented systems, which is not our case. We will thus work on estimating the satisfaction of the user for the interaction from objective measures. The system will be evaluated with patients from medical facilities and students.

Acknowledgements This work is funded by the French ANR (http://projet_armen.byethost4.com). The authors wish to thank the association APPROCHE for their help during the data collection and the SME Voxler, member of the project.

References

1. Bernsen, N., et al.: Wizard of Oz Prototyping: How and when? CCI Working Papers in Cognitive Science and HCI, WPCS-94-1. Center for Cognitive Science, Roskilde University (1994)
2. Bickmore, T., Caruso, L., Clough-Gorr, K., Heeren, T.: It's just like you talk to a friend - Relational agents for older adults. Interact. Comput. **17**, 711–735 (2005)
3. Brendel, M., Zaccarelli, R., Devillers, L.: Building a system for emotions detection from speech to control an affective avatar. In: Proceedings of the 7th International Conference on Language Resources and Evaluation (2010)
4. Chang, C.C., Lin, C.J.: LIBSVM : a library for support vector machines. ACM Trans. Intell. Syst. Tech. **2**, 27:1–27:27 (2011)
5. Courgeon, M., Martin, J.-C., Jacquemin, C.: MARC: a Multimodal Affective and Reactive Character. In: Proceedings of the 1st Workshop on Affective Interaction in Natural Environments (AFFINE). Chania, Crete (2008)
6. Delaborde, A., Tahon, M., Barras, C., Devillers, L.: A wizard-of-Oz game for collecting emotional audio data in a children-robot interaction. In: Proceedings of the International Workshop on Affective-Aware Virtual Agents and Social Robots, ICMI-MLMI. Boston, USA (2009)
7. Devillers, L., Vidrascu, L., Layachi, O.: Automatic detection of emotion from vocal expression. In: Scherer, K., Bänziger, T., Roach, E. (eds.) A Blueprint for an Affectively Competent Agent, Cross-Fertilization Between Emotion Psychology, Affective Neuroscience, and Affective Computing, pp. 232–244. Oxford University Press, Oxford (2009)
8. Graf, B., Hans, M., Kubacki, J., Schraft, R.: Robotic home assistant care-o-bot II. In: Proceedings of the Joint EMBS/BMES Conference, vol. 3, pp. 2343–2344. Houston, TX, USA (2002)
9. Han, J.G., Gilmartin, E., De Looze, C., Vaughan, B., Campbell, N.: Speech and multimodal resources - The Herme database of spontaneous multimodal human-robot dialogues. In: Proceedings of LREC 2012, pp. 1328–1331. Istanbul, Turkey (2012)
10. Lee, C.M., Narayanan, S.S.: Toward detecting emotions in spoken dialogs. IEEE Trans. Speech Audio Process. **13**(2), 293–303 (2005)
11. McQuiggan, S., Rowe, J., Lester, J.: The effects of empathetic virtual characters on presence in narrative-centered learning environments. In: Proceedings of the ACM Conference on Human Factors in Computing Systems, pp. 1511–1520 (2008)

12. Oertel, C., Scherer, S., and Campbell, N.: On the use of multimodal cues for the prediction of degrees of involvement in spontaneous conversation. In: INTERSPEECH 2011, pp. 1541–1544 (2011)
13. Robins, B., Dautenhahn, K., Boekhorst, R., Billard, A.: Robotic assistants in therapy and education of children with autism: Can a small humanoid robot help encourage social interaction skills? Universal Access in the Information Society (2005)
14. Schröder, M., Bevacqua, E., Cowie, R., Eyben, F., Gunes, H., Heylen, D., ter Maat, M., McKeown, G., Pammi, S., Pantic, M., Pelachaud, C., Schuller, B., de Sevin, E., Valstar, M., Wöllmer, M.: Building autonomous sensitive artificial listeners. IEEE Trans. Affective Comput. **99** 134–146 (2011)
15. Schuller, B., Müller, R., Eyben, F., Gast, J., Hörnler, B., Wöllmer, M., Rigoll, G., Höthker, A., Konosu, H.: Being Bored? Recognizing natural interest by extensive audiovisual integration for real-life application. Image Vision Comput. J. Special Issue Vis. Multimodal Anal. Human Spontaneous Behav. **27**, 1760–1774 (2009)
16. Schuller, B., Steidl, S., Batliner, A.: The Interspeech 2009 emotion challenge. In: Proceedings of the 10th Annual Conference of the International Speech Communication Association. Brighton, UK (2009)
17. Walker, M.A., Littman, D.J., Kamm, C.A., Abella, A.: PARADISE: A framework for evaluation of spoken dialog agents. In: Proceedings of the 35th Annual Meeting of the Association for Computational Linguistics. Madrid, Spain (1997)

Chapter 19
Multimodal Open-Domain Conversations with the Nao Robot

Kristiina Jokinen and Graham Wilcock

Abstract In this paper we discuss the design of human-robot interaction focussing especially on social robot communication and multimodal information presentation. As a starting point we use the WikiTalk application, an open-domain conversational system which has been previously developed using a robotics simulator. We describe how it can be implemented on the Nao robot platform, enabling Nao to make informative spoken contributions on a wide range of topics during conversation. Spoken interaction is further combined with gesturing in order to support Nao's presentation by natural multimodal capabilities, and to enhance and explore natural communication between human users and robots.

19.1 Introduction

In recent years, human-robot interaction has been an active research area resulting in development of integrated platforms for interactive applications with various input and output technologies, as well as opportunities to study natural human-machine interaction with rich communicative capabilities. It is envisaged that future interactive systems will operate in ubiquitous computing contexts and smart spaces and will be characterised by a flexible use of modalities, such as speech, gesturing, touch, gaze, and movement, so that users can readily understand how to get their task completed without needing to think about how the interaction should take place. Human-robot interactions should thus afford natural communicative capabilities and offer the user more intuitive and conversational ways for interaction [5].

Novel technology enables robotic agents not only to record, analyse, and react to the changing environment but also to be sensitive to users' presence, their communicative needs, and social norms. This is useful in different applications

K. Jokinen (✉) • G. Wilcock
University of Helsinki, Helsinki, Finland
e-mail: kristiina.jokinen@helsinki.fi; graham.wilcock@helsinki.fi

J. Mariani et al. (eds.), *Natural Interaction with Robots, Knowbots and Smartphones: Putting Spoken Dialog Systems into Practice*, DOI 10.1007/978-1-4614-8280-2_19,

Fig. 19.1 Human-robot interaction at eNTERFACE 2012

and can be especially deployed in the realm of social robotics [3] which focuses on communicating robots, capable of interacting and cooperating with humans and exhibiting relevant social behaviours in the context of human society and culture. Moreover, social robots exemplify conversational interfaces which do not necessarily aim at the most economical interaction in terms of numbers of contributions and turn-exchanges, but at interaction that allows associative and open-domain conversations which emphasise mutual rapport and construction of social bonds.

In this paper we discuss the design of human-robot interaction exemplifying aspects of social robot communication related to multimodal information presentation. The robot gets information about topics from Wikipedia and presents it to the user in speech. We explore how to present the new information to the user and how to detect human contact and the partner's interest in continuing. In particular, multimodal observations are important in this context, assuming that much of the feedback behaviour in conversations is conducted using gaze and gesture signals rather than explicit speech.

As a starting point for speech interaction we use WikiTalk [17], a system that supports open-domain conversations using Wikipedia as a knowledge source, previously developed using a robotics simulator [8, 9]. We show how to extend this model to a situated agent that is capable of multimodal communication and refer to a prototype implementation on the Nao robot made at the eNTERFACE 2012 Summer Workshop in Metz (Fig. 19.1).

The structure of the paper is as follows. We give an overview of the dialogue modelling framework and its application to the Nao robot in Sect. 19.2. In Sect. 19.3 we discuss two important research issues concerning interaction with the Nao robot: the flow of information and the use of gestures in information presentation. The implementation of the Nao WikiTalk prototype and its evaluation at the eNTERFACE 2012 Summer Workshop are briefly described in Sect. 19.4. Conclusions and future work are discussed in Sect. 19.5.

19.2 Communicative Enablements in the Context of Human-Robot Interaction

The constructive dialogue model (CDM) [5] assumes that the speakers are rational agents who interact in a cooperative manner. The speakers coordinate and control their interaction by providing feedback on the basic enablements of communication: contact, perception, understanding (CPU), and reaction. The enablements, set out by [1], are independent requirements for the communication to take place in the first place, and they operate on different levels. They can be said to model the agent's awareness of the communication as well as their involvement in the communicative interaction.

Contact refers to the fact that the agents need to be in a suitable proximity so that hearing and seeing, and in some cases also touching the partner, are possible, while perception refers to the agent's conscious perception that the partner is sending meaningful symbols which are intended to be understood and evaluated as a communicative message in the current context. The enablements of Understanding and Reaction concern more intentional functioning of the agent: finding a semantic interpretation for the partner's action and producing one's own behaviour as a reaction to it.

Successful communication requires that the enablements are fulfilled, and the agents thus monitor each other and the communicative situation so they can react to problems and proactively avoid possible errors and misunderstandings. For instance, if the agent has lost contact, is not interested, or does not understand the partner's contribution, there is an obligation to make these CPU problems known to the partner, who, analogously, is obliged to adjust their communication to the level that addresses the problem situation.

The Aldebaran (http://www.aldebaran-robotics.com) Nao robot has many sensors of different types that it uses to perceive its environment. For instance, it has four microphones and two cameras in its head that provide sound and vision about the environment, sonar sensors to check the distance of objects in its vicinity and tactile sensors on its head and body which are triggered when touched. By mapping the robot's sensor technology and its general processing capability to the basic enablements for communication, we can implement the concepts of a general communicative theory with the help of the robot's physical devices. Figure 19.2 depicts how CDM and the basic enablements are interpreted in the context of human-robot interaction.

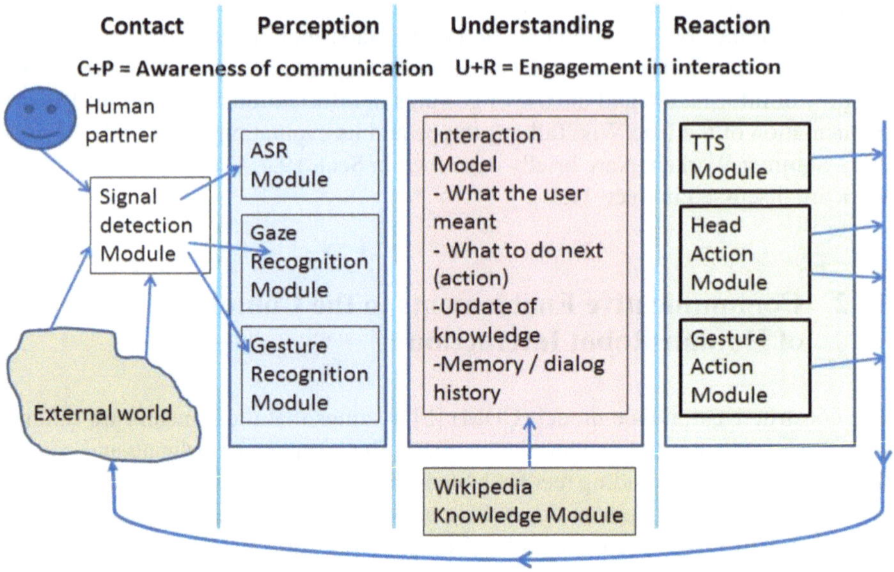

Fig. 19.2 Overall view of the CDM architecture in human-robot interaction

Contact and perception form the basis of the robot's awareness of the communicative situation. Contact, or the detection of the presence of another agent, can be operationalised through the robot's signal detection mechanism: the perception of a visual or auditive signal is a sign of another agent or an object being in contact range. Perception, on the other hand, can be implemented through various modality recognition engines such as speech, gaze and gesture recognition modules, which produce the first, signal-level analysis of the perceived auditive and visual signals.

The agent's engagement in the interaction requires conscious action that indicates the agent's active participation in the situation and willingness to cooperate with the partner. This includes Understanding, or further analysis of the signals through which the pragmatic meaning of the partner's contribution is constructed with respect to the agent's intentions and the current task, and Reaction, or the production of an appropriate response to the partner's message in the communicative situation. The Understanding component corresponds to the traditional dialogue manager that creates a semantic interpretation for the user's contribution and makes the decisions about what to do next. In the current implementation Understanding is a state-based module which coordinates WikiTalk conversations and the topic management with respect to the Wikipedia articles. The Reaction components refer to the robot's speech and motor control engines through which the execution of the robot's response takes place. They also render the communicative action with respect to the available modalities: speech and gesturing.

The agent's behaviour affects the external world, and the changes as well as the partner's reaction to them will function as a new input to the agent. The robot gets information about the changes via its sensors and starts a new analysis and reaction according to the new input. Through the communicative cycle of action and observation, the agent can learn how its actions change the world and formulate suitable interaction strategies to be used in different communicative situations. If the robot's functionalities include a learning algorithm, it is possible to implement adaptative behaviour.

19.3 Dialogue Modelling with New Information

Dialogues are modelled as a series of dialogue states, representing the agent's beliefs of the external world and the current state of a communicative situation. Each transition from one state to another (or looping in the same state) corresponds to the communicative cycle described in Fig. 19.2 and is characterised by the new information that is conveyed by the agent's action and used to update the shared dialogue context. An important aspect of the interaction management is the flow of new information that is exchanged between the partners: it is important to show continuation with the topic or mark awkward shifts so as to maintain coherence of the overall dialogue.

One of our goals is to study how to present information in speech so as to help the partner to understand what the new information is. One of the most obvious mechanisms is to use prosody and mark the new information by prosodic cues [16]. However, in this paper, the main focus is on non-verbal communication and how much dialogue control information can be conveyed by visual cues, i.e. gesturing, nodding, and body posture. The different research issues related to coherence and NewInfo presentation are briefly discussed in the subsections below.

19.3.1 Topic Management and Coherence of Presentation

We use Wikipedia as the knowledge base and follow the WikiTalk approach to enable open-domain conversations with the Nao robot (see [9, 17] for more details about WikiTalk). The Wikipedia articles are considered as possible topics that the robot can talk about, while each link in the article is treated as new information that the user can shift their attention to and ask for more information about. The paragraphs and sentences in the article are considered propositional chunks or pieces of information that structure the topic into subtopics and form the minimal units for presentation, i.e. they can be presented in one utterance by the robot. The challenge in presenting Wikipedia information is how to convey its structure to the users so that they can readily understand which are the new information links and how to navigate in the topic structure smoothly. In other words, dialogue modelling should support interaction that affords natural information flow [5].

In dialogue management, topics are often managed by a stack, which is a convenient last-in-first-out mechanism to handle topics that have been recently talked about. However, stacks are a rather rigid means to describe the information flow in cases where the dialogues are more conversational and do not follow any particular task structure. We prefer topic trees [13], which enable more flexible management of the topics. The trees can be traversed in whatever order, while the distance of the jumps determines the manner of presentation of the information. Moreover, we use the concepts of Topic and NewInfo [5], where Topic refers to the particular issue (Wikipedia article) that the speakers are talking about, and NewInfos are the parts of the message (the hyperlinks) that are new in the context of the current Topic.

It must be emphasised that dialogue coherence, or the discourse relations between consecutive utterances that enable the listener to infer what their connection is, becomes rather straightforward: we can rely on the structure of Wikipedia to provide coherence for us. As the articles have already been written as coherent texts, and in particular the hyperlinks between the articles have been inserted so that they form coherent hypertexts, we can assume that the content of the topics and the NewInfo links that will be presented to the user form a coherent discourse. Interaction is then driven by the user's interests based on the content of the presentation rather than a particular task structure that would constrain the suitable topics. However, what is important in WikiTalk is to capture the partner's attentional state in such a way that the partner can focus attention on the available NewInfos.

19.3.2 Gesturing and Presentation of NewInfo

Gestures and body movement play important roles in human communicative behaviour and can be directly related to the information flow of interactions [11]. Hand movements are often used to visually explain the topic, e.g. the size or shape of an object, while beat gestures are usually associated with emphasis and rhythm of the speech and deictic gestures with pointing or attention catching. We have experimented with various gestures to make the robot's presentation more expressive, especially to mark the NewInfo and to structure the WikiTalk presentation. Such gestures indicate to the partner how the conversation is to be understood and divided into communicatively important segments [6]. They are distinguished from iconic gestures and beats in that they aim to catch the partner's attention, and thereby they also control the dialogue flow. Kendon [11] calls them meta-discursive gestures.

Much research concerns the use of gestures in interaction [6, 11]. Following Quek [15], we classify hand and arm movements into unintentional movements and gestures and divide the latter into manipulative commands and communicative gestures. Manipulative commands are gestures that the user issues to control the robot. Sophisticated gesture recognition would require a robust recognition

Function	Body	Speed	Movement	Frequency	Orientation
Greeting	Hand	Fast	Complex (wave)	Repeated	Palm vertical
Start presentation	Both hands	Slow	Front, side	Single	Palm-up
Start topic	Left	Fast	Front, side	Repeated	Palm-up
Give feedback	Head	Fast	Down	Single/ Repeated	Nod
Elicit feedback	Head	Slow	Down	Single	Nod
Emphasise	Left	fast	Up-down	Single	Palm vertical
Beat	Left	fast	Up-down	Repeated	Palm vertical
Surprise	Head	slow	Up	Single	Nod

Fig. 19.3 Gesture taxonomy and parameters

algorithm, so we experimented with only one manipulative command: a stop gesture with fingers and palm vertical, that the user can use to stop Nao talking. However, hand recognition was too much dependent on the lighting and background, so in practice we did not include it in the interaction repertoire, but used tapping on the robot's head instead. See [4] for a review of the experiments on hand recognition.

For communicative gestures we use the idea of gesture families, following Kendon [11]. Gesture families consist of gestures which have similar form and meaning, such as Palm Down (Open Hand Prone), Palm Up (Open Hand Supine), Palm Sideways (Open Hand Vertical), and Index Finger Extended. They are associated with a semantic theme, for example, gestures of the Palm Down family are often used in contexts where something is being denied, negated, interrupted, or stopped, while those in the Palm Up family are used in contexts where the speaker is offering, giving, or showing something or requesting the reception of something.

Figure 19.3 shows the gesture taxonomy and parameters in the Nao WikiTalk prototype. Gestures are organised into a gesture library, a collection of behaviours that Nao can choose from. The library consists of gesture families, each linked to a particular semantic theme. Selection of a gesture in a particular communicative context is based on the dialogue situation (give/elicit feedback, inform, greet) and on the task (continue/change/stop the topic). More detailed description of the robot's gesturing can be found in [14].

We used Nao's Choregraphe tool to model a set of gestures which were then exported as Python code to be run by the dialogue manager. Gestures are parameterized, which helps to constrain the choice to a certain gesture family, and alternatives in the same gesture family are selected in a loop.

One of the main issues in gesture studies is the synchrony of gestures and speech. Speakers coordinate speaking and gesturing in an accurate manner so that the peak

Fig. 19.4 Some key poses: standing, speaking, start presentation, emphasis, surprise

of the gesture is aligned with the speech and visually contributes to the information content of the message [11]. Theoretical questions are related to how simultaneous cognitive processing and planning of speech and gestures take place. Although mental modelling of communication is outside the scope of our research, the correct timing of beat gestures with the NewInfo is important for the natural presentation capabilities of the robot: gesturing provides "silent" feedback to the partner and prepares them to have the right stance to interpret the forthcoming message in the intended way.

We followed Kendon's three phases (preparation, nucleus and retraction) for temporal segmentation of our gestures and experimented with methods for counting utterance words and timing their pronunciation. However, often gesturing is too slow, and more accurate synchrony would require access to speech synthesis timings in order to synchronise it with the motorics that implement robot gestures. Further research is needed to provide more accurate methods for the timing and synchrony of gesturing and speech.

Besides gesture and speech coordination, the whole posture of the robot is relevant in providing natural and intuitive presentation. Figure 19.4 presents some specific Nao key poses. The speaking posture with hands half way up differs from the standing position with hands by the side, and the robot has a particular palm-up presentation gesture, used at the beginning and end of presentations. Emphasis is expressed by single hand beat gestures, while surprise, e.g. after the user interruption, is expressed by a head nod up. In the current version we cannot use pointing gestures due to the anatomy of Nao's hand: Nao can only have palm closed or palm open.

19.4 Implementation and Evaluation

The implementation of the WikiTalk system on Nao and the development of the multimodal interaction modules were done during the eNTERFACE 2012 Summer Workshop in Metz. Further details are given in [2].

Technologies	Nao behaviours
Wikitalk	– Knowledge from Wikipedia – Segmentation of articles into paragraphs, sentences, and NewInfos
Face recognition	– Recognition of an object in the vision field – If human face, start to follow – Recognition of human contact and interest
Gesturing	1) User commands: Stop! 2) Nao's own communication gestures: – Presentation of information with palm-up – Elicit of user feedback with nod down – Surprise at interuption with nod up – Emphasis of NewInfo by a beat
Dialogue management	– Start a conversation – Continue with the same topic – Change to a new topic

Fig. 19.5 Basic technologies and behaviours implemented at eNTERFACE 2012

The basic technologies, including WikiTalk, face recognition, gesturing and dialogue management, listed in Fig. 19.5, were integrated into the Nao system. The robot's functionalities include segmenting articles into paragraphs, sentences and NewInfos, detecting a human face, using communicative gestures, as well as managing the dialogue. For instance, to track the human face and thus draw conclusions about the user's interest, a simple face tracker was implemented. At the start of the interaction, when Nao detects a face, it makes contact by saying "Hello, I can tell you many interesting things, what would you like to hear?". If the face disappears for more than a few seconds at any time during the dialogue, Nao pauses and explicitly asks if the user wishes to hear more. If there is no answer, Nao stops and waits. More details about using non-verbal cues in human-robot interaction are given in [4].

The Nao Wikitalk prototype was evaluated with 12 students and staff at the eNTERFACE 2012 workshop. The evaluation is reported in [2, 14].

Following the evaluation scheme proposed in [7], the users were asked to fill in a questionnaire twice, first to capture their expectations of the system before interacting with it, and then to measure their experience of the system after their interaction with it. The subjects tested three different versions of the system: with gaze-tracking only (no hand gestures), with small gesturing (head nods and presentation postures), and with full gesturing, along the dimensions of Interface, Responsiveness, Expressiveness, Usability, and Overall Experience. Averaged results of some of these aspects between user expectations and experience are shown in Fig. 19.6.

Version 2 (small gesturing) exceeded user expectations in appearing more lively and gesturing more naturally than expected. However, in all versions, the behaviour was not as expressive as expected and the timing of nodding was not suitable. Perhaps "expressiveness" requires more facial expressions than is possible with

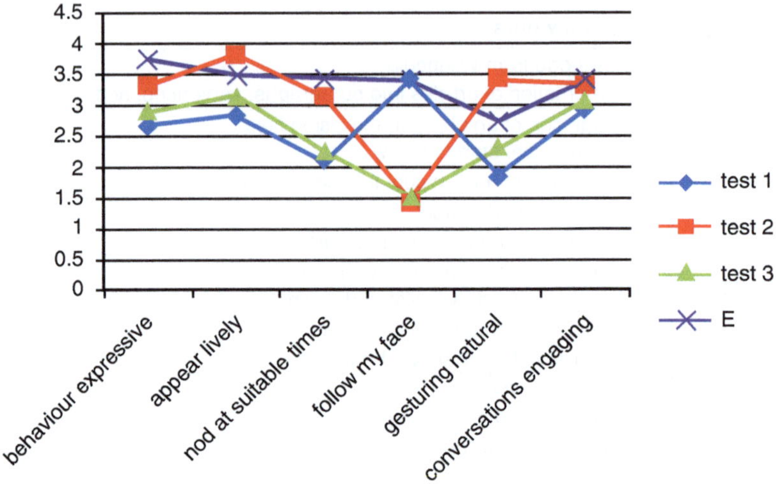

Fig. 19.6 User expectations (-X-) and experience on some aspects of the three system versions

Nao, and the scores were thus low. It is also likely that asynchrony of gesturing was the reason why Version 3 with "big" visible gestures was rated less natural than Version 2 with "small", less visible gesturing. Concerning engagement in the interaction, however, Version 2 seemed to fulfill user expectations, and the other versions also ranked highly, which supports the view that interactions with robots are generally engaging.

The evaluation data is available for further research purposes by request from the authors. The data is unique in that it is a fairly large, systematic collection of open-domain multimodal human-robot interactions.

19.5 Discussion and Future Work

The paper describes interaction with the Nao robot from the point of view of constructive dialogue modelling and demonstrates how the framework can be applied to the Nao WikiTalk application. Nao's interaction capabilities are greatly extended by the multimodal aspects related to gesturing and by enabling it to make informative spoken contributions on a wide range of topics during conversations. As far as we know, this is the first multimodal human-robot conversational interaction system that is open-domain.

The evaluation of the prototype system implemented at the eNTERFACE 2012 Summer Workshop in Metz shows that the system can engage humans in interaction which is lively and fairly natural. The combination of speech communication with gesturing supports natural interaction between human users and robots and enhances possibilities for successful application of the technology to various other types of tasks such as educational applications, tourist guiding, and game interfaces.

At the summer workshop we also explored other multimodal features, in particular gaze-tracking and motion capture. Gaze-tracking is important in order to manage smooth turn-taking [10, 12] and to get feedback about the partner's interest in the topic. As humans direct their gaze towards objects of interest, it is useful if the robot can infer where the partner's attention is focussed, and if they are still interested in what it is presenting. Further experiments are planned to model agents' awareness and focus of attention and to explore the notion of conversational engagement. Finally, we also experimented with motion capture technology using Kinect as one of the robot's inputs. Preliminary work on this is presented in [2].

Acknowledgements We would like to thank Adam Csapo, Emer Gilmartin, Jonathan Grizou, JingGuang Han, Raveesh Meena and Dimitra Anastasiou for implementing and evaluating Nao WikiTalk and the multimodal interaction capabilities on the Nao robot at eNTERFACE 2012 in Metz in July 2012.

References

1. Allwood, J.: Linguistic Communication as Action and Cooperation: A Study in Pragmatics. Gothenburg Monographs in Linguistics 2. University of Gothenburg, Gothenburg (1976)
2. Csapo, A., Gilmartin, E., Grizou, J., Han, J., Meena, R., Anastasiou, D., Jokinen, K., Wilcock, G.: Multimodal conversational interaction with a humanoid robot. In: Proceedings of 3rd IEEE International Conference on Cognitive Infocommunications (CogInfoCom 2012). Kosice (2012)
3. Fong, T., Nourbaksh, I., Dautenhahn, K.: A survey of socially interactive robots. Robot. Auton. Syst. **42**, 143–166 (2003)
4. Han, J., Campbell, N., Jokinen, K., Wilcock, G.: Investigating the use of non-verbal cues in human-robot interaction with a Nao robot. In: Proceedings of 3rd IEEE International Conference on Cognitive Infocommunications (CogInfoCom 2012). Kosice (2012)
5. Jokinen, K.: Constructive Dialogue Modelling: Speech Interaction and Rational Agents. Wiley, Chichester (2009)
6. Jokinen, K.: Pointing gestures and synchronous communication management. In: Esposito, A., Campbell, N., Vogel, C., Hussein, A., Nijholt, A. (eds.) Development of Multimodal Interfaces: Active Listening and Synchrony, pp. 33–49. Springer, Heidelberg (2010)
7. Jokinen, K., Hurtig, T.: User expectations and real experience on a multimodal interactive system. In: Proceedings of 9th International Conference on Spoken Language Processing (Interspeech 2006). Pittsburgh, USA (2006)
8. Jokinen, K., Wilcock, G.: Emergent verbal behaviour in human-robot interaction. In: Proceedings of 2nd International Conference on Cognitive Infocommunications (CogInfoCom 2011). Budapest (2011)
9. Jokinen, K., Wilcock, G.: Constructive interaction for talking about interesting topics. In: Proceedings of Eighth International Conference on Language Resources and Evaluation (LREC 2012). Istanbul (2012)
10. Jokinen, K., Harada, K., Nishida, M., Yamamoto, S.: Turn-alignment using eye-gaze and speech in conversational interaction. In: Proceedings of 11th International Conference on Spoken Language Processing (Interspeech 2010). Makuhari, Japan (2010)
11. Kendon, A.: Gesture: Visible Action as Utterance. Cambridge University Press, Cambridge (2004)

12. Levitski, A., Radun, J., Jokinen, K.: Visual interaction and conversational activity. In: Proceedings of the 4th Workshop on Eye Gaze in Intelligent Human Machine Interaction: Eye Gaze and Multimodality. Santa Monica, USA (2012)
13. McCoy, K.F., Cheng, J.: Focus of attention: Constraining what can be said next. In: Paris, C., Swartout, W., Mann, W. (eds.) Natural Language Generation in Artificial Intelligence and Computational Linguistics, pp. 103–124. Kluwer Academic Publishers, Boston (1991)
14. Meena, R., Jokinen, K., Wilcock, G.: Integration of gestures and speech in human-robot interaction. In: Proceedings of 3rd IEEE International Conference on Cognitive Infocommunications (CogInfoCom 2012). Kosice (2012)
15. Quek, F.: Toward a vision-based hand gesture interface. In: Proceedings of the Virtual Reality System Technology Conference, pp. 17–29. Singapore (1994)
16. Swerts, M., Geluykens, R.: Prosody as a marker of information flow in spoken discourse. Lang. Speech **37**, 21–43 (1994)
17. Wilcock, G.: WikiTalk: A spoken Wikipedia-based open-domain knowledge access system. In: Proceedings of the COLING 2012 Workshop on Question Answering for Complex Domains, pp. 57–69. Mumbai, India (2012)

Chapter 20
Component Pluggable Dialogue Framework and Its Application to Social Robots

Ridong Jiang, Yeow Kee Tan, Dilip Kumar Limbu, Tran Anh Dung, and Haizhou Li

Abstract This paper is concerned with the design and development of a component pluggable event-driven dialogue framework for service robots. We abstract standard dialogue functions and encapsulate them into different types of components or plug-ins. A component can be a hardware device, a software module, an algorithm or a database connection. The framework is empowered by a multipurpose XML-based dialogue engine, which is capable for pipeline information flow construction, event mediation, multi-topic dialogue modeling and different types of knowledge representation. The framework is domain-independent, cross-platform, and multilingual. Experiments on various service robots in our social robotics laboratory showed that the same framework works for all the robots that need speech interface. The development cycle for new dialogue system is greatly shortened while the system robustness, reliability, and maintainability are significantly improved.

20.1 Introduction

Spoken-dialogue system is an intuitive, flexible, and natural mean of communication between human and computer. It possesses several advantages over other types of interface such as natural, inexpensive, and easy to use. With the advances of speech technology, language processing, dialogue modeling, as well as the emergence of faster and more powerful computers, spoken-dialogue systems have obtained an increasingly widespread use in a wide variety of applications: voice-operated cell phones, car navigation systems, commercial information retrieval, gaming, education, healthcare and talking agents, etc. [1]. Especially in recent years, with

R. Jiang (✉) • Y.K. Tan • D.K. Limbu • T.A. Dung • H. Li
Institute for Infocomm Research, 1 Fusionopolis Way #21-01 Connexis, Singapore 138632, Singapore
e-mail: rjiang@i2r.a-star.edu.sg; yktan@i2r.a-star.edu.sg; dklimbu@i2r.a-star.edu.sg; tanhdung@i2r.a-star.edu.sg; hli@i2r.a-star.edu.sg

J. Mariani et al. (eds.), *Natural Interaction with Robots, Knowbots and Smartphones: Putting Spoken Dialog Systems into Practice*, DOI 10.1007/978-1-4614-8280-2_20,

more and more social robots making their way into the world of our everyday life, for instance, bank tellers, tour guides, elder care nurses, home maids, and receptionists, there is an increasing need for spoken-dialogue applications which can be equipped on various robots for natural and intuitive human-robot interaction. However, building a spoken dialogue for a robot is a great challenge for the engineers working on robots. Especially when they are developing multiple different robots, things become even worse because the hardware configurations and software platforms of these robots may vary from one to another. For instance, some robots need close-talk microphone while others may request far-talk recording. Some robots need Windows operation system while some may run on Linux or Macintosh. In the meantime, any hardware or software upgrading may also affect the existing spoken-dialogue application. There are cases where the robots must be built with the capability to communicate in different spoken languages. All these facts pose a great challenge to us: how to design a spoken-dialogue platform that works for as many robots as possible without repeating effort?

The aforementioned challenges and our past experience working on different social robots motivate us to develop a component pluggable event-driven spoken-dialogue framework which targets for a broad spectrum of domains. The framework must be component based. Different dialogue systems can be assembled and configured based on the existing dialogue components with minimum effort. To make the system easy to use, there should be no high-level programming as well as complex third-party software installation and configuration in developing new dialogue applications.

In this paper, we first review some related spoken-dialogue frameworks in Sect. 20.2. Then the pluggable system architecture is presented in Sect. 20.3. After that, detailed component design is described in Sect. 20.4. Section 20.5 presents XML-based dialogue model which allows finite state automata and frame-based representation. Finally, we conclude our work on the component pluggable spoken-dialogue platform and future enhancement.

20.2 Related Work

Much effort has been made by researchers from both industry and academia on the research of spoken-dialogue systems. Industrial research puts more weight on pragmatic and reusable systems while academic research focuses more on the natural and flexible dialogue applications [2]. Among these researches, some projects work on the application framework for reducing the cost of new dialogue application development. A notable project is American DARPA communicator project [3, 4]. The Galaxy Communicator software infrastructure is a distributed, message-based, hub-and-spoke infrastructure optimized for constructing spoken-dialogue systems. Many spoken-dialogue systems were developed on top of Galaxy II communicator, for instance, Olympus dialogue system (with its own dialogue management framework) [5] and AT&T spoken-dialogue system [6]. Galaxy II-based systems treat every component as a separate server and the routing logic of Galaxy hub is described via simple configuration script. To be different from the Galaxy-based system, our proposed framework allows every dialogue component to be remote server or local plug-in. Message routing and business logic are controlled by extensible XML with the unified syntax. With the integrated control on message routing and dialogue logic, multiple engines of the same type (e.g., language understanding) can work in parallel or be switched dynamically based on current dialogue status. Another widely used dialogue framework is systems based on information state update (ISU). ISU approach provides the user with a dialogue modeling framework using information state, update rules, and static plans. Trindikit [7] is a toolbox for building dialogue managers based on an information state and dialogue move engine. DIPPER [8] is implemented on top of Open Agent Architecture and it comes with its own dialogue management component, which is similar to Trindikit. ISU-based systems require Open Agent Architecture for communication and non-free dialect of the programming language Prolog for information state update and dialogue control [9]. VoiceXML is designed for automated telephone services with its major goal of bringing the advantages of Web-based development and content delivery to interactive voice response applications. It is widely used by leading organizations from the speech, telecommunications, mobile and wireless, as well as information technology industries. It is a frame-based system and support mixed-initiative dialogue management technique. Compared to VoiceXML, our framework shows some similarities in terms of XML support, dialogue control strategy, etc. However, VoiceXML lacks features on low-level message routing, database access, programming logic control, multi-topic management, and modeling of returning dialogue flow [10]. It is also not able to persistently store and describe a specific state of a dialogue [11].

20.3 System Architecture

In order to make the dialogue framework reusable, we employ object-oriented approach, loose-component coupling, event-driven paradigm and plug-and-play strategy to design and develop the proposed dialogue framework. The overall system architecture is shown in Fig. 20.1. The dialogue system is functionally divided into dialogue manager and a number of standard dialogue components—plug-ins. A programmable message center is designed to facilitate the communication between internal components, as well as message routing to graphic user interface (GUI) or middleware interface for the communication with external module through TCP/IP protocol. All messages from different sources are represented in a unified form and can be dispatched and handled in the same way. A message may come from a local or remote dialogue component, XML script command or a user command issued by graphic user interface. With the developed database plug-in, rule engine plug-in and information retrieval plug-in, the dialogue framework is able to access rich information from the backend (database, knowledge base and Web). A detailed introduction about the system architecture of the framework can be found in [12].

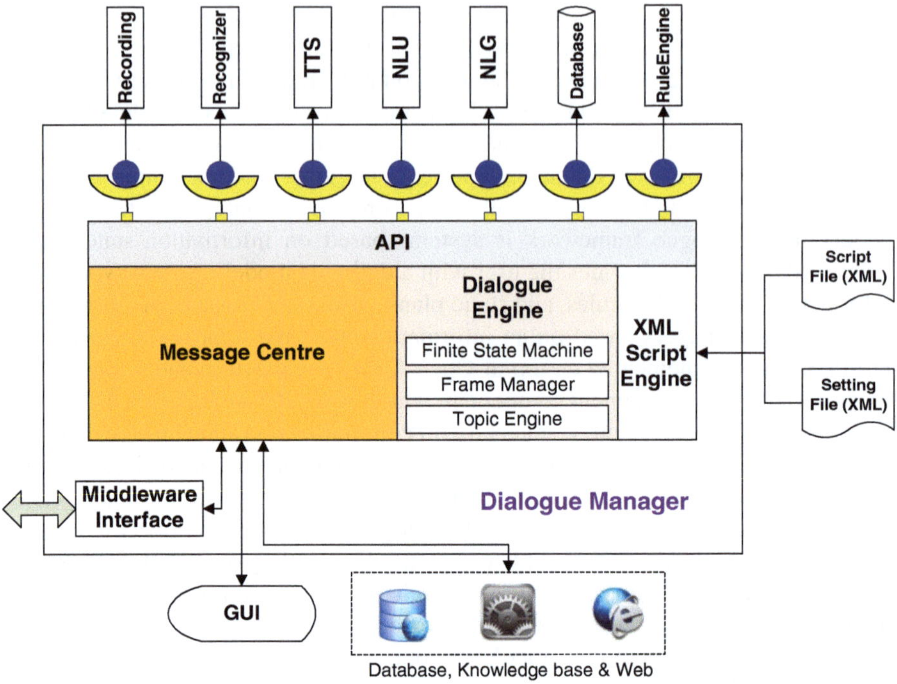

Fig. 20.1 System architecture of component-based spoken dialogue framework

20.4 Component Design

A component is an independent functional module which is designed to perform one or more specific functions. A component can be reused and provides service to a dialogue system which needs its functions. From the perspective of software engineering, a component often takes the form of object or class and is developed by object-oriented paradigm for code reuse, better encapsulation and extensibility. In the proposed dialogue framework, all components can be broadly classified into two types: Generic Components and Standard Dialogue Components. Generic Components are components whose functions are not standard dialogue functions such as *push-to-talk, database access, and rule engine.* Generic Components only provide fundamental interface to work with the framework. Standard Dialogue Components are referring to components that perform one of the standard dialogue functions, for instance, *sound recording, voice activity detection, speech recognition, spoken language understanding, nature language generation, and speech synthesis*. Besides providing fundamental interface, these components support their own specific interface in order to carry out their specific functions, for instance, speech recognition and utterance generation.

20.4.1 Interface Design for Generic Component

To work with the proposed dialogue framework, the minimum requirement for a component is that it must be able to communicate with the dialogue manager, any other component in the framework, or remote agent through middleware interface. Hence, following interfaces are fundamental:

- OnMessage—this is the interface for a component to handle all incoming messages from external sources. The message can come from a remote agent, other components in the framework, or the framework itself, which fires the message in XML script or through its graphic user interface.
- FireEvent—interface to fire an event by the particular component. This event can be captured by event hub and handled in XML script or relayed to one or many other components.
- SendMessage—this is the interface to communicate with remote agent through the platform. This creates a way for new plug-ins to directly communicate with remote agents or dialogue components. For instance, a dialogue plug-in can directly send a message to the robot to perform certain actions or even request information from a particular sensor.
- Print—interface to print whatever message to the console of the dialogue framework, for instance, normal component output and warning errors.

20.4.2 Interface Design for Standard Component

Besides the fundamental interface for Generic Component, a Standard Component must provide its specific interface to perform its correspondent functions. For instance, *sound recording* must support commands for "start recording," "stop recording," "select recording device," "set recording format (frequency, channel, etc.)," "return recorded buffer," etc., regardless of whether the microphone is close-talk or far-talk. Similarly, a *speech recognition* component should provide basic interfaces such as "do recognition on buffer," "do recognition on file," "set grammar," and "return recognition result." In the proposed dialogue framework, the common functions for all standard dialogue components are abstracted and correspondent component interfaces are designed and developed.

With the designed interfaces of standard dialogue components and their plug-ins, various spoken-dialogue systems can be quickly configured.

20.4.3 API for New Component Development

One of the advantages of this component pluggable dialogue framework lies in its extensibility through various plug-in developments. Currently the framework

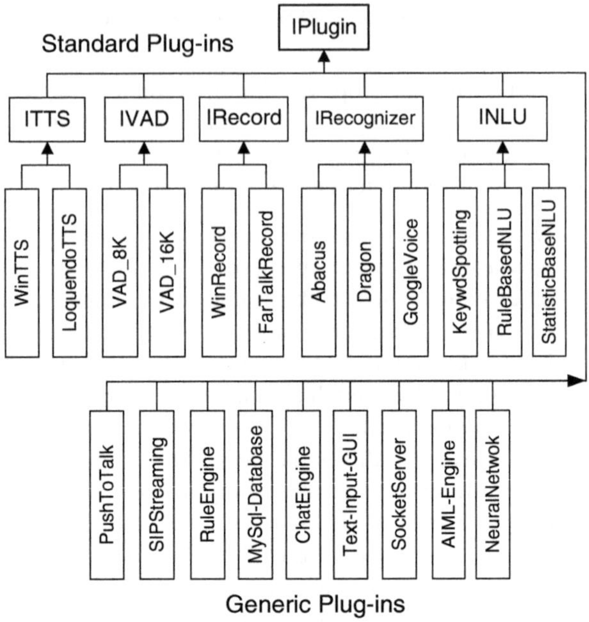

Fig. 20.2 Object-oriented interface and its hierarchical relationship with plug-ins

provides C++ application programming interface (API) for both Generic Component and Standard Dialogue Component development. The APIs are encapsulated into different objects with inheritance and worked as adaptors for new plug-in development. With the API provided, third-party algorithms and dialogue components can be quickly integrated into the framework. Figure 20.2 illustrates the generic interface and standard dialogue interface as well as their inheritance hierarchy with some of the developed plug-ins. Currently we have created a pool of dialogue plug-ins that are sufficient enough to assemble different dialogue systems. For speech recognition, we have our own in-house developed speaker-independent recognition engine-Abacus (for both English and Chinese) [13]. We also integrate commercial Dragon speech recognition engine and Web-based Google speech recognition engine (multiple languages). For speech synthesis, Microsoft Speech API and commercial Loquendo text-to-speech are integrated.

20.4.4 Component Management and Communication

A component can be configured to be loaded during system initialization. In the meantime, it can also be dynamically loaded and unloaded by XML script or a load/unload message based on the current dialogue status. Once a component is loaded into the framework, it is ready to provide services to any module through different events. For instance, a database component is able to query

```
<Plugins>
    <Plugin>SR_WinTTS</Plugin>
    <Plugin>SR_SpeechEnhancement</Plugin>
    <Plugin>SR_Dragon</Plugin>
    <Plugin>SR_StatisticNLU</Plugin>
    <Plugin>SR_NLG</Plugin>
    <Plugin>MySQLDatabase</Plugin>
</Plugins>
```

Fig. 20.3 Configuration of plug-ins for a typical spoken-dialogue system

```
   Module: IRecord
   Command: BufferFull
   Action: <post  module ="ASR "
command ="DoRecognition  "/>
-----------------------------------------------------
   Module: IRecognizer
   Command: RecognitionResult
   Action:  <post  module ="NLU "  command ="DoNLU "
param ="_LPARAM1_  "/>
-----------------------------------------------------
   Module : INLU
   Command: FrameInfo
   Action:  <object  name ="nluRes "  class ="list "/>
        <getlist  name ="nluRes "/>
        <fire  param ="nluRes "/>
-----------------------------------------------------
   Module: INLG
   Command: Speak
   Action:  <speak  param ="_LPARAM1_  "/>
```

Fig. 20.4 Message control for typical pipeline information flow

specified database and return the results to XML script, graphic user interface, a natural language understanding plug-in, or a remote agent through middleware communication. In the meantime, it can also share all the resources and services within the framework.

To build a new spoken-dialogue application, what we need to do is only to assemble the selected dialogue components through a configuration file and then control the communication between these components. Figure 20.3 shows XML snippet from a dialogue setting file which specifies the dialogue components required for a dialogue application. Figure 20.4 shows one type of message control which targets to form a popular pipeline information flow.

In addition, the framework supports multiple components of the same type. These multiple components can be used in parallel as shown in Fig. 20.5a or only one suitable component was chosen at one time based on the scenario as shown in Fig. 20.5b. In this way, more advanced dialogue systems can be quickly configured. For instance, multiple speech recognition engines can be used and then results are fused for higher recognition accuracy. Rule-based natural language understanding and statistical natural language understanding components can be selectively used based on the scenario for better understanding performance.

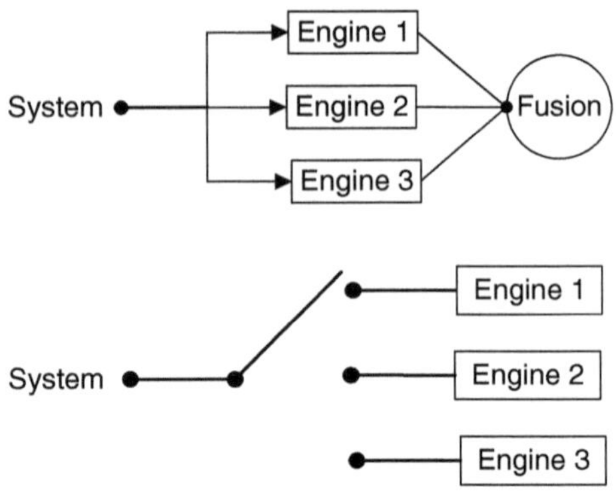

Fig. 20.5 Diagram of multiple engines (the same type): (**a**) work in parallel for improved results and (**b**) dynamically switched based on scenarios

20.5 Dialogue Model

The framework supports several types of task representation and dialogue management: finite state automata, object-oriented frame-based representation and the combination of these two representations.

20.5.1 Finite State Automata

An XML-based finite state machine was built into this framework to support simple dialogue task representation. First, task is decomposed into predetermined steps or stages [14]. In every state, expected events are processed. Then transitions are designed based on the dialogue control logic. The structural XML representation of a state is shown in Table 20.1. *Messages* is the field for handling all the messages. When the state is active, all messages are directed to this state and messages are handled based on the definition of the *Messages* field.

The transition is triggered by certain events or after execution of a piece of XML script. The advantage of this task representation is easy to build, possesses better task success rate. However, it inhibits user's opportunity to take initiative, usually requires explicit confirmation and longer dialogue; hence it is not efficient and less natural.

20.5.2 Object-Oriented Frame-Based Representation

We propose an object-oriented dialogue model for the development of natural human-machine communication. The model employs object-oriented technology to represent conversational topics and their internal relationships in a structured XML script language. Complex topics can be modeled as topic tree through hierarchical

Table 20.1 Field configuration of a dialogue state

Field	Representation
Entrance	Define actions for the first entry of the state. It is usually used for state initialization
Exit	Define actions when the dialogue exits from this state
Messages	Define actions responding to different messages
Online	Define message handler which responses to the event when a remote agent is online
Offline	Define message handler which responses to the event when a remote agent is offline
Functions	A place holder for the definition of script functions

Table 20.2 Slot field configuration and representation

Field	Representation
Help	Define help message when the system is automatically seeking information about this slot
Confirm	Define confirmation message when the slot value needs confirmation
Cardinality	Define cardinality constraints for this slot
Filled	Define actions when the slot value is filled. This is different from the "filled" field at the object level

decomposition. Discourse attributes, system intentions, goals, beliefs and dialogue history, context, etc., can be encapsulated in an XML-based class. This integrated and comprehensive representation facilitates the implementation of frame-based dialogue control strategy. Multi-topic, mixed-initiative spoken-dialogue system can be realized by defining multiple objects and embedding rules and knowledge into the representation. The proposed approach takes the advantage of object-oriented technology in object modeling and makes use of the power of XML for data and knowledge representation. Codes can be reused and extended by class instantiation and inheritance. Most importantly, this approach preserves modularity, reusability and extensibility within the dialogue model.

In this framed-based model, the slot representation includes many attributes and a number of fields to facilitate the task representation and dialogue control. Table 20.2 illustrates some of the fields used for the slot representation.

Table 20.3 illustrates a sample "FacilityName" slot for a FACILITY object with cardinality and filled fields. It is a self-contained automated dialogue representation for question asking, answer checking and information query.

20.5.3 Hybrid Task Representation

Finite state automata and frame-based representation can be combined in this framework for the better representation of dialogue task and logic control as well as

Table 20.3 Sample slot representation in XML

```
<slot name=''FacilityName''
      type=''Functional''
      expr=''unknown''
      alias=""
      question=''What's the facility name?''>
  <cardinality type=''set''>
    <value>Swimming pool</value>
    <value>Sky garden</value>
    <value>Auditorium</value>
    <value>Star home</value>
  </cardinality>
  <filled>
    <querydb param='''select * from facility where
                   name=\''+this.FacilityName+'\''''
              return=''dbRes''/>
    <post module=''NLG'' command=''Generate'' param=''dbRes[6]''/>
</filled>
</slot>
```

higher dialogue success rate. This representation takes the advantages of both finite state automata and frame-based representation. For instance, when the dialogue needs simple confirmation or selection from a flat menu, then state representation will be used. Otherwise, frame-based representation is used.

20.6 Conclusion

This paper presents our work on the design and development of component pluggable event-driven spoken-dialogue framework. The motivation of this research comes from the fact that most of the dialogue frameworks are not configurable; components are coupled with other modules or closely tied with specific domain and cannot be reused; dialogue modeling and backend information access need expert knowledge or high-level programming experience such as C++ or Java; systems can only support single spoken language and are not cross-platform; both the framework and dialogue script are not open, they are not extensible and very difficult to maintain; systems have steep learning curve—they are very complicated and come with a bunch of third-party software; all these defects make the development of spoken-dialogue interface extremely difficult for normal users such as robotics engineers. Keeping all these flaws in mind, we strive to come out a new dialogue framework which will ultimately avoid these drawbacks.

The proposed framework employs object-oriented approach, loose-component coupling, event-driven paradigm, hub-and-spoke topology and service-oriented architecture to facilitate the easy construction of robust and efficient spoken-dialogue applications. The framework supports state-based and frame-based dialogue control strategies, user-initiative, system-initiative and mixed-initiative dialogue techniques, multi-topic and topic-changing management, confirmation, database query, etc. Compared with other XML-based dialogue systems, the current dialogue framework is empowered by a full-fledged XML script engine which supports scoped variables, arrays, list, string handling, expression, if/else statement, for loop, functions, timer, socket communication, file I/O, as well as pipeline information flow programming. In addition, the platform is domain independent, lightweight and does not require any other supporting software or environment setting. Most importantly, it can be enhanced by new plug-in development with API provided by the framework. So far, we have created quite a number of plug-ins which include Dragon and Google Voice engine for speech recognition, Microsoft SAPI and Loquendo engine for speech synthesis, MySql engine for database access, and Artificial Intelligence Markup Language (AIML) engine for chat services. Based on this pool of generic and standard components as well as dialogue framework, we have successfully built spoken-dialogue interface for all our service robots which need speech interface: robotic butler, Fusionbot (home robot helper), Mika (coffee robot), Lucas (information Kiosk & butler), Olivia (receptionist robot), etc. By component reuse, every speech interface only needs two XML files: one configuration file specifying components used and one script file describing dialogue model and message routing.

Our practice on developing these new spoken-dialogue interfaces showed that the development cycle for a new dialogue system can be greatly reduced by its nature of configurability and component reusability as well as the powerful XML-based dialogue engine. The developed systems are robust, reliable, and easy to maintain.

In the future, we will continue to increase the number of standard dialogue components and generic components. In the meantime, to ease the XML scripting for dialogue model, an authoring tool is needed.

Acknowledgements The research described in this paper is the result of collaboration with colleagues in the Lab of Agency for Science, Technology and Research (A*STAR) Social Robotics (ASORO) and colleagues from Human Language Technology Department of Institute for Infocomm Research; their contributions are gratefully acknowledged.

References

1. McTear, M.F.: Spoken dialogue technology: enabling the conversational user interface. ACM Comput. Surv. **34**(1), 90–169 (2002)
2. Pieraccini, R., Huerta, J.: Where do we go from here? Research and commercial spoken dialog systems. In: SixthSIGdial Workshop on Discourse and Dialogue. Lisbon, Portugal (September 2–3, 2005)
3. Galaxy Communicator Documentation: http://communicator.sourceforge.net/sites/MITRE/distributions/GalaxyCommunicator/docs/manual/index.html
4. Seneff, S., Hurley, E., Lau, R., Pao, C., Schmid, P., Zue, V.: Galaxy-II: A reference architecture for conversational system development. In: Proceedings of the Fifth International Conference on Spoken Language Processing (ICSLP '98), pp. 931–934. Sydney (December, 1998)
5. Bohus, D., Raux, A., Harris, T.K., Eskenazi, M., Rudnicky, A.I.: Olympus: An open-source framework for conversational spoken language interface research. In: Proceedings of HLT-NAACL 2007 Workshop on Bridging the Gap: Academic and Industrial Research in Dialog Technology, Rochester (2007)
6. Levin, E., Narayanan, S., Pieraccini, R., Biatov, K., Bocchieri, E., Di Fabbrizio, G., Eckert, W., Lee, S., Pokrovsky, A., Rahim, M., Ruscitti, P., Walker, M.: The AT&T-DARPA: Communicator mixed-initiative spoken dialog system. In: Proceedings of the Sixth International Conference on Spoken Language Processing (ICSLP'2000), pp. 122–125. Beijing (2000)
7. Larsson, S., Traum, D.R.: Information state and dialogue management in the TRINDI dialogue move engine toolkit. Nat. Lang. Eng. **6**(3–4), 323–340 (2000)
8. Bos, J., Klein, E., Lemon, O., Oka, T.: DIPPER: Description and formalisation of an information-state update dialogue system architecture. In: Proceedings of the Fourth SIGdial Workshop of Discourse and Dialogue, Sapporo (2003)
9. Ljunglöf, P.: Trindikit.py: An open-source Python library for developing ISU-based dialogue systems. In: Proceedings of IWSDS'09, First International Workshop on Spoken Dialogue Systems Technology Workshop. Kloster Irsee, Germany (2009)
10. Hamerich, S.W., Wang, Y.H., Schubert, V., Schless, V., Igel, S.: XML-based dialogue descriptions in the GEMINI project. In: Proceedings of the Berliner XML-Tage 2003, pp. 404–412. Germany (2003)
11. Heinroth T., Denich, D.: Spoken interaction within the computed world: evaluation of a multitasking adaptive spoken dialogue system. In: 35th Annual IEEE International Computer Software and Applications Conference (COMPSAC 2011), Munich. IEEE (2011)

12. Jiang, R.D., Tan, Y.K, Limbu, L.M., Tung, A.T., Li, H.Z.: A configurable dialogue platform for ASORO robots. In: Asia Pacific Signal and Information Processing Association Annual Summit and Conference, APSIPA ASC 2011. Xi'an, China (2011)
13. Li, H., Ma, B., Lee, C.H.: A vector space modeling approach to spoken language identification. IEEE Trans. Audi. Speech Lang. Process. **15**(1) (2007)
14. McTear, M.F.: Spoken Dialogue Technology. Springer, Berlin (2004)

Part V
Spoken Dialog Systems Components

Chapter 21
Visual Contribution to Word Prominence Detection in a Playful Interaction Setting

Martin Heckmann

Abstract This paper investigates how prominent words can be distinguished from non-prominent ones in a setting where a user was interacting in a small game, designed as a Wizard-of-Oz experiment, with a computer. Misunderstandings of the system were triggered and the user was asked to correct them naturally, i. e. using prosodic cues. Consequently, the corrected word is expected to be highly prominent. Audio-visual recordings with a distant microphone and without visual markers were made. As acoustic features relative energy, duration and fundamental frequency were calculated. From the visual channel rigid head movements and image transformation-based features from the mouth region were extracted. Different feature combinations are evaluated regarding their power to discriminate the prominent from the non-prominent words using a SVM. Depending on the features accuracies of approximately 70%–80% are achieved. Thereby the visual features are in particular beneficial when the acoustic features are weaker.

21.1 Introduction

Current commercial spoken dialog systems are insensitive to prosodic characteristics of speech even though it is well known that prosodic cues play a very important role in human communication [20]. Nevertheless, quite a few research systems included such prosodic cues in a human-machine dialog [15, 18, 21]. In general the inclusion of prosodic cues is quite difficult as they show not only a large variability from speaker to speaker but are also difficult to extract from the speech signal. The latter problem could possibly be reduced by including visual information. Information on the movements of the speaker's mouth and face notably improves

M. Heckmann (✉)
Honda Research Institute Europe GmbH, D-63073 Offenbach/Main, Germany
e-mail: martin.heckmann@honda-ri.de

J. Mariani et al. (eds.), *Natural Interaction with Robots, Knowbots and Smartphones: Putting Spoken Dialog Systems into Practice*, DOI 10.1007/978-1-4614-8280-2_21,

the accuracies of automatic speech recognition particular in difficult situations [11, 16, 19, 23]. Humans are also able to use such visual information to extract prosodic cues [1, 2, 9, 17, 22]. Studies quantifying these visual prosodic cues have shown that they are mainly manifested in larger jaw opening, lip spreading and protrusion and to some extent to head movements [7, 8].

In [10] we showed that the discrimination of prominent from non-prominent words can be improved by visual features extracted from the speaker's face without the use of additional visual markers. As visual features we used image transformations calculated on the mouth region of the speaker. For this paper we improved the feature extraction and investigate how rigid head movements and dynamic features can support this discrimination.

In the next section an overview on the recording of the data will be given. After that Sect. 21.3 describes the different features extracted from the acoustic and visual channel. Following this Sect. 21.4 will present the results of the classification experiments. In the last section we will discuss the results.

21.2 Dataset

For the recording of the data the subjects interacted via speech in a Wizard-of-Oz experiment with a computer in a small game where they would move tiles to uncover a cartoon. With this playful setting we expected to obtain more natural speech, in particular regarding the prosody. This game yielded utterances of the form "put green in B one". Occasionally, a misunderstanding of one word of the sequence was triggered and the corresponding word highlighted, verbally and visually. Verbal feedback was based on the FESTIVAL speech synthesis system [3]. The subjects were told to repeat in these cases the phrase as they would do with a human, i. e. emphasizing the previously misunderstood word. However, they were not allowed to deviate from the sentence grammar e. g. via beginning with "No". This was expected to create a narrow focus condition (in contrast to the broad focus condition of the original utterance) and thereby making the corrected word highly prominent.

In total three subjects, one female and two males, either speaking British English as their sole native language or being bilingual British English/German were recorded. The audio signal was originally sampled at 48 kHz and later downsampled to 16 kHz. For the video images a CCD camera with a resolution of $1,280 \times 1,024$ pixels and a frame rate of 25 Hz was used.

Due to the strong resemblance of the recorded speech in grammar and vocabulary to that of the Grid Corpus [6] a speech recognition system trained on that corpus could be used to perform a forced alignment on the acquired data. For the alignment HTK and a combination of RASTA-PLP and spectro-temporal HIST features [12] were used as this gave upon visual inspection better results than either of the feature sets alone or MFCC features. In particular, we first performed a speaker adaptation with a maximum likelihood linear regression (MLLR) step followed by a maximum A-Posteriori (MAP) step, both using HTK [24].

For further processing those turns where the original utterance and a correction were available were selected. This yielded 137 turn pairs (original utterance + correction) for subject A, 146 for subject B and 94 for subject C. From these the word which was emphasized in the correction was determined. Then it was extracted as well in the original utterance as in the correction. This yields a dataset with each individual word taken from a broad and a narrow focus condition. An analysis of acoustic features related to word prominence in [10] showed that the words in the narrow focus condition were notably more prominent than in the broad focus condition.

21.3 Features

In the following experiments the features described in Table 21.1 which have previously been proposed to capture word prominence were used. From these features (except for duration) the mean value for each word was calculated and used in the subsequent analysis. The beginning and end of the word were taken from the forced alignment.

For the visual modality, the openCV library [4] was applied to first detect the face in each image frame and then determine the nose position. As the nose moves only slightly in relation to the skull during articulation it yields information on the rigid head movements and hence also on the current position of the mouth region. For determining the mouth region from the images a fixed and for all speakers identical offset from the nose was used and also the size of the mouth region was kept identical. After downsampling by a factor of 2, this yields an image of 80×80 pixels of the mouth region (compare Fig. 21.1). On these images either a two-dimensional fast Fourier transform (FFT) or discrete cosine transform (DCT) was calculated. In case of the FFT and DCT out of the 6,400 coefficients per image the 50 with the highest energy were selected. This was done by calculating for each speaker separately the mean energy of all 6,400 coefficients on a randomly selected subset of 10% of the data. As FFT coefficients are complex, we only used their magnitude in all steps. Consequently we obtain for FFT and DCT 50 coefficients per frame to capture the mouth shape. All visual features, i. e. for the nose and the mouth shape, were smoothed along the time axis with a 5th order FIR low-pass filter with a cut-off frequency of 5 Hz. Furthermore, first and second derivatives (Δ and $\Delta\Delta$) were calculated.

Table 21.1 Description of the different features

Acoustic	
dur	Duration of the word
en	Energy relative to the mean of the utterance
f0	Mean fundamental frequency (extracted according to [13, 14])
Visual	
y	Nose y position relative to the mean of the utterance
d, dd	First and second derivative

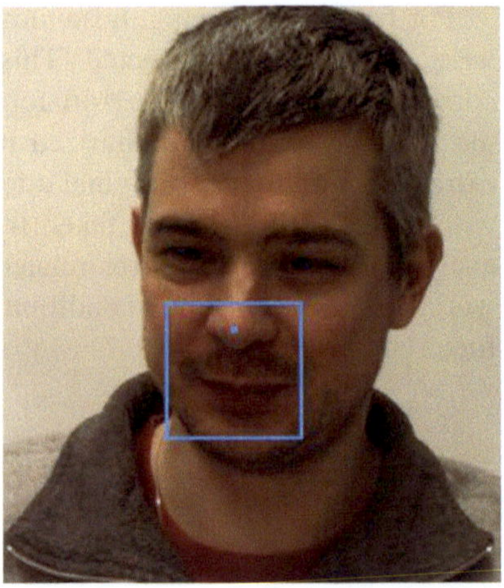

Fig. 21.1 Image from recording after cropping to face region, nose detection, downsampling and highlighting of the mouth region

21.4 Results

To discriminate prominent from non-prominent words, an SVM with a Radial Basis Function Kernel was trained using LibSVM [5]. For each feature combination a grid search for C, the penalty parameter of the error term, and γ, the variance scaling factor of the basis function, were performed using the whole dataset. Prior to the grid search the data was normalized to the range $[-1 \ldots 1]$. With the found optimal parameters, an SVM was trained on 75% of the data and tested on the remaining 25%. Hereby a 30-fold cross-validation in which the data set was always split such that an identical number of elements is taken from both classes was run. To establish the 30 sets, a sampling with replacement strategy was applied. This process was performed individually for each speaker. In the following all results are averaged over all speakers.

When taking only the acoustic cues into account, we see that the relative energy achieves a performance of 63%, duration of 59% and f0, a notably stronger cue, 70% correct. The combination of all acoustic cues achieves 80% correct.

If we now look at the results for the visual cues in Fig. 21.2, we see that FFT achieves 64% and DCT 67% correct. Adding the first derivative to the FFT improves its performance to 66% correct, almost the level of the DCT. Adding the second derivative has a negative effect. For the DCT all derivatives have a negative effect. Combining FFT and DCT features does not yield to superior performance than DCT alone. Regarding the nose y movement, only the second derivative, i. e. the acceleration of the nose, yields accuracies notably above chance level.

In Fig. 21.3 a combination of the f0 feature, the strongest acoustic feature, and different visual features is depicted. Adding the nose y acceleration to the f0 feature leads to a small improvement. When combining either FFT, DCT, or both with f0 the improvement is from 70% to 78%. Including either derivatives of FFT or DCT or the nose y acceleration does not further improve the results.

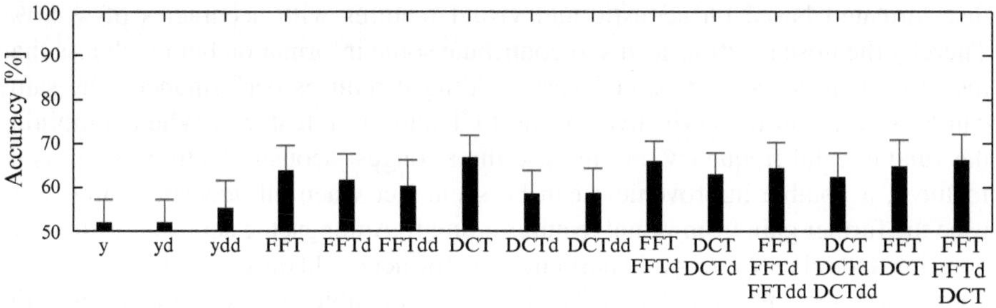

Fig. 21.2 Discrimination accuracies for different visual feature combinations. The horizontal lines indicate the standard deviation of the 30-fold cross-validation. See Table 21.1 for an explanation of the abbreviations

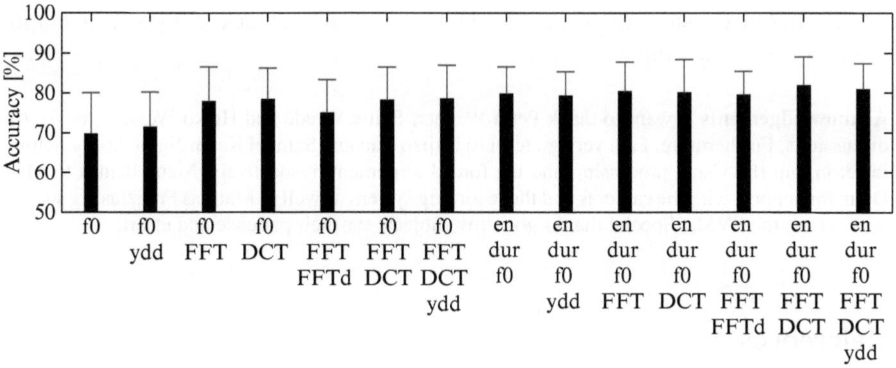

Fig. 21.3 Discrimination accuracies for different acoustic and visual feature combinations. The horizontal lines indicate the standard deviation of the 30-fold cross-validation. See Table 21.1 for an explanation of the abbreviations

Finally Fig. 21.3 also shows a combination of all acoustic features and different visual features. The performance of the acoustic features alone is 80% correct. Adding either FFT or DCT alone does not change this. However, when both are added 82% correct are obtained. When adding also the nose *y* acceleration it is reduced to 81% correct.

21.5 Conclusion

We set up a Wizard-of-Oz experiment where subjects were interacting in a small game with a computer. The experiment was designed to elicit prominent and non-prominent versions of a word. The results showed that these two classes can be discriminated based on acoustic and visual features with accuracies of $\approx 80\%$. Thereby the nose position seems to contribute some information but much less than the other features and in several cases adding it reduces performance. The same can be said about the derivatives of the FFT and DCT features. When combining the fundamental frequency, in our case the strongest acoustic feature, with visual features, a notable improvement can be seen. Yet when all acoustic features are used the further gain from visual features is small. Amongst the acoustic features are duration and relative energy. In particular the former could only be reliably extracted due to the forced alignment used. The latter is vulnerable to varying distances to the microphone. Hence these would be available with much less precision in a real world system. In these cases the additional visual features can be very beneficial. Therefore, we think that adding visual features renders the extraction of prosodic cues more reliable and hence can be helpful to develop more natural spoken dialog systems which can deal meaningfully with prosodic variations. Evaluation of the whole utterance instead of only the prominent word to include e. g. hypo-articulation effects of the following words [8].

Acknowledgements I want to thank Petra Wagner, Britta Wrede and Heiko Wersing for fruitful discussions. Furthermore, I am very grateful to Rujiao Yan and Samuel Kevin Ngouoko for helping in setting up the visual processing and the forced alignment, respectively. Many thanks to Mark Dunn for support with the cameras and the recording system as well to Mathias Franzius for support with tuning the SVMs. Special thanks go to my subjects for their patience and effort.

References

1. Al Moubayed, S., Beskow, J.: Effects of visual prominence cues on speech intelligibility. In: Proceedings of the International Conference on Auditory Visual Speech Process. (AVSP), vol. 9, p. 16. ISCA, Austin (2009)
2. Beskow, J., Granström, B., House, D.: Visual correlates to prominence in several expressive modes. In: Proceedings of INTERSPEECH, pp. 1272–1275. ISCA (2006)
3. Black, A., Taylor, P., Caley, R.: The festival speech synthesis system. Tech. rep. (1998)
4. Bradski, G., Kaehler, A.: Learning OpenCV: Computer vision with the OpenCV library O'reilly (2008)
5. Chang, C.C., Lin, C.J.: LIBSVM: A library for support vector machines. ACM Transactions on Intelligent Systems and Technology **2**, 27:1–27:27 (2011). Software available at http://www.csie.ntu.edu.tw/~cjlin/libsvm
6. Cooke, M., Barker, J., Cunningham, S., Shao, X.: An audio-visual corpus for speech perception and automatic speech recognition. J. Acoust. Soc. Am. **120**, 2421 (2006)
7. Cvejic, E., Kim, J., Davis, C., Gibert, G.: Prosody for the eyes: Quantifying visual prosody using guided principal component analysis. In: Proceedings of INTERSPEECH. ISCA (2010)
8. Dohen, M., Lœvenbruck, H., Harold, H., et al.: Visual correlates of prosodic contrastive focus in french: Description and inter-speaker variability. In: Speech Prosody. Dresden, Germany (2006)
9. Graf, H., Cosatto, E., Strom, V., Huang, F.: Visual prosody: Facial movements accompanying speech. In: International Conference on Automatic Face and Gesture Recognition, pp. 396–401. IEEE (2002)
10. Heckmann, M.: Audio-visual evaluation and detection of word prominence in a human-machine interaction scenario. In: Proceedings of INTERSPEECH. ISCA, Portland, OR (2012)

11. Heckmann, M., Berthommier, F., Kroschel, K.: Noise adaptive stream weighting in audio-visual speech recognition. EURASIP J. Applied Signal Process. **11**, 1260–1273 (2002)
12. Heckmann, M., Domont, X., Joublin, F., Goerick, C.: A hierarchical framework for spectro-temporal feature extraction. Speech Comm. **53**(5), 736–752 (2011). DOI: 10.1016/j.specom.2010.08.006. Perceptual and Statistical Audition
13. Heckmann, M., Gläser, C., Vaz, M., Rodemann, T., Joublin, F., Goerick, C.: Listen to the parrot: Demonstrating the quality of online pitch and formant extraction via feature-based resynthesis. In: Proceedings of IEEE/RSJ International Conference on Intelligent Robots and Systems (IROS). Nice (2008)
14. Heckmann, M., Joublin, F., Goerick, C.: Combining rate and place information for robust pitch extraction. In: Proceedings of INTERSPEECH, pp. 2765–2768. Antwerp (2007)
15. Hirschberg, J., Litman, D., Swerts, M.: Prosodic and other cues to speech recognition failures. Speech Communication **43**(1-2), 155–175 (2004)
16. Kolossa, D., Zeiler, S., Vorwerk, A., Orglmeister, R.: Audiovisual speech recognition with missing or unreliable data. In: Proceedings of International Conference on Auditory Visual Speech Processing (AVSP) (2009)
17. Munhall, K., Jones, J., Callan, D., Kuratate, T., Vatikiotis-Bateson, E.: Visual prosody and speech intelligibility. Psychol. Sci. **15**(2), 133 (2004)
18. Nöth, E., Batliner, A., Kießling, A., Kompe, R., Niemann, H.: Verbmobil: The use of prosody in the linguistic components of a speech understanding system. IEEE Trans. Speech and Audio Process. **8**(5), 519–532 (2000)
19. Potamianos, G., Neti, C., Gravier, G., Garg, A., Senior, A.: Recent advances in the automatic recognition of audiovisual speech. Proc. IEEE **91**(9), 1306–1326 (2003)
20. Shriberg, E.: Spontaneous speech: How people really talk and why engineers should care. In: Proceedings of EUROSPEECH, ISCA (2005)
21. Stolcke, A., Ries, K., Coccaro, N., Shriberg, E., Bates, R., Jurafsky, D., Taylor, P., Martin, R., Ess-Dykema, C., Meteer, M.: Dialogue act modeling for automatic tagging and recognition of conversational speech. Comput. ling. **26**(3), 339–373 (2000)
22. Swerts, M., Krahmer, E.: Facial expression and prosodic prominence: Effects of modality and facial area. J. Phonetics **36**(2), 219–238 (2008)
23. Yoshida, T., Nakadai, K., Okuno, H.: Automatic speech recognition improved by two-layered audio-visual integration for robot audition. In: Proceedings of 9th IEEE-RAS International Conference on Humanoid Robots, pp. 604–609. IEEE (2009)
24. Young, S., Odell, J., Ollason, D., Valtchev, V., Woodland, P.: The HTK Book. Cambridge University, Cambridge, United Kingdom (1995)

Chapter 22
Label Noise Robustness and Learning Speed in a Self-Learning Vocal User Interface

Bart Ons, Jort F. Gemmeke, and Hugo Van hamme

Abstract A self-learning vocal user interface learns to map user-defined spoken commands to intended actions. The voice user interface is trained by mining the speech input and the provoked action on a device. Although this generic procedure allows a great deal of flexibility, it comes at a cost. Two requirements are important to create a user-friendly learning environment. First, the self-learning interface should be robust against typical errors that occur in the interaction between a non-expert user and the system. For instance, the user gives a wrong learning example to the system by commanding "Turn on the television!" and pushing a power button on the wrong remote control. The spoken command is then supervised by a wrong action and we refer to these errors as label noise. Secondly, the mapping between voice commands and intended actions should happen fast, i.e. require few examples. To meet these requirements, we implemented learning through supervised NMF. We tested keyword recognition accuracy for different levels of label noise and different sizes of training sets. Our learning approach is robust against label noise, but some improvement regarding fast mapping is desirable.

22.1 Introduction

In *command-and-control* speech recognition, the user usually speaks a phrase from a set of predefined grammars and words. Large transcribed speech data sets are used to train the acoustic model beforehand and the language model is often written by hand in the form of context-free grammars. Such models fit well for users that a

---

B. Ons (✉) • J.F. Gemmeke • H. Van hamme
Department ESAT-PSI, KU Leuven, Leuven, Belgium
e-mail: Bart.ons@esat.kuleuven.be; Jort.gemmeke@esat.kuleuven.be; hugo.vanhamme@esat.kuleuven.be

J. Mariani et al. (eds.), *Natural Interaction with Robots, Knowbots and Smartphones: Putting Spoken Dialog Systems into Practice*, DOI 10.1007/978-1-4614-8280-2_22,

developer in a lab had in mind. Although these models suit the average speaker very well, they are inadequate for speakers with deviant speech. There is a renewed interest in dialogue systems that allow for more freedom in the interaction between man and machine by means of adaptation ([1, 2]). However, these systems are still based on predefined language and acoustic models that adapt through interactions to real-life situations.

Contrarily, we aim to design a vocal user interface that learns to understand normal or deviant speech by associating the spoken commands and their related actions during its usage (see [3]). The vocal user interface is trained by the end user by mining the speech input and the changes that are provoked on the device. The end user is able to specify his own commands and trains the system by giving examples to the system. For instance, the user might say "Please, turn on the television" and turns on the television with the remote control. The learning problem is a machine learning problem where the user has to demonstrate the intended action to the vocal user interface and by doing so, he provides supervision to the machine learning process. The vocal user interface should learn the association between the vocal command and the intended action, and it should control the intended action in future command calls by the user.

In a command-and-control speech application, some words are meaningful while others are not. For instance, for the command "Please, turn on the television", the informative parts are "turn on" and "television", but the polite introduction "please" is not informative. The required information needed to control a device is solely determined by the device. Informative parts lead to the identification of the desired action and we call these parts "keywords", while the others are referred to as "filler words".

When we assign a label to each keyword, the learning problem can be redefined: the vocal user interface should learn the association between the spoken keywords and the labels associated with the appropriate action. Because supervision is solely depending on the examples provided by the end user, correct demonstrations of consistent spoken commands will be more effective for learning all necessary associations between spoken keywords and labels. However, instead of imposing consistency on the user, we would rather prefer to design a user-friendly system where natural variation in communications with the system is allowed. For instance, when the user stands in front of the television, the user might say "Turn on, please!" and turns on the television. The provoked action is associated with the keyword labels "turn on" and "television" while the last keyword is actually missing in the acoustic input. Another kind of error would occur if the user says, "Turn on the television" but mistakenly pushes the power button on the remote control of the audio system. Obviously, such errors occur easily in the natural interaction between man and machine, and the learning algorithm should be robust against them. We refer to these errors with the term *label noise* (see [4] for a discussion of the impact of label noise in clustering methods).

Since the end user has to provide some effort to train the device, it is desirable that mappings should be learned fast. Label noise robustness and fast learning are two requirements that we investigate in the current study. In fact, the two may be related because the speed of learning might slow down when the system is not robust against it. Conversely, the system might be less robust against label noise when the system is able to learn mappings from only a few examples, as the number of examples might be too low to obtain a representative sample for future data.

We have chosen a supervised NMF approach [5–9] to learn the mappings between the spoken keywords and the keyword labels. NMF is a method to learn the underlying patterns in data like speech, images [10], documents [11] and many other types of data. Supervised NMF allows to discover latent patterns in data signalling the occurrence of meaningful parts like for instance the keywords in spoken commands and dialogues. The strength of the NMF approach is that the acoustic representation of a keyword can be found through weak supervision. We only have to specify which keywords are spoken in which utterance. NMF is able to extract the acoustic representation of a keyword from the examples of multiple weakly supervised spoken utterances. The goal of the current study is to test the two previously mentioned requirements, i.e. label noise robustness and fast learning, within the context of a supervised NMF framework.

The rest of the text is organized as follows. In Sect. 22.2, we briefly explain supervised NMF learning. In Sect. 22.3, we list four different types of errors that we expect to occur in the user's environment and we explain the effect of these errors for the NMF framework. In Sect. 22.4, we discuss the experimental design and we provide all technical details that are specific to the current experiments. Finally, in Sects. 22.5 we show and discuss the results.

22.2 Supervised Word Learning

NMF is a machine learning approach aiming at the discovery of latent structure in data based on the decomposition of a larger data matrix **V** into the product of two matrices **W** and **H** of lower dimensionality:

$$\mathbf{V} \approx \mathbf{W}\mathbf{H} \tag{22.1}$$

The columns in **W** represent the latent structure (recurring patterns) of the columns in **V** and the columns in **H** indicate which patterns are combined to approximate the columns in **V**. Iterative update rules for minimizing a distance measure between **V** and (**WH**) can be found in [6,9,10]. In the current NMF keyword learning approach, we have chosen the Kullback-Leibler divergence as the distance measure.

In our approach, the matrices **V**, **W** and **H** can be interpreted as follows. The nth utterance in the training set is represented by the column $\mathbf{v}_n$ in **V** ($n = 1 \ldots N$). Thus, the columns in **V** comprise the leaning examples (vectorized in a column vector) provided by the user. Supervised NMF learning leads to columns in **W** corresponding to keywords while columns in **H** indicate which columns in **W** are combined to compose the keywords in the utterance of the spoken command $\mathbf{v}_n$.

In supervised NMF learning (see [6] and [9]), the observation data of the nth utterance $\mathbf{v}_n$ in the learning phase consists actually of two parts: the acoustic representation of the command spoken by the user and the keyword labels in the action handled by the machine. We denote the acoustic part of $\mathbf{V}$ by $\mathbf{V}_a$ and the part indicating the presence of keywords by $\mathbf{V}_l$. For each to-be-learned keyword label, there is one row foreseen in $\mathbf{V}_l$ and its entries represent the number of times that the respective keyword was uttered in the nth utterance. In the matrix $\mathbf{W}$, an equal number of rows for the keyword labelling part are added to the acoustic part of $\mathbf{W}$. Keyword labels in $\mathbf{W}$ consist of 1 on row k for the kth keyword and 0 elsewhere. The purpose of the supervised NMF learning is to find the latent acoustic representations of the keywords (the acoustic part of $\mathbf{W}$). When we denote the labelling part of $\mathbf{W}$ by $\mathbf{W}_l$ and the acoustic part by $\mathbf{W}_a$, Eq. 22.1 (see [6] and [9]) is extended to

$$\begin{bmatrix} \mathbf{V}_l \\ \mathbf{V}_a \end{bmatrix} \approx \begin{bmatrix} \mathbf{W}_l \\ \mathbf{W}_a \end{bmatrix} \mathbf{H} \tag{22.2}$$

When the total set of vocal commands contains L keywords, $\mathbf{W}$ should count at least L columns, but in practice, some extra D columns are added to $\mathbf{W}$ to model the filler words.

The representation of the keywords in ($\mathbf{W}_a$) can be found by minimizing the Kullback-Leibler divergence between both sides of Eq. 22.2:

$$(\mathbf{H}^*, \mathbf{W}_a^*, \mathbf{W}_l^*) = \arg \min_{(\mathbf{H}, \mathbf{W}_a, \mathbf{W}_l)} D_{KL}\left(\begin{bmatrix} \mathbf{V}_l \\ \mathbf{V}_a \end{bmatrix} || \begin{bmatrix} \mathbf{W}_l \\ \mathbf{W}_a \end{bmatrix} \mathbf{H} \right) \tag{22.3}$$

When the acoustic representations of keywords ($\mathbf{W}_a^*$) are learned, keyword recognition can be tested on a test set consisting of unseen (unlabelled) utterances. We denote the data matrix $\mathbf{V}$ of the unseen utterances in the test phase by $\mathbf{V}_t$ and it only contains the acoustic representations of the spoken utterances. The matrix $\mathbf{H}$ in the test phase is denoted by $\mathbf{H}_t$. To recognize the keywords in $\mathbf{V}_t$, $\mathbf{H}_t$ is optimized in order to minimize the distance between $\mathbf{V}_t$ and ($\mathbf{W}_a^*\mathbf{H}_t$):

$$\mathbf{H}_t^* = \arg \min_{\mathbf{H}_t} D_{KL}(\mathbf{V}_t || \mathbf{W}_a^* \mathbf{H}_t) \tag{22.4}$$

The obtained matrix $\mathbf{H}_t^*$ is used to provide the keyword activation matrix $\mathbf{A}$:

$$\mathbf{A} = \mathbf{W}_l^* \mathbf{H}_t^* \tag{22.5}$$

$\mathbf{A}$ is a $(L \times N)$ matrix and each column in $\mathbf{A}$ corresponds to the respective column in $\mathbf{V}_t$. The higher the score in the rows of $\mathbf{A}$, the more likely that the respective keyword has appeared in the spoken test utterances.

22.3 Label Noise

Although supervised NMF approaches ([5–9]) have been studied frequently within the general scope of machine learning (see Sect. 22.1), to the best of our knowledge, no attention has ever been directed towards the robustness of the approach against label noise. When data is manually annotated, labels are expected to be correct. However, there are numerous ways to end up with labelling errors when a self-learning user interface is trained by the end user. We consider here four types of label noise: insertion, deletion, substitution and command substitution.

Insertion The command of the user can be underspecified: the command lacks part of the information needed to uniquely determine the intended action of the user. For instance, when the user says "Turn on" and then turns on the television, the executed action is associated with two keyword labels, "turn on" and "television", while the acoustic input only contains one spoken keyword. There is one additional keyword in the label input compared to the acoustic input. Omitting a spoken keyword in a command is equivalent to adding a wrong keyword label to a correctly labelled utterance. Because it is difficult to adapt a spoken utterance in an existing speech corpus, we simulate this error by activating a keyword label in $\mathbf{V}_l$ that was not present in the transcription of the spoken utterance. In different words, we increase the frequency count by one in $\mathbf{V}_l$ for a keyword that we simulate to be missing in the spoken command. We call this error an insertion.

Deletion A second kind of error is the over-specification of a spoken command. For instance, the user says "Turn on, the radio, uh no, the television". The acoustic input contains one keyword more than the label input, i.e. "radio". Decreasing a non-zero entry in $\mathbf{V}_l$ by one allows us to simulate the occurrence of one extra keyword in the spoken command. We call this error a deletion.

Substitution The user might mistakenly say "Turn on the radio" but then turns on the television. As a consequence, the label part and the acoustic part of $\mathbf{V}$ are sharing one keyword, i.e. "Turn on", but they are not sharing the second keyword. This error is simulated by one deletion followed by one insertion in the same column of the label matrix $\mathbf{V}_l$, i.e. in a correctly transcripted utterance. We call this error a substitution.

Command substitution Finally, the user might push the wrong button on a manual user interface. For instance, the end user might ask his/her partner to switch off the lights because he/she would like to watch television. In the meantime, the user might turn on the television. A total mismatch between the voice command "Switch off the lights" (the acoustic input) and the executed command "Turn on, television" (the keyword labels) is then expected. We simulate this error by taking a correctly annotated utterance, and then, changing the complete column in $\mathbf{V}_l$ to a label vector consisting of a few randomly selected keywords. We call this error a command substitution.

22.4 Experimental Set-Up

22.4.1 Speech Data

The speech data was selected from the English corpus constructed in the second year of the ACORNS project [12]. The database consists of 13,160 utterances, produced by ten speakers. Each utterance consists of 1 to 4 different keywords embedded in a carrier sentence with unrelated filler words. In total, there are 50 unique predefined keywords. An example of a sentence is presented in Fig. 22.1 and the keywords are underlined. Note that, although the speech does not resemble a command and control task, this makes no difference for the purpose of evaluating NMF learning since the size and complexity of the data is similar to, for example, a home automation task. Similar to [6], 9,821 utterances were randomly selected to compose the training set and 3,268 utterances were selected to compose the test set. The training set and the test set contained utterances of all ten different speakers.

22.4.2 Feature Extraction

The feature extraction (see Fig. 22.1) was done with the Hamming window of 25 ms size and frame shift of 10 ms. Mel-band spectral magnitudes were converted into a 39-dimensional feature vector: 12 Mel Frequency Cepstral Coefficients (MFCC's) plus the frame's log energy and the respective velocity (Δ) and acceleration ($\Delta\Delta$) vectors. The MFCC features were mean and variance normalized. In an intermediate step, the frame-based features of each utterance were transformed into a single histogram represented by one column vector in $\mathbf{V}_a$. To vectorize the acoustics of the utterances into successive columns, K-means clustering ([13]) was performed on the frames and the cluster centres were used as a vector quantization (VQ) codebook of size K. The frame-based features of the whole utterances were then converted into a sequence of VQ labels. The co-occurrence of all pairs of VQ labels was counted and these counts were ordered to form the Histogram of Acoustic Co-occurrence (HAC, [6, 9]). More precisely, HAC representations are built by counting the co-occurrence of two VQ labels in different frames over a particular time offset τ between frames. In the toy example of Fig. 22.1, each feature vector is clustered and the letters' 'D","F""M", and "K" represent the VQ labels of the clusters. The co-occurrence of the VQ labels in each utterance is counted over a time delay $\tau = 2$. In the experiments, the utterance-based feature vectors consisted of co-occurrence counts of VQ labels for three different codebook sizes: $K = 20, 100$ and 400. Additionally, three different time offsets were used: $\tau = 20, 50$ and 90 ms. Co-occurrence counts for codebook sizes $K = 20, 100$ and 400 for three time offsets resulted in $3 \times (20^2 + 100^2 + 400^2) = 511,200$ features in each column vector of $\mathbf{V}_a$.

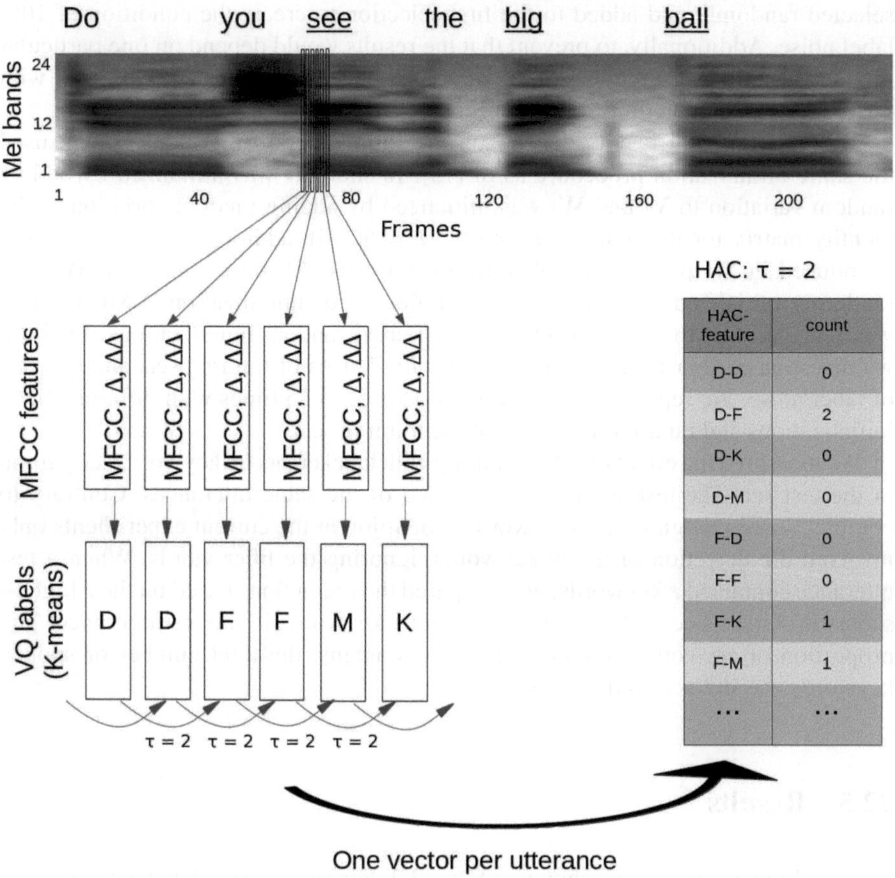

Fig. 22.1 The preprocessing and the feature extraction method

22.4.3 Experiment

We tested label noise robustness for different sizes of the training sets: $N = 100$, 200, 500, 1,000, 2,000, 4,000 and 9,821 utterances. Each utterance is considered one training example. For each type of label noise, we had label noise affecting 0, 10, 30, 50, 70 and 90 percents of the utterances in the training sets.

We took some precautions to limit the variation of the experimental results due to the random selection of utterances. First, the smaller training sets were nested in the larger training sets. For instance, the first 100 utterances were selected randomly to compose the training set of $N = 100$ utterances, then, 100 more utterances were selected randomly and added to the first 100 utterances to compose the training set of $N = 200$ utterances. Second, a similar procedure was followed for adding label noise. We first selected 5% of the utterances randomly, then, an additional 5% was selected randomly and added to the first selection to create the condition of 10% label noise. Additionally, to prevent that the results would depend on one particular random selection of utterances, the whole procedure was repeated five times with different random selections of utterances.

The experimental results depend on the initialization of **W** and **H**. We used the same initialization procedure as in [13]. In short, **H** was initialized by adding random variation to $\mathbf{V}_l$ and $\mathbf{W}_l$ was initialized by adding random variation to the identity matrix for the first $L \times L$ entries ($L = 50$). In addition, 25 columns were introduced in **W** to represent labelless filler words. All other entries in **W** were randomly initialized. To control for the influence of initialization, NMF training was repeated five times, each time with different random initializations. In short, we investigated four types of label errors with 7 sizes of training sets and 6 levels of label noise. We repeated all experiments ($5 \times 5 =$) 25 times with different NMF initializations and random selections of the training sets.

We measured the effect of NMF training with label errors on keyword recognition in the test set. The test set always consisted of the same utterances. Contrary to common word recognition tasks, word recognition in the current experiments only involved the detection of a few keywords, ignoring the filler words. When a test utterance contained r keywords, we compared the predictions based on the r highest scores in **A** (see Sect. 22.2) with the correct r keywords in the test utterance. The proportion of correctly recognized keywords against the total number of spoken keywords was defined as the accuracy.

22.5 Results

The resulting accuracies are shown in Fig. 22.2. For each type of label error, there is one graph showing the mean recognition accuracy as a function of the percentage of utterances affected with label noise. For each training set size, there is one trend line and the training set size is indicated at the end or on top of the trend line. The error bars denote the standard errors. In a system that is robust against label noise, the performance is not degrading too much as a function of label noise. Accordingly, we can observe that the proposed method is very robust against label insertions, deletions and substitutions since the lines are nearly horizontal over the whole range. For the largest training set size and without label noise, the performance was 95.6% correctly identified keywords. In the case of 90% utterances affected by label noise, the accuracies were only slightly lower, 95.2%, 93.7% and 93.4% respectively, for the insertions, deletions and substitutions.

However, the curves are more rapidly descending for command substitution errors. Given the definition of the four error types, this difference can be expected because command substitution is affecting all keyword labels in an utterance and not just one. For instance, 90% of the utterances with one insertion, deletion or substitution error corresponds with 31.3% of keyword labels that were affected in the training sets. Contrary to insertion, deletion and substitution, 90% of the

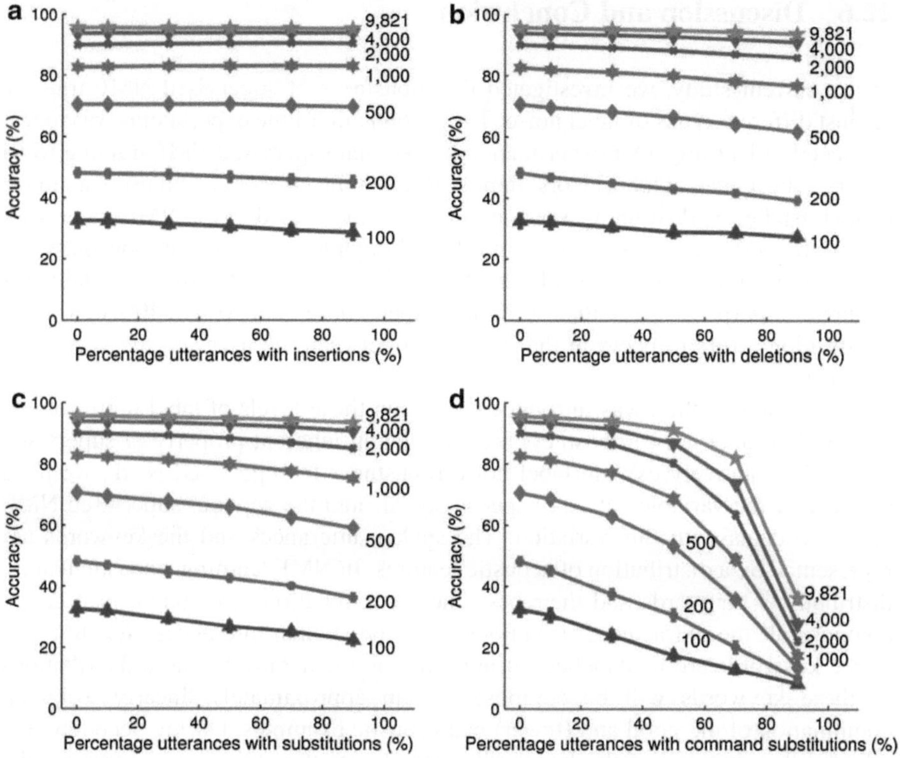

Fig. 22.2 Mean recognition accuracy as a function of training set size and percentage of affected utterances for each label error type: (**a**) insertions, (**b**) deletions, (**c**) substitutions, (**d**) command substitutions

utterances with command substitution errors corresponds to 90% of keyword labels that were affected in the training sets. The difference with the former types of label noise demonstrates that the effect of label noise depends on the number of wrong keyword labels rather than on the number of affected utterances.

The second issue of interest is the speed of learning. The higher the accuracies for small training sets, the faster the system picks up the keywords. After 100 correctly labelled utterances of training, which corresponds to an average of 5.6 examples per keyword, the self-learning vocal user interface picked up the associations with an average accuracy rate of 34%. However, 2,000 utterances, which corresponds to an average of 112 examples per keyword in the training set, were needed to reach a performance close to 90%.

22.6 Discussion and Conclusion

In the current study, we investigated the robustness of supervised NMF training against different types of label noise. The second aim of the experiments was to test the speed of learning. Our experiments showed that supervised NMF training [6, 9] is very robust against label errors. In practice, that means that an end user can make many mistakes while training a self-learning interface based on the NMF framework before the performance starts to degrade. Although less label noise robustness has been demonstrated for command substitutions in Sect. 22.5, it is still acceptable for application purposes given that we do not expect more than 30% to 40% command substitution errors to occur in the interaction between the user and the vocal user interface.

No explicit method was introduced to obtain these levels of label noise robustness, rather label noise robustness seems to be an inherent property of supervised NMF. To tentatively explain label noise robustness in supervised NMF, we need to consider the variable nature of speech signals and the way our supervised NMF approach deals with this variation. The spoken utterances and the keywords are represented by a distribution of acoustic features. In NMF learning, acoustic feature distributions for words and their best linear combinations are sought in order to compose all the extracted distributions from the spoken utterances. When a part of the keywords are mislabelled during training, the acoustic feature distributions of these keywords will be composed of an approximately linearly weighted combination of the good and (fewer) bad acoustic examples. During decoding, the affected distributions of the keywords might lower the activation scores, but the highest activations are possibly still corresponding to the spoken keywords in the utterances.

The end user can permit to make many label errors before the performance starts to degrade, but he still has to put some effort in training the device. When end users have to demonstrate the meaning of 2,000 commands to reach a performance of 90%, it looks like a very demanding task. It is difficult to determine an acceptable lower bound for the learning speed of a vocal user interface. Nevertheless, the faster the better, as more users will be prepared to keep on training the device until some comfortable level of service is experienced by the user. Therefore, there is still room for improvement concerning the speed of learning.

There are some suggestions that are helpful to improve overall accuracy rates and speed up the learning curve. First, and most importantly, in the applied set-up, the keywords are learned individually without using common units. Unlike human and contemporary automatic speech recognition systems, there is no phone-level acoustic model that is shared across words. Such a model is required to be able to learn compact lexical-type word descriptions, which would be possible from small training sets. Secondly, it should be noted that the speech data was produced by ten different speakers, males and females. Since the targeted vocal user interface, however, is self-learning, it aims at learning the speech of only a single user. Keyword recognition accuracies are higher when the experiments are conducted on the data of the individual speakers.

To conclude, the occurrence of label noise is not an issue in a self-learning command-and-control application built on supervised NMF. The first requirement for creating a user-friendly learning environment is therefore met. However, it would be desirable to have a higher learning speed. More attention to speed up the learning curve is therefore required in future research.

Acknowledgements This work is funded by IWT-SBO project 100049 (ALADIN).

References

1. Heinroth, T., Grotz, M., Nothdurft, F., Minker, W.: Adaptive speech understanding for intuitive model-based spoken dialogues. In: Proceedings of LREC, pp. 1281–1288 (2012)
2. Taguchi, R., Iwahashi, N., Funakoshi, K., Nakano, M., Nose, T., Nitta, T.: Learning physically grounded lexicons from spoken utterances. In: Inaki, M.(ed.) Human Machine Interaction–Getting Closer, pp. 69–84 (2012). URL http://www.intechopen.com/books/human-machine-interaction-getting-closer/learning-physically-grounded-lexicons-from-spoken-utterances
3. van de Loo, J., Gemmeke, J.F., De Pauw, G., Driesen, J., Van hamme, H., Daelemans, W.: Towards a self-learning assistive vocal interface: Vocabulary and grammar learning. In: Proceedings of the workshop Speech and Multimodal Interaction in Assistive Environments (SMIAE) (2012)
4. Bootkrajang, J.: Learning with labeling errors. Tech. Rep. CSR-11-07, School of Computer Science, University of Birmingham (2011)
5. Driesen, J., ten Bosch, L., Van hamme, H.: Adaptive non-negative matrix factorization in a computational model of language acquisition. In: Proceedings of the Interspeech, pp. 1711–1714. Brighton, UK (2009)
6. Driesen, J., Gemmeke, J., Van hamme, H.: Weakly supervised keyword learning using sparse representations of speech. In: Proceedings ICASSP, pp. 5145–5148. Kyoto (2012)
7. Driesen, J., Van hamme, H.: Modelling vocabulary acquisition, adaptation, and generalization in infants using adaptive bayesian plsa. Neurocomputing **74**, 1874–1882 (2011)
8. Lee, H., Yoo, J., choi, S.: Semi-supervised nonnegative matrix factorization. IEEE Signal Process. Lett. **17**, 4–7 (2009)
9. Van hamme, H.: Hac-models: a novel approach to continuous speech recognition. In: Proceeding of Interspeech, pp. 255–258. Brisbane (2008)
10. Lee, D., Seung, H.: Learning the parts of objects by nonnegative matrix factorization. Nature **401**, 788–791 (1999)
11. Xu, W., Liu, X., Gong, Y.: Document clustering based on non-negative matrix factorization. In: Proceedings of ACM SIGIR Conference on Research and Development in Information RetrievalInterspeech. Toronto (2003)
12. Boves, L., ten Bosch, L., Moore, R.: Acorns-towards computational modeling of communication and recognition skills. In: Proceedings IEEE International Conference On Cognitive informatics, pp. 349–355. California (2007)
13. Driesen, J.: Discovering words in speech using matrix factorization. Ph.D. thesis, K.U.Leuven, ESAT (2012)

Chapter 23
Topic Classification of Spoken Inquiries Using Transductive Support Vector Machine

Rafael Torres, Hiromichi Kawanami, Tomoko Matsui, Hiroshi Saruwatari, and Kiyohiro Shikano

Abstract In this work, we address the topic classification of spoken inquiries in Japanese that are received by a guidance system operating in a real environment, with a semi-supervised learning approach based on a transductive support vector machine (TSVM). Manual data labeling, which is required for supervised learning, is a costly process, and unlabeled data are usually abundant and cheap to obtain. TSVM allows to treat partially labeled data for semi-supervised learning, including labeled and unlabeled samples in the training set. We are interested in evaluating the influence of including unlabeled samples in the training of the topic classification models, as well as the amount of them that could be necessary for improving performance. Experimental results show that this approach can be useful for taking advantage of unlabeled samples, especially when using larger unlabeled datasets. In particular, we found gains in classification performance for specific topics, such as city information, with a 6.30% F-measure improvement in the case of children's inquiries and 7.63% for access information in the case of adults' inquiries.

23.1 Introduction

The interest of this work is to improve topic classification performance of spoken inquiries in Japanese, received by a speech-oriented guidance system operating in a real environment. In previous work, we evaluated the classification performance of three supervised methods: a support vector machine (SVM) with a radial basis

R. Torres (✉) • H. Kawanami • H. Saruwatari • K. Shikano
Graduate School of Information Science, Nara Institute of Science and Technology, Nara, Japan
e-mail: rafael-t@is.naist.jp; kawanami@is.naist.jp; sawatari@is.naist.jp; shikano@is.naist.jp

T. Matsui
Department of Statistical Modeling, The Institute of Statistical Mathematics, Tokyo, Japan
e-mail: tmatsui@ism.ac.jp

J. Mariani et al. (eds.), *Natural Interaction with Robots, Knowbots and Smartphones: Putting Spoken Dialog Systems into Practice*, DOI 10.1007/978-1-4614-8280-2_23,

function (RBF) kernel, PrefixSpan boosting (pboost), and maximum entropy (ME) [5]. We have also evaluated a stacked generalization scheme to combine the predictions of these three classifiers, improving predictive accuracy compared with the performance of individual classifiers [4].

Manual data labeling, which is required for supervised learning, is a costly process, and unlabeled data are usually abundant and cheap to obtain. Because of this, it is desirable to be able to use unlabeled samples to improve topic classification performance and minimize the generalization error of the classifiers. In the present work, we address the topic classification of spoken inquiries in Japanese received by a guidance system, with a semi-supervised learning approach based on a TSVM, which extends a regular SVM to treat partially labeled data, including labeled and unlabeled samples in the training set.

TSVMs were proposed by Vapnik in 1998 and were introduced by Joachims [2] for text classification. TSVMs use labeled samples to find optimal hyperplanes that maximize the separation margin of two classes of data and then use unlabeled samples to adjust that margin.

Our task, topic classification of spoken inquiries, shares some similarities with text classification; however, classification of spontaneous speech includes automatic speech recognition (ASR) errors. In this work we evaluate the viability of using a TSVM for semi-supervised learning in this task, as well as the amount of unlabeled data that would be necessary for improving classification performance.

23.2 Speech-Oriented Guidance System *Takemaru-kun*

The *Takemaru-kun* system [3] (Fig. 23.1) is a real-environment speech-oriented guidance system placed inside the entrance hall of the Ikoma City North Community Center in Nara, Japan, and it has been operating daily since November 2002.

The system uses a one-question-to-one-response strategy for interaction, which fits the purpose of responding simple questions to a large number of users.

Fig. 23.1 Speech-oriented guidance system *Takemaru-kun*

It provides information about the center facilities and services, local sightseeing, weather forecast, and news, among other.

Since the *Takemaru-kun* system started operating, the received utterances have been recorded. Utterances from Nov. 2002 to Oct. 2004 and from Dec. 2004 to Mar. 2005 have been manually transcribed and labeled. However, because this is a very costly process, there is still a vast amount of data that remains unlabeled.

23.3 Transductive Support Vector Machine

A transductive support vector machine (TSVM) extends a regular SVM to treat partially labeled data for semi-supervised learning, including labeled and unlabeled samples in the training set. In this work we use SVMLight [1] to implement it.

In TSVM, the primal optimization problem follows the form

$$\min_{\mathbf{w},b,\xi,\xi^*} \frac{1}{2}\|\mathbf{w}\|^2 + C\sum_{i=1}^{n}\xi_i + C^*_-\sum_{\{j:y^*_j=-1\}}\xi^*_j + C^*_+\sum_{\{j:y^*_j=+1\}}\xi^*_j$$

$$\text{sb.t.}\quad \forall^n_{i=1}: y_i[\mathbf{w}\cdot\mathbf{x}_i + b] \geq 1 - \xi_i$$

$$\forall^k_{j=1}: y^*_j[\mathbf{w}\cdot\mathbf{x}_j + b] \geq 1 - \xi^*_j \tag{23.1}$$

where $\mathbf{x}_i$ represents a labeled training sample and $\mathbf{x}_j$ an unlabeled training sample and $y_i \in \{1,-1\}$ and $y^*_j \in \{1,-1\}$ a class for labeled and unlabeled samples, respectively. The hyperparameters C, C^*_- and C^*_+ penalize the sum of the slack variables ξ_i and ξ^*_j to allow soft-margin, where * is used to denote unlabeled samples.

The TSVM algorithm [2] begins with labeling unlabeled samples based on the classification of a regular SVM trained with only labeled samples. Then, it retrains the model using all samples and improves the solution by switching the labels of the newly labeled samples so that the objective function decreases. The label switching part of the algorithm consists of two embedded loops:

- An external loop uniformly increases the influence of the newly labeled samples by incrementing C^*_- and C^*_+, which are initialized with a very low value, up to a defined value C^*. Very low values of C^*_- and C^*_+ mean that these samples are almost ignored when finding the classification margin, because these are still considered not reliable. As the reliability of the newly labeled samples improves, the values of C^*_- and C^*_+ are increased.
- An internal loop identifies two newly labeled samples for which switching the labels leads to a decrease in the current objective function and switches the labels if this condition is met. For this, it identifies two samples with opposite labels and checks if the value of ξ^*_j, which measures classification error, is greater than a predefined value, which indicates that the samples may be mislabeled, and then it switches both labels. In each iteration, the optimization problem is solved again.

In our approach, we use labeled and unlabeled samples to train a model using a TSVM and use the resultant model to classify test data, which differs from the approach of Joachims [2], where unlabeled samples are the test data.

We use bag-of-words (BOW) to represent utterances as vectors and use character unigrams, bigrams, and trigrams as features, as it was previously shown to improve classification performance in comparison to words [5]. We also use a RBF kernel and follow a one-vs-rest approach, constructing one binary classifier for each topic, as it showed better performance in preliminary experiments.

23.4 Experiments

We evaluated the performance of a TSVM against a regular SVM used as baseline. For this, we classified ASR results of inquiries in Japanese received by the speech-oriented guidance system *Takemaru-kun* in topics. We performed experiments with separate datasets for children and adults. Classification performance was evaluated using the F-measure, which was calculated individually for each topic and then averaged by frequency of samples. Optimal hyperparameter values were obtained experimentally using a grid search strategy and were set a posteriori.

23.4.1 Characteristics of the Datasets

The labeled data correspond to the utterances collected by *Takemaru-kun* in the period from Nov. 2002 to Oct. 2004 and from Dec. 2004 to Mar. 2005. Julius was used as ASR engine. Acoustic models (AMs) and language models (LMs) were separately prepared for children and adults. The AMs were trained using the samples collected by the system from Nov. 2002 to Oct. 2004, and the LMs were constructed using the transcriptions of the samples of the same period. Samples corresponding to the months of Aug. 2003 and from Dec. 2004 to Mar. 2005 were used for testing and were not included in the training sets. For these experiments we selected the 15 topics with most training samples. Table 23.1 shows the amount of samples and word recognition accuracy of the ASR engine in the labeled datasets.

The unlabeled data correspond to the utterances collected by *Takemaru-kun* in the period from Apr. 2005 to Dec. 2007. Julius was also used as ASR engine, and

Table 23.1 Amount of samples and ASR word recognition accuracy in the labeled datasets

(Labeled datasets)	Children training	Children test	Adults training	Adults test
Amount of samples	43,494	15,524	14,431	3,085
ASR word recognition acc.	72.95%	66.77%	88.42%	81.60%

Table 23.2 Amount of samples in the unlabeled datasets

(Unlabeled datasets)	Children training	Adults training
Unlabeled dataset #1 (2005.04 to 2005.12)	119,322	110,537
Unlabeled dataset #2 (2005.04 to 2006.12)	271,744	252,428
Unlabeled dataset #3 (2005.04 to 2007.12)	413,144	385,165

Table 23.3 Averaged F-measure results per training dataset combination (open test)

Training dataset combination	Children(%)	Adults(%)
Labeled only (SVM)	83.54	**93.03**
Labeled dataset + unlabeled dataset #1 (TSVM)	83.02	91.75
Labeled dataset + unlabeled dataset #2 (TSVM)	84.17	92.86
Labeled dataset + unlabeled dataset #3 (TSVM)	**84.28**	92.81

we used the same AMs and LMs that were used to recognize the labeled data. We created three datasets, incrementing the size of them. Table 23.2 shows the amount of samples in the unlabeled datasets.

23.4.2 Experiment Results

Table 23.3 presents the averaged topic classification performance per training dataset combination in the open test, for children and adults. In the case of the children's datasets, the TSVM outperformed the baseline when using the two largest unlabeled datasets, with an improvement of 0.74% when using the largest one. However, in the case of the adults' datasets, the averaged performance of the TSVM was not better than the baseline.

We can observe that the topic classification performance with children's datasets is lower in comparison to adults', which leaves more room for improvement. The main reason for this is the lower ASR accuracy for children. The results obtained with the TSVM suggest that the inclusion of unlabeled samples in the training of the topic classification models can help to deal with the influence of ASR errors. We can also observe a tendency to obtain better performance with the TSVM when using larger unlabeled datasets.

Table 23.4 presents the classification performance per topic. We can observe that most of the topics presented improvements in the case of children's data, while more than half of the topics were improved for adults' data. The best gain in performance for children's inquiries was presented by the *info-city* topic, with 6.30% F-measure improvement and 7.63% by *info-access* for adults' inquiries.

Table 23.4 F-measure results per topic (open test)

Topic	Children SVM(%)	Children TSVM(%)	Adults SVM(%)	Adults TSVM(%)
Chat-compliments	64.24	**66.91**	**86.35**	81.64
Info-services	58.06	**59.04**	**87.65**	87.12
Info-news	88.89	**92.47**	95.52	**96.30**
Info-local	56.71	**59.50**	83.08	**84.44**
Info-facility	**82.70**	82.15	**89.36**	**89.36**
Info-city	67.06	**73.37**	84.34	**88.27**
Info-weather	83.89	**85.40**	95.46	**95.74**
Info-time	89.67	**90.83**	95.56	**96.83**
Info-sightseeing	74.74	**75.34**	91.39	**92.21**
Info-access	44.58	**44.89**	84.39	**92.02**
Greeting-end	**84.87**	83.81	**93.48**	92.17
Greeting-start	91.64	**91.88**	97.76	**97.83**
Agent-name	79.15	**80.84**	**92.15**	89.50
Agent-likings	89.75	**90.62**	**93.30**	91.64
Agent-age	89.26	**89.69**	**95.71**	95.68
Averaged	83.54	**84.28**	**93.03**	92.81

23.5 Conclusions

This work evaluated the topic classification of spoken inquiries received by a guidance system with a semi-supervised learning approach based on a TSVM. Experimental results with children's data show an overall improvement of 0.74% with the TSVM in comparison to a regular SVM. In particular, we found gains in classification performance for specific topics, such as city information, with 6.30% F-measure improvement for children's inquiries and 7.63% for access information in the case of adults' inquiries. A tendency to obtain better performance when using larger unlabeled datasets was observed. Future work will focus on the evaluation and improvement of other semi-supervised learning approaches.

References

1. Joachims, T.: Making large-scale support vector machine learning practical. In: Schölkopf, B., Burges, C.J.C., Smola, A.J. (eds.) Advances in Kernel Methods, pp. 169–184. MIT Press (1999). Software available at http://svmlight.joachims.org/
2. Joachims, T.: Transductive inference for text classification using support vector machines. In: ICML 1999 Proceedings, pp. 200–209 (1999)
3. Nisimura, R., Lee, A., Saruwatari, H., Shikano, K.: Public speech-oriented guidance system with adult and child discrimination capability. In: ICASSP 2004 Proceedings, pp. 433–436 (2004)

4. Torres, R., Kawanami, H., Matsui, T., Saruwatari, H., Shikano, K.: Topic classification of spoken inquiries based on stacked generalization. In: APSIPA 2011 Proceedings (2011)
5. Torres, R., Takeuchi, S., Kawanami, H., Matsui, T., Saruwatari, H., Shikano, K.: Comparison of methods for topic classification in a speech-oriented guidance system. In: Interspeech 2010 Proceedings, pp. 1261–1264 (2010)

Chapter 24
Frame-Level Selective Decoding Using Native and Non-native Acoustic Models for Robust Speech Recognition to Native and Non-native Speech

Yoo Rhee Oh, Hoon Chung, Jeom-ja Kang, and Yun Keun Lee

Abstract This paper proposes a frame-level selective-decoding method by using both native acoustic models (AMs) and non-native AMs in order to construct a robust speech recognition system for non-native speech as well as native speech. To this end, we use two kinds of well-trained AMs: (a) AMs trained with a large amount of native speech (*native AMs*) and (b) AMs trained with a plenty amount of non-native speech (*non-native AMs*). First, each speech feature vector is decoded using *native AMs* and *non-native AMs* in parallel. And, we select proper AMs by comparing the likelihoods of the two AMs. Then, the next M frames of speech feature vectors are decoded by using the selected AMs, where M is a pre-defined parameter. The selection and the decoding procedures are repeated until an end of an utterance is encountered. From automatic speech recognition (ASR) experiments for English spoken by Korean speakers, it is shown that an ASR system employing the proposed method reduces an average word error rate (WER) by 16.6% and 41.3% for English spoken by Koreans and native English, respectively, when compared to an ASR system employing an utterance-level selective-decoding method.

24.1 Introduction

According to many applications adopting an automatic speech recognition (ASR) system [1,2], non-native speech is increasingly exposed to an ASR system; however, non-native speech poorly degrades the performance of speech recognition since there are many different characteristics on non-native speech when compared to

Y.R. Oh (✉) • H. Chung • J.-ja Kang • Y.K. Lee
Spoken Language Processing Team, Electronics and Telecommunications Research Institute (ETRI) 138 Gajeongno, Yuseong-gu, Daejeon 305-700, Korea
e-mail: yroh@etri.re.kr

J. Mariani et al. (eds.), *Natural Interaction with Robots, Knowbots and Smartphones: Putting Spoken Dialog Systems into Practice*, DOI 10.1007/978-1-4614-8280-2_24,

native speech [3]. Therefore, this paper focuses on a robust speech recognition to non-native speech as well as native speech.

There have been considerable researches on non-native ASR reported. First, pronunciation variants are investigated and applied to pronunciation models for non-native speech [4–10]. Second, the effects of non-native speech are compensated by adapting acoustic models [4, 11–13]. Third, the speaking styles or grammatical effects of non-native speech are considered to language models[14]. Fourth, speech feature vectors are transformed to cope with non-native speech [15]. Finally, a hybrid approach combines several approaches in order to improve further on non-native ASR [16,17]. Especially, while most researches on non-native ASR are based on the use of a small amount of non-native speech data, we attempt to improve the performance of non-native ASR using a relatively large amount of non-native speech.

The organization of this paper is as follows. In Sect. 24.2, we propose a frame-level selective-decoding method using native AMs and non-native AMs for a robust speech recognition to non-native speech as well as native speech. Next, we show the performance comparison of several ASR systems for English spoken by Korean speakers in Sect. 24.3. Finally, we conclude our findings in Sect. 24.4.

24.2 Proposed Frame-Level Selective Decoding for Robust Speech Recognition to Native and Non-native Speech

This section proposes a frame-level selective-decoding method by using both *native AMs* that is well trained with a large amount of native speech data and *non-native AMs* that is well trained with a plenty amount of non-native speech data for a non-native robust recognition system having a considerable performance on native speech. In this paper, the term "frame" refers to a unit of a speech recognition feature vector.

As shown in Fig. 24.1, the main idea of the proposed method is that proper AMs are first selected among *native AMs* and *non-native AMs* for every M-th frame and then the selected AMs are decoded during next M frames, where M is a pre-defined parameter. In other words, every M-th frame of an input utterance is used to select proper AMs by decoding *native AMs* and *non-native AMs* in parallel and by comparing the log-likelihoods of the two different AMs. Second, the next M frames are decoded by using the selected AMs. The two steps are repeated until an end of speech is encountered.

In an ideal case that proper AMs are correctly selected throughout an utterance, it would be expected that the performance for non-native speech would be similar to the performance when employing *non-native AMs* whereas the performance for native speech would be similar to the performance when employing *native AMs*.

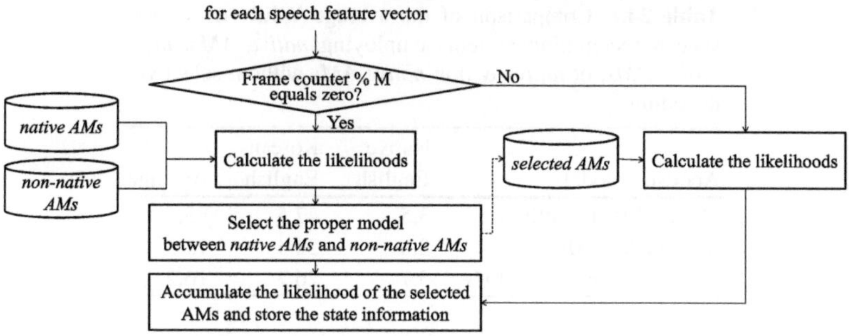

Fig. 24.1 The main procedure of the proposed frame-level decoding using *native AMs* and *non-native AMs* for non-native speech recognition, where a selection of proper AMs is performed per *M*-frames

24.3 Experiments

This section shows the performance comparison of several ASR systems for non-native speech recognition. For the speech recognition experiments, we select English as a target language and Korean speakers as non-native speakers.

24.3.1 Speech Database and Baseline ASR System

We used two kinds of speech databases: (a) native English speech databases and (b) Korean-spoken English speech databases. For native English speech databases, we used a subset of the Wall Street Journal (WSJ1)[18], the native English speech corpus that was supported by Speech Information Technology and Industry Promotion Center (SiTEC), and the native English speech corpus collected by Electronics and Telecommunications Research Institute (ETRI). For non-native English speech databases, we used a subset of the Korean-spoken English speech corpus collected by ETRI. For a training set, we used 331,527 utterances (280 h long in total) and 205,879 utterances (180 h long in total) for English spoken by native speakers and Korean speakers, respectively. Moreover, for an evaluation set, we used 4,878 utterances of a subset of WSJ1 for native English and 3,109 utterances of a subset of ETRI corpus for Korean-spoken English.

As a recognition feature, we used a 39-dimensional feature vector by extracting 12 mel-frequency cepstral coefficients (MFCCs) with logarithmic energy for every 10 ms analysis frame and by concatenating their first and second derivatives. Moreover, each feature vector was normalized using a Cepstral mean subtraction (CMS) method [19]. The acoustic models were based on 3-state left-to-right, 8-mixture, and cross-word triphone-based hidden Markov models (HMMs). In other words, by using the training set of native English, the monophone-based HMMs were expanded to the triphone-based HMMs, and then the states of the triphone-

Table 24.1 Comparison of the average WERs of several speech recognition systems employing *native AMs*, *non-native AMs*, or *native & non-native AMs* without selective-decoding

Acoustic models	Native English	Korean English	Average
Native AMs (baseline)	3.5	47.5	25.5
Non-native AMs	8.6	22.4	15.5
Native & non-native AMs	5.7	26.7	16.2

based HMMs were clustered by employing a decision tree. As a result, the acoustic models were composed of 24,560 physical triphones and 7,019 states. In addition, two bigram language models were used for native English and for Korean-spoken English, respectively. And, the native English pronunciation for each word was generated using grapheme-to-phoneme engine of ETRI. As a speech recognizer, we used a large vocabulary automatic speech recognizer of ETRI [20].

The average word error rates (WERs) of the baseline ASR system were 3.5% and 47.5% for English spoken by native speakers and Korean speakers, respectively. It was shown that non-native speech severely degrades the performance of an ASR system that was trained with native speech.

24.3.2 Performance Comparison of ASR Systems

First, we construct three kinds of AMs: *native AMs* using native speech, *non-native AMs* using non-native speech, and *native & non-native AMs* using both native and non-native speech. It is shown from Table 24.1 that *non-native AMs* achieves the WER reduction of 52.8% for Korean-spoken English when compared to *native AMs*; but, it degrades the WER by 146.4% for native English. On the other hand, *native & non-native AMs* achieved the WER reduction of 43.8% for Korean-spoken English, whereas it degrades the WER by 62.2% for native English. Even though *non-native AMs* has better performance than *native & non-native AMs* for non-native speech, *native & non-native AMs* are commonly used for non-native speech recognition since it maintains the performance for native speech than *non-native AMs*.

Second, the proposed frame-level selective-decoding method is applied to an ASR system using *native AMs* and *non-native AMs* by varying the parameter M from 1 to 5, which are referred to as $ASR_{proposed,M=1}$, $ASR_{proposed,M=2}$, ..., $ASR_{proposed,M=5}$. It is shown from the first row of Table 24.2 that $ASR_{proposed,M=1}$ achieves the WER reductions by 49.7% and 5.4% for Korean-spoken English and native English, respectively, when compared to an ASR system employing *native AMs*. In other words, the WERs of $ASR_{proposed,M=1}$ for Korean-spoken English and native English are comparable to an ASR system employing *non-native AMs* and an ASR system employing *native AMs*, respectively. Moreover, it is shown from the table that $ASR_{proposed,M}$ has higher WERs when the parameter M is increased.

Table 24.2 Comparison of the average WERs of several speech recognition systems by using *native AMs* and *non-native AMs* with selective-decoding

Selective-decoding	Acoustic models	Native English	Korean English	Average
Frame-level, M=1	*Native AMs, non-native AMs*	3.3	23.9	13.6
Frame-level, M=2	*Native AMs, non-native AMs*	3.4	24.0	13.7
Frame-level, M=3	*Native AMs, non-native AMs*	3.5	24.2	13.9
Frame-level, M=4	*Native AMs, non-native AMs*	3.6	24.6	14.1
Frame-level, M=5	*Native AMs, non-native AMs*	3.9	25.4	14.6
Utterance-level ($M=\infty$)	*Native AMs, non-native AMs*	5.6	28.6	17.1

Third, as a comparison of the proposed method, we apply an "utterance-level selective-decoding" method by performing speech recognition using *native AMs* and *non-native AMs* in parallel and selecting more probable recognition sequence at the end of each decoding. It is shown from the last row of Table 24.2 that the utterance-level selective-decoding achieves the WER reduction of 39.8% for Korean-spoken English when compared to an ASR system employing *native AMs*; but, it degrades the WER by 61.0% for native English.

From the speech recognition experiments, it can be concluded that the proposed frame-level selective-decoding method provides best performance when the selection of the proper AMs is performed for each frame.

24.4 Conclusion

In this paper, we proposed a frame-level selective-decoding method by using both *native AMs* and *non-native AMs* in order to provide a robust ASR system for both native speech and non-native speech. The proposed method used two well-trained AMs: *native AMs* using a large amount of native speech data and (b) *non-native AMs* using a plenty amount of non-native speech data. For every M-th frame of speech feature vectors where M is a pre-defined parameter, the feature vector was decoded using *native AMs* and *non-native AMs* in parallel, and then proper AMs were selected by comparing the likelihoods of the two AMs. Then, the selected AMs were used to decode the next M frames of feature vectors. The selection and decoding procedures were repeated until an utterance ends. It was shown from the speech recognition experiments for Korean-spoken English that an ASR system employing the proposed method reduced an average WER by 16.6% and 41.3% for English spoken by Koreans and native English, respectively, when compared to an ASR system employing an utterance-level selective-decoding method. Moreover, it was noted that the proposed method provided lower performance when the parameter M was increased.

Acknowledgements This work was supported by the Industrial Strategic technology development program, 10035252, Development of dialog-based spontaneous speech interface technology on mobile platform funded by the Ministry of Knowledge Economy(MKE, Korea)

References

1. Kreger, TechNavio Insights : Global automatic speech recognition applications market, 2011–2015 (2012)
2. Global Industry Analysts, Inc. : Voice/speech recognition systems: a global outlook. (2012)
3. Lawson, A.D., Harris, D.M., Grieco, J.J.: Effect of foreign accent on speech recognition in the NATO N-4 corpus. In: Proceedings Eurospeech, pp. 1505–1508. Geneva (2003)
4. Gruhn, R., Markov, K., Nakamura, S.: A statistical lexicon for non-native speech recognition. In: Proceedings ICSLP, pp. 1497–1500. Jeju Island (2004)
5. Raux, A.: Automated lexical adaptation and speaker clustering based on pronunciation habits for non-native speech recognition. In: Proceedings ICSLP, pp. 613–616. Jeju Island (2004)
6. Strik, H., Cucchiarini, C.: Modeling pronunciation variation for ASR: A survey of the literature. Speech Commun. **29**(2–4), 225–246 (1999)
7. Fosler-Lussier, E.: Multi-level decision trees for static and dynamic pronunciation models. In: Proceedings Eurospeech, pp. 463–466. Budapest (1999)
8. Amdal, I., Korkmazasky, F., Suredan, A. C.: Data-driven pronunciation modelling for non-native speakers using association strength between phones. In: Proceedings Automatic Speech Recognition, pp. 85–90. Paris (2000)
9. Goronzy, S., Rapp, S., Kompe, R.: Generating non-native pronunciation variants for lexicon adaptation. Speech Commun. **42**(1), 109–123 (2004)
10. Kim, M., Oh, Y. R., Kim, H. K.: Non-native pronunciation variation modeling using an indirect data driven method. In: Proceeding of ASRU, pp. 231–236. Kyoto (2007)
11. Stemmer, G., Steidl, S., Hacker, C., Noth, E.: Adaptation in the pronunciation space for non-native speech recognition. In: Proceedings of ICSLP, pp. 2901–2904. Jeju Island (2004)
12. Morgan, J.J.: Making a speech recognizer tolerate non-native speech through Gaussian mixture merging. In: Proceedings InSTIL/ICALL Symposium on Computer Assisted Learning, pp. 213–216. Venice (2004)
13. Oh, Y.R., Yoon, J.S., Kim, H.K.: Acoustic model adaptation based on pronunciation variability analysis for non-native speech recognition. Speech Commun. **49**(1), 59–70 (2007)
14. Bellegarda, J.R.: An overview of statistical language model adaptation. In: Proceedings ITRW on Adaptation Methods for Speech Recognition, pp. 165–174. Sophia Antipolis (2001)
15. Oh, Y.R., Kim, H.K.: On the use of feature-space MLLR adaptation for non-native speech recognition. In: Proceedings ICASSP, pp. 4314–4317. Texas (2010)
16. Bouselmi, G., Fohr, D., Illina, I.: Combined acoustic and pronunciation modelling for non-native speech recognition. In: Proceedings Interspeech, pp. 1449–1452. Antwerp (2007)
17. Oh, Y.R., Kim, M., Kim, H.K.: Acoustic and pronunciation model adaptation for context-independent and context-dependent pronunciation variability for non-native speech. In: Proceedings ICASSP, pp. 4281–4284. Las Vegas (2008)
18. Paul, D.B., Baker, J.M.: The design for the Wall Street Journal-based CSR corpus. In: Proceedings of ICSLP, pp. 899–902. Springer, Banff (1992)
19. Lee, S.J., Kang, B.O., Jung, H.Y.: Statistical model-based noise reduction approach for car interior applications to speech recognition. ETRI J. **32**(5), 801–809 (2010)
20. Chung, H., Park, J., Jeon, H., Lee, Y.: fast speech recognition for voice destination entry in a car navigation system, In: Proceeding of Interspeech, pp. 975–978. Brington (2009)

Chapter 25
Analysis of Speech Under Stress and Cognitive Load in USAR Operations

Marcela Charfuelan and Geert-Jan Kruijff

Abstract This paper presents ongoing work on analysis of speech under stress and cognitive load in speech recordings of Urban Search and Rescue (USAR) training operations. During the training operations several team members communicate with other members on the field and members on the control command using only one radio channel. The type of stress encountered in the USAR domain, more specifically on the human team communication, includes both physical or psychological stress and cognitive task load. Physical stress due to the real situation and cognitive task load due to tele-operation of robots and equipment. We were able to annotate and identify the acoustic correlates of these two types of stress on the recordings. Traditional prosody features and acoustic features extracted at sub-band level probed to be robust to discriminate among the different types of stress and neutral data.

25.1 Introduction

For several years the applications of stress detection in speech were mainly related to improve speech recognition, speaker recognition, or to improve the naturalness of synthetic speech [3]. Nowadays applications including detection of speech under stress and/or cognitive load span many fields. In human-computer interaction (HCI) and human-machine interaction (HMI) there is an increasing interest in analyzing stress in speech. For example, Jameson et al. [5] explored the prospects of exploiting

M. Charfuelan (✉)
DFKI GmbH, Language Technology Lab, Berlin, Germany
e-mail: marcela.charfuelan@dfki.de

G.-J. Kruijff
DFKI GmbH, Language Technology Lab, Saarbrücken, Germany
e-mail: gj@dfki.de

J. Mariani et al. (eds.), *Natural Interaction with Robots, Knowbots and Smartphones: Putting Spoken Dialog Systems into Practice*, DOI 10.1007/978-1-4614-8280-2_25,

the user's speech as a source of evidence for the recognition of resource limitation. Models of cognitive task load (CTL) as well as models of affective task load (ATL) and performance level are proposed in Looije et al. [7] to recognize critical states, with the objective of enhancing geo-collaboration on teamwork. The type of stress encountered in the Urban Search and Rescue (USAR) domain, more specifically on the human team communication, includes both physical or psychological stress and cognitive task load. Physical stress due to real situation and cognitive task load due to tele-operation of robots and equipment. The expectation is that collaborative teamwork will benefit from the automatic detection of critical affective states (stress). For example, in an application involving multiple sources of information, the control command might decide to adapt or limit the information presented to the team members when different stress conditions are detected.

One approach that has been shown to be robust to analyze speech under stress in real situations is the multiband processing of speech. Hansen et al. [3] have developed an acoustic feature based on multiband nonlinear processing of speech: the autocorrelation envelope of the critical band filtered Teager Energy Operator (TEO-CB-AutoEnv). This feature has been used to recognize simulated and actual speech under stress from the SUSAS database [4]. In our study we have used traditional prosody features extracted at full-band level, and TEO-AutoEnv, spectral, and voicing strength features extracted at sub-band level.

This paper is organized as follows: first we briefly describe our experience collecting and annotating speech data from USAR training sessions (Sect. 25.2). Then we briefly describe the acoustic features (Sect. 25.3), how we use them to identify acoustic correlates of the annotated stress and preliminary stress classification results (Sect. 25.4). Conclusions and future work are presented in Sect. 25.5.

25.2 Data Collection and Annotation

The speech database analyzed in this paper corresponds to the recordings of the NIFTi Join Exercises 2011 on human-robot-teaming (NJEx2011) [6]. The NIFTi Join Exercises took place in a constructed, complex environment where four different teams performed several missions in two days. On the first day (0706) each team had two missions: in mission 1 the teams traversed a complex arena with an unmanned ground vehicle (UGV), helped by an unmanned aerial vehicle (UAV); each team got 45 min. In mission 2 the teams explored two floors on the Red Building searching for victims; each team got 75 min. On the second day of exercises (0707) the teams went into the Red Building again but this time under more severe circumstances: smoke, fire, more floors to explore, and in less time. Each team explored three floors of the Red Building searching for victims; each team got 90 min. In all the exercises UGV operation was remote, UAV was Line Of Sight (LOS), and the communication was done via open voice loop only. Seven sessions (missions) were recorded during the first day and 4 during the second day. Different team players (persons) participate in each session.

Table 25.1 NJEx2011 distribution of turns per day and speaker

	Day	
Speaker	0706	0707
Missiondirector	161	272
Safetydirector	817	324
Teamrole	47	25
UAVpilot	31	48
UGVpilot	343	197
Whitecommand	53	36
Total time	410 min.	315 min.

Table 25.2 NJEx2011 distribution of turns per speaker type and annotated stress level, where the annotators agree

Speaker	Higher	Medium	Neutral
Missiondirector	0	13	375
Safetydirector	24	188	629
Teamrole	0	4	63
UAVpilot	0	1	74
UGVpilot	0	16	437
Whitecommand	0	4	79
Total	24	226	1,657
Percentage	1.2%	11.8%	86.8%

The recordings of each session were segmented per turn and annotated according to the speakers, or team players, that participate on the mission. Table 25.1 shows the distribution of turns (utterances) per day and speaker.

The segmented sessions were further annotated according to three levels of stress: (1) unstress: normal or neutral speech, happy, and relax; (2) stress: speech is nervous, there is tension in the voice, there is more speed, and there are hesitations; and (3) very stressed: there are shouts, anger, and despair. Two people annotated these three levels of stress on each utterance of all sessions. The distribution of data according to speakers and three stress categories, higher (stress level 3), medium (stress level 2), and neutral (stress level 1), is presented in Table 25.2. According to this table, there is very small number of higher and medium stress turns, and in particular higher stress is only exhibited by the Safetydirector speaker of the sessions. The inter-rater agreement is presented in Table 25.3. The number of observed agreements is 1,908 (81.02% of the observations) and the number of agreements expected by chance is 1,553.1 (65.95% of the observations). The Kappa value is 0.443 with 95% confidence interval: from 0.401 to 0.484. The strength of agreement is considered to be "moderate," although as reported by Burkhardt et al. [1], Kappa values between 0.4 and 0.7 are usually regarded as fair agreement in annotations of this type of expressive speech data. For the analysis of stress in this data, we have selected the turns where the two annotators agree.

Table 25.3 NJEx2011 stress annotation: two annotators inter-rater agreement, Kappa = 0.443

Stress level	Neutral	Medium	Higher	Total turns
Neutral	**1,658**	287	2	1,947
Medium	118	**226**	14	358
Higher	3	23	**24**	50
Total turns	1,779	536	40	2,355

Turns where the two annotators agree (diagonal) are indicated in bold.

25.3 Acoustic Features

Standard prosodic features and TEO sub-band features reported in the literature as good correlates of stress were extracted from the data; these and other sub-band features were extracted with snack [11] and are described below:

(a) Standard prosodic features: Fundamental frequency or pitch (f0); maximum, minimum, and range of f0; duration of the utterance in seconds; voicing rate calculated as the number of voiced frames (frames for which $f0 > 0$) per time unit; and log power calculated as the logarithm of the averaged short-term energy: log_pow= $\log(\frac{1}{N}\sum s^2)$ where N is the length of the window frame. Prosodic features are extracted frame based and at full band.

(b) Teager energy operator—autocorrelation envelope (TEO-AutoEnv): This is a measure that has been used to detect and classify speech under stress (emotional stress, task load stress, and Lombard effect) in the SUSAS database. The Teager operator for a discrete-time signal s is defined as [12]

$$\Psi[s(n)] = s^2(n) - s(n+1)s(n-1)$$

Similarly to the TEO-AutoEnv measure proposed in [12], we have implemented five band-pass filters with pass-bands: 0–1 kHz, 1–2 kHz, 2–4 kHz, 4–6 kHz, and 6–8 kHz. In our implementation of the TEO-AutoEnv, we apply the TEO operator to the five filtered signals, then the autocorrelation from each TEO band is calculated, and the area under the autocorrelation envelope is calculated and normalized over the window lag.

(c) Voicing strengths (STR): Estimated with peak normalized cross correlation of the input signal. The correlation coefficient for a signal s and delay t is defined by

$$c_t = \frac{\sum_{n=0}^{N-1} s(n)s(n+1)}{\sqrt{\sum_{n=0}^{N-1} s^2(n) \sum_{n=0}^{N-1} s^2(n+t)}}$$

In a previous work [2], we have found that voicing strengths are correlated with vocal effort of dominant speech, so it is expected that these features are correlated as well with some type of stressed speech (shouting, angry speech, etc.).

(d) Spectral entropy (SPE): Is a kind of "peakiness" of the spectrum that has been used in speech endpoint detection and in classification of emotions. This feature is calculated as follows [8]: the spectrum X is converted into a probability mass function (PMF) normalizing it by

$$x_i = \frac{X_i}{\sum_{i=1}^{N} X_i} \qquad i = 1 : N$$

where X_i is the energy of the ith frequency component of the spectrum, x is the PMF of the spectrum, and N is the number of points in the spectrum. Entropy for each frame is calculated by

$$H(x) = -\sum_{x \in X} x_i * log_2 x_i$$

25.4 Acoustic Correlates of Higher and Medium Stress Types

One of the objectives in this work is to get a better understanding of the acoustic characteristics of the annotated data. Analysis of variance (AOV) of the acoustic features described in Sect. 25.3 was performed in order to establish the main acoustic correlates of the two types of annotated stress. The idea is to find out which features are significantly different among the sets of data: higher (H), medium (M), and neutral (N). Results are presented in Table 25.4. We have analyzed which features are significantly different among the three classes (H/M/N), between the medium level and neutral (M/N) and between the higher level of stress and the medium and neutral levels together (H/(M&N)). We can observe in this table that most of the features, except voicing strengths in some bands, are significantly different among the three classes (H/M/N). Prosody, TEO-AutoEnv, and spectral features in higher bands are significantly different between medium and neutral data (M/N). In average f0 for medium stress is greater than f0 for neutral speech; log_pow is also in average greater for medium than for neutral, and the spectral entropy values in average are smaller for medium than for neutral, which indicates an increase in the proportion of energy in higher frequencies. According to Scherer et al. [10] these are characteristics of cognitive load or stress due to task load/engagement. On the other hand, significantly different features between higher stress and medium and neutral speech data (H/(M&N)) are mainly f0 and TEO features. In average f0 for higher stress is greater than f0 for medium and neutral data together. Taking into account the studies in Patil et al. and Scherer et al. [9, 10], we can conclude that indeed our annotated higher stress corresponds to physical or emotional stress.

Table 25.4 NJEx2011 AOV: Analysis of variance of acoustic features between different levels of stress: higher (H), medium (M), and neutral speech (N). Signif. codes: ***< 0.001, **< 0.01, *< 0.05, • < 0.1, – < 1. Preliminary classification results are presented for the different sets

Acoustic features			Stress types and neutral		
			H / M / N	M / N	H / (M & N)
Full-band	(a) Prosody	f0	***	***	***
		max_f0	**	**	–
		min_f0	***	*	***
		range_f0	•	*	–
		dur_seconds	***	***	**
		voicing_rate	•	*	–
		log_pow	***	***	*
Sub-band	(b) Voicing strengths	str1	**	–	***
		str2	*	–	*
		str3	–	–	–
		str4	–	–	–
		str5	•	*	–
	(c) TEO-AutoEnv	teo1	–	–	–
		teo2	***	***	–
		teo3	***	***	***
		teo4	***	***	***
		teo5	***	***	***
	(d) Spectral entropy	se1	***	–	***
		se2	***	***	–
		se3	***	***	•
		se4	**	**	–
		se5	***	***	*
SVM classification accuracy (avg)			75%	76%	83%
Classification per class %			**H:**43 **M:**66 **N:**76	**M:**75 **N:**76	**H:**71 **(M&N):**83

Preliminary classification results of neutral speech and two levels of stressed speech are presented in Table 25.4. Three classifiers are trained with different sets of features, one for classifying three classes H/M/N and two for classifying two classes M/N and H/(M&N). Since the data is very unbalanced, a weighted support vector machine (SVM) classifier is used; weight values are determined by the proportion of data in each class. Twenty repetitions of stratified sampling are performed, where 2/3 of the data in each class are randomly selected to train the models and the other 1/3 is used for testing. The preliminary results indicate that the detection of higher and medium levels of stress is improved when the classifiers are trained with different sets of features. For example, the detection of higher stress improved from 43% to 71% when using the H/(M&N) classifier.

25.5 Conclusions and Future Work

In this paper we have presented ongoing work on analysis of speech under stress and cognitive load in speech recordings of USAR training operations. In contrast to most of the analysis of speech under stress and/or cognitive load reported in the literature, we have analyzed speech recordings of real situations under very noisy conditions. The stress levels in this data were determined by manual annotation and not by the recording condition or experimental setting. We were able to annotate and identify the acoustic correlates of two types of stress on the recordings: physical stress and cognitive load. Traditional prosody features and sub-band acoustic features probed to be robust to discriminate among the different types of stress and neutral data. Our future work is to design appropriate classifiers of stress for the USAR domain that can cope with the very unbalanced data; when designing the classifiers, we will take into account that the acoustic correlates of the two types of stress are very different, so the classifier/detector of physical stress should not be trained with the same features as the classifier/detector of cognitive load.

Acknowledgements The work reported in this paper has received funding from the EU-FP7 ICT 247870 NIFTi project. We would like to thank Holmer Hemsen for assistance with data annotation.

References

1. Burkhardt, F., Polzehl, T., Stegmann, J., Metze, F., Metze, F., Huber, R.: Detecting real life anger. In: IEEE International Conference on ICASSP, Taipei (2009)
2. Charfuelan, M., Schröder, M.: The vocal effort of dominance in scenario meetings. In: Interspeech. Florence (2011)
3. Hansen, J., Patil, S.: Speech under stress: Analysis, modeling and recognition. In: Speaker Classification I, Lecture Notes in Computer Science, vol. 4343, pp. 108–137. Springer, Berlin (2007)
4. Hansen, J.H.L., Bou-Ghazale, S.E.: Getting started with susas: a speech under simulated and actual stress database. In: Eurospeech. Rhodes (1997)
5. Jameson, A., Kiefer, J., Müller, C., Gromann-Hutter, B., Wittig, F., Rummer, R.: Assessment of a user's time pressure and cognitive load on the basis of features of speech. In: Resource-Adaptive Cognitive Processes, Cognitive Technologies. Springer, Berlin (2010)
6. Kruijff, G.: Proceedings of NJEx 2011, NID 2011 (2012). DFKI internal report
7. Looije, R., te Brake, G., Neerincx, M.: Geo-collaboration under stress. In: Workshop on Mobile HCI for Emergencies. Singapore (2007)
8. Misra, H., Ikbal, S., Sivadas, S., Bourlard, H.: Multi-resolution spectral entropy feature for robust ASR. In: IEEE International Conference ICASSP. Philadelphia (2005)
9. Patil, S.A., Hansen, J.H.L.: Detection of speech under physical stress: Model development, sensor selection, and feature fusion. In: Interspeech. Brisbane (2008)
10. Scherer, K.R., Grandjean, D., Johnstone, T., Klasmeyer, G., Bänziger, T.: Acoustic correlates of task load and stress. In: ICSLP2002–Interspeech 2002. Denver (2002)
11. Sjölander, K.: The Snack Sound Toolkit. http://www.speech.kth.se/snack (2012)
12. Zhou, G., Hansen, J.H.L., Kaiser, J.F.: Nonlinear feature based classification of speech under stress. IEEE Trans. Speech Audio Process. **9**(3), 201–216 (2001)

Part VI
Dialog Management

Chapter 26
Does Personality Matter? Expressive Generation for Dialogue Interaction

Marilyn A. Walker, Jennifer Sawyer, Grace Lin, and Sam Wing

Abstract This paper summarizes our recent work on developing the technical capabilities needed to automatically generate dialogue utterances that express either a personality or the persona of a dramatic character. In previous work, we developed a personality-based generation engine, PERSONAGE, that produces dialogic restaurant recommendations that varied according to the speakers personality. More recently we have been exploring three issues: (1) how to coordinate verbal expression of personality or character with nonverbal expression through facial or body animation parameters; (2) whether we can express character models that we learn from film dialogue with the existing parameters of the PERSONAGE engine; and (3) whether we can show experimentally that expressive generation is useful in a range of tasks. Our long-term goal is to create off-the-shelf tools to support the creation of spoken dialogue agents with their own persona and personality, for a broad range of types of dialogue agents in task-oriented applications or in interactive stories and games.

26.1 Introduction

The last twenty years have seen tremendous progress in spoken dialogue systems for a large range of applications, from smartphone personal assistants to human-robot interaction. Much of this progress has been driven by large corpora of training data, which has vastly improved the state of the art [5, 15, 41, 50]. However, to date, there has only been limited work on adaptation of dialogue agents to a particular

M.A. Walker (✉) • J. Sawyer • G. Lin • S. Wing
Natural Language and Dialogue Systems Lab, Baskin School of Engineering, University of California, Santa Cruz, CA, USA
e-mail: maw@soe.ucsc.edu; jsawyer@soe.ucsc.edu; glin@soe.ucsc.edu; sampwing@soe.ucsc.edu

J. Mariani et al. (eds.), *Natural Interaction with Robots, Knowbots and Smartphones: Putting Spoken Dialog Systems into Practice*, DOI 10.1007/978-1-4614-8280-2_26,

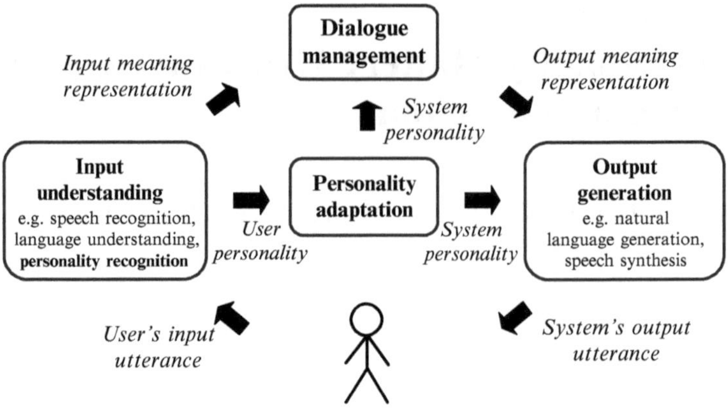

Fig. 26.1 High-level architecture of a dialogue system with personality-based user adaptation

user, either within the context of a single interaction or over time. How might such adaptation be enabled and what evidence is there that it would be useful?

Figure 26.1 depicts a general model for user adaptation, namely the adaptation of a dialogue agent to the user's personality. Any type of adaptation to the user requires progress on three research problems: (1) USER MODEL: acquiring relevant user traits (a model of the user that drives adaption), (2) ADAPTATION MODEL: deciding how to modify system responses to orient toward the relevant user traits, and (3) GENERATION: producing a consistent response matching those traits. We believe that progress in user adaptation has been slow because software has not been available that would allow one to easily control and integrate the generation of verbal and nonverbal dialogue agent behavior. Without such software, it is not possible to properly test whether a particular USER MODEL or ADAPTATION MODEL is really functional. What is required to make progress in this area is a flexible and parameterizable generation component that can be dynamically controlled in real time. Our recent work has focused on developing and testing parameterizable generation capabilities for a range of dialogue applications as we discuss in more detail below.

Why do we think user response adaptation will be useful? First of all, dialogue generation is one of the few remaining modules of spoken dialogue systems that are completely handcrafted; this is a bottleneck for building systems of all types. In addition, the last 15 years have seen research by the SDS community on mixed-initiative dialogue. Mixed-initiative dialogue, however, not only makes demands on the ASR, SLU, and DM; it also requires a much more flexible method for generating responses. Moreover, it is widely known that in natural conversation, humans dynamically adapt to one another in many different ways [10, 11, 18, 25, 36, 37, 40]. And there are a relatively large number of experimental results that strongly suggest that a dialogue agent would be more effective if it could adapt its responses to the user [2, 9, 12, 16, 17, 20, 33, 46, 48, 52, 54]. For example, experimental results have shown that when users interact with a system that adapts to them, they will spend more time on their medically recommended exercises, or their perception of the system's intelligence or competence increases, or they learn more [24,28,43,46,54,55,61]. Finally, intuitively it seems that people are interested in social aspects of interaction with SDS. For example, the recent interest in Siri seems to be at least partly driven by the perception that Siri has a personality. There also seems to be an explosion in the range of voices supported by commercial TTS engines, including dialectical variations and celebrity voices [4,5,56].

This paper (1) briefly describes our framework for adaptive response generation (Sec 26.2) and (2) describes four different projects on adaptive response generation ranging from our work on a travel planning agent for DARPA Communicator (Sec. 26.3) to our recent work on dialogue generation for nonplayer characters (NPCs) for interactive story systems and games (Sec. 26.4).

26.2 Spoken Language Generation Framework

Our work on expressive and adaptive generation for dialogue interaction starts with the standard NLG pipeline architecture [13,22,32,45,47,51]. Figure 26.2 shows how we integrated the standard NLG pipeline architecture into AT&T Communicator [53,59,60]. The standard NLG architecture consists of the following modules:

- Content (text) planning: refine communicative goals; select and structure content.
- Sentence planning: choose linguistic resources (lexicon, syntax) to achieve goals.
- Realization: use grammar (syntax, morphology) to generate surface utterances.

This framework provides a basic modular architecture that supports adaptation in either content selected (what is said) or in the way the content is realized (how it is said).

As part of our work in this area, we have extended this standard architecture in several ways. First, we developed several novel methods for statistical language generation that can be trained to adapt to a user or to aspects of a particular domain. Secondly, we have introduced finer grained distinctions in these modules in order to support expressive generation. Below we will discuss in detail how our PERSONAGE

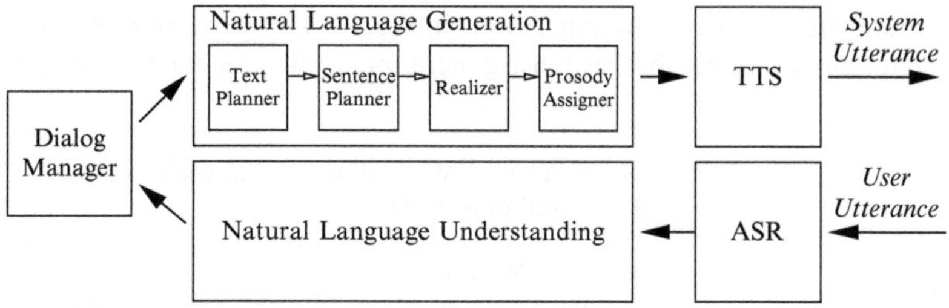

Fig. 26.2 Architecture of a dialogue system with natural language generation

generator introduces finer distinctions in the generation process and new modules in order to generate more expressive language, in comparison with this basic generation for dialogue architecture. Third, we have begun to explore integration of verbal and nonverbal behavior using the same model of highly parameterizable behaviors. Finally, we are working on several different applications that we expect will show the utility of this framework.

26.3 Generation for Task-Oriented Dialogue

We started our work on dynamic language generation in the context of the AT&T Communicator project [26, 42]. Here we describe our experimental work on evaluating generation for task-oriented dialogue, different ways of constructing generators, and our evaluation comparisons to more traditional methods of response generation such as template-based generation. Because both AT&T Communicator and the MATCH system had a template-based generator available that had been developed first, we were able to directly compare our statistical generation approach to template-based generation on a number of metrics as we were developing it. For example, consider a mixed-initiative dialogue system for travel planning, with a dialogue strategy that allows the user to provide any type of information in any turn and which aims to combine confirming the information the user has provided with requests for additional information needed to complete the task, as illustrated in the dialogue in Fig. 26.3.

Utterance S1 in Fig. 26.3 requests information about the caller's departure airport, but in U2, the caller takes the initiative to provide information about her destination rather than answering the question that the system asked. In S3, AT&T Communicator's dialogue strategy is to *implicitly confirm* the destination (because of the possibility of error in the speech recognition component) and *request information* (for the second time) of the caller's departure airport. In U4, the caller provides this information but **also** provides the month and day of travel. The system's dialogue strategy for turn S5 is to *implicitly confirm* all the information that the user has provided so far, i.e., the departure and destination cities and the month and day information, as well as to *request information* about the time of travel that has not yet been provided by the user. The system's representation of its communicative goals for utterance S5 is in Fig. 26.4. In general, mixed-initiative dialogue requires very flexible response generation. If the user can take the initiative

Fig. 26.3 D1: Mixed-initiative dialogue from DARPA Communicator

S1: Welcome.... What airport would you like to fly out of?
U2: I need to go to Dallas.
S3: Flying to Dallas. What departure airport was that?
U4: from Newark on September the 1st.
S5: What time would you like to travel on September the 1st to Dallas from Newark?

Fig. 26.4 The communicative goals for utterance S5 in dialogue D1

implicit-confirm(orig-city:NEWARK)
implicit-confirm(dest-city:DALLAS)
implicit-confirm(month:9)
implicit-confirm(day-number:1)
request(depart-time)

System	Realization
TEMPLATE	Flying from Newark to Dallas, Leaving on the 1st of September, And what time did you want to leave?
Spot Alt1	What time would you like to travel on September the 1st to Dallas from Newark?
Spot Alt2	Leaving on September the 1st. What time would you like to travel from Newark to Dallas?
RANDOM	Leaving in September. Leaving on the 1st. What time would you, traveling from Newark to Dallas, like to leave?
NOAGG	Leaving on the 1st. Leaving in September. Going to Dallas. Leaving from Newark. What time would you like to leave?

Fig. 26.5 Sample outputs for System5 of dialogue D1 for each type of generation system used in the evaluation experiment

Table 26.1 Summary of results. Score out of 5

System	Mean score	S.D.
TEMPLATE	3.94	1.11
SPoT	3.88	1.27
No aggregation	3.01	1.22
Random	2.66	1.45

to provide any of N items of information, then there are 2^N possibilities for different combinations of information that must be confirmed in a system turn, as illustrated by the example dialogue goals in Fig. 26.4.

Figure 26.5 shows some of the different ways of producing system responses for utterance S5 that we experimented with, including two alternatives for the dynamic language generation system SPoT that we developed for AT&T Communicator. SPoT was a trainable sentence planner (Sentence Planner Trainable) that provided 12 parameters to control how concepts were combined into sentences. Multiple parameters could apply in a single turn; the parameter selection was controlled with a probability distribution over operators, allowing us to generate many different N-best outputs for each set of dialogic goals. By using conceptual representations for dialogue goals, as illustrated in Fig. 26.4, and general linguistic representations and rules, we developed SPoT as a trainable overgenerate and rank statistical generator [23, 44, 58]. Thus SPoT's N-best outputs were represented by features, feedback on the quality of the outputs was elicited in a training phase, and a ranker was trained to rank all the generated possibilities. In addition, because SPoT was implemented for AT&T's DARPA Communicator system, we were able to directly compare SPoT to the handcrafted, template-based prompt generation first implemented for AT&T Communicator, as well as to several BASELINE generators. See Fig. 26.5 and Table 26.1. One baseline, which we call NO AGGREGATION, produces one sentence for each communicative goal. Note that the NoAgg system represents a template-based generation system without significant handcrafting for mixed-initiative dialogue systems. In the AT&T Communicator system, one template was handcrafted to combine departure and arrival airports, and another template combined date and month information, as shown in the TEMPLATE-labeled utterance in Fig. 26.5. But of course templates had not been handcrafted for all of the 2^N combinations, and therefore there was no way to achieve all of the dialogue goals shown in Fig. 26.4 with either of the alternatives shown as SPoT possibilities in Fig. 26.5. SPoT produces these combinations dynamically by using deep representations that can be combined with general rules. Parameter settings control which combination operations are preferred for expressing a particular style. The RANDOM baseline generation strategy in Fig. 26.5, with results in Table 26.1, randomly makes decisions about how to combine communicative goals into sentences, using the same underlying mechanisms as SPoT but without the trained ranker.

We compared these different approaches by having 60 human subjects rank the outputs of these different generators in the context of a spoken dialogue. As Table 26.1 indicates, some system outputs received more consistent scores than others, e.g., the standard deviation for TEMPLATE was much smaller than RANDOM. The ranking of the systems by average score is TEMPLATE, SPoT, NoAgg, and RANDOM. Post hoc comparisons of the scores of individual pairs of systems using the adjusted Bonferroni statistic revealed several different groupings. The highest ranking systems were TEMPLATE and SPoT, whose ratings were not statistically significantly different from one another. This shows that it is possible to match the quality of a handcrafted system with a trainable one, which should be more portable, more general, and require less overall engineering effort. Along with the TEMPLATE system, SPoT also scored better than the baseline systems NoAgg and RANDOM, as we would have expected.

Our next generation system, SPaRKy, implemented several additional ways of parameterizing the dialogue agent's output. SPaRKy had a mechanism for using a model of the user to customize the content selection parameters so as to include the subset of available information that the user was most interested in when making a decision about a restaurant. SPaRKy also introduced parameters for manipulating the rhetorical and argument structure of a recommendation. Figure 26.6 shows several alternative recommendations that SPaRKy could produce that vary the ordering of information, and sentence length, and use of cues like *since*. Figure 26.6 also shows how individual users A and B actually vary in their preferences for stylistic utterance variations. SPaRKy was implemented for MATCH, a multimodal system on a first generation mobile device, that provided local information about restaurant and entertainment options [21]. This project demonstrated that (1) we could extend our approach to providing information that potentially had a more complex rhetorical structure; (2) we could use it to train individualized sentence planners. Table 26.2 provides the results of an evaluation study comparing the performance of the trained rankers for users A and B, when A's ranker was used for B and vice versa, as opposed to a ranker trained on the average of A and B's

Alt	Realization	A	B	SPR_A	SPR_B	SPR_{AVG}
6	Chanpen Thai has the best overall quality among the selected restaurants since it is a Thai restaurant, with good service, its price is 24 dollars, and it has good food quality.	1	4	0.16	0.65	0.58
2	Chanpen Thai has the best overall quality among the selected restaurants. Its price is 24 dollars, and it is a Thai restaurant. It has good food quality and good service.	4	4	0.75	0.50	0.74
8	Chanpen Thai is a Thai restaurant, with good food quality. It has good service. Its price is 24 dollars. It has the best overall quality among the selected restaurants.	4	2	0.81	0.29	0.73

Fig. 26.6 Some alternatives for recommending Chanpen Thai, with feedback from users A and B (1=worst and 5=best) and rankings from the trained generation rankers for users A and B and mean(A,B) ($[0,1]$)

Table 26.2 Ranking error for various configurations with the RECOMMEND strategy

RECOMMEND	A's model	B's model	AVG model
A's test data	0.17	0.52	0.29
B's test data	0.52	0.17	0.27
AVG's test data	0.31	0.31	0.20

rankings. Figure 26.6 shows how the rankers for user A (SPR_A) and user B (SPR_B) vary in the ranking score that they give different possible utterance realizations. Our work on SPaRKy illustrated for the first time that people have quite specific individual preferences regarding information ordering, sentence aggregation, and use of discourse cues. Furthermore, our results showed that a trainable sentence planner can model these individual preferences and that in some cases the individualized sentence planners are better than, or statistically indistinguishable from, the template-based generator.

In sum, our work on task-oriented dialogue systems demonstrated (1) that we can develop trainable methods that are as good or better than traditional methods; (2) that we can develop the architecture we started with for requesting information and easily extend it to providing information with rhetorical structure; and (3) that the training methods allow us to automatically train dialogue agents that are adaptive for individual users.

26.4 Expressive Generation

We then turned to the question of producing expressive variation that could target individual preferences in terms of expressivity, rather than simply in terms of information provided and information structure. An obvious initial question is what types of parameters are needed to produce a wide range of expressive variation. To answer this question, we turned to social psychology theories of individual variation, namely the *Big Five* Theory of Personality. The Big Five model is based on the observation that, when talking about a close friend, one can usually produce a large number of descriptive adjectives [1]. This observation is described as the

Table 26.3 Trait adjectives associated with the Big Five traits

Trait	High	Low
Extraversion	Warm, assertive, excitement seeking, active, spontaneous, optimistic, talkative	Shy, quiet, reserved, passive, solitary, moody, joyless
Emotional stability	Calm, even-tempered, reliable, peaceful, confident	Neurotic, anxious, depressed, self-conscious, oversensitive, vulnerable
Agreeableness	Trustworthy, friendly, considerate, generous, helpful, altruistic	Unfriendly, selfish, suspicious, uncooperative, malicious
Conscientiousness	Competent, disciplined, dutiful, achievement striving, deliberate, careful, orderly	Disorganized, impulsive, unreliable, careless, forgetful
Openness to experience	Creative, intellectual, imaginative, curious, cultured, complex	Narrow-minded, conservative, ignorant, simple

Lexical Hypothesis, i.e., that any trait important for describing human behavior has a corresponding lexical token, which is typically an adjective, such as *trustworthy* or *friendly*. The Lexical Hypothesis, proposed in 1936, has, over the subsequent years, led to a consensus that there are essential traits, known as the Big Five personality traits [8, 14, 19, 38, 39, 49]. The Big Five traits and some of their corresponding trait adjectives are shown in Table 26.3.

Our first goal was to determine whether studies in psychology that documented correlations between Big Five traits and verbal and nonverbal external behaviors could provide a specification of the parameters needed to express a broad range of individual variation in character and personality. We first organized the psychological findings in terms of the modules in the NLG reference architecture that was shown in Fig. 26.2. The upper part of Fig. 26.7 shows how this resulted in an exploded architecture, with new modules introduced in order to compartmentalize the 67 new parameters we implemented into a pipeline architecture. Note that what was previously called Sentence Planning in Fig. 26.2 now consists of 4 modules: syntactic template selection, aggregation, pragmatic marker insertion, and lexical choice. The lower part of Fig. 26.7 specifies some of the new parameters that were added to those new architectural components.

We first applied this extended expressive architecture to our previous application of restaurant recommendations, with the goal of determining whether the dialogue agent's personality would be perceived by humans in the way that the agent intended. In other words if parameters had been set with the goal of expressing the personality of an extraverted, conscientious agent, would people perceive the agent as extraverted and conscientious? In terms of the adaptation model in Fig. 26.1, these experiments aim to establish that the generation module can actually achieve the goals that the adaptation module gives it.

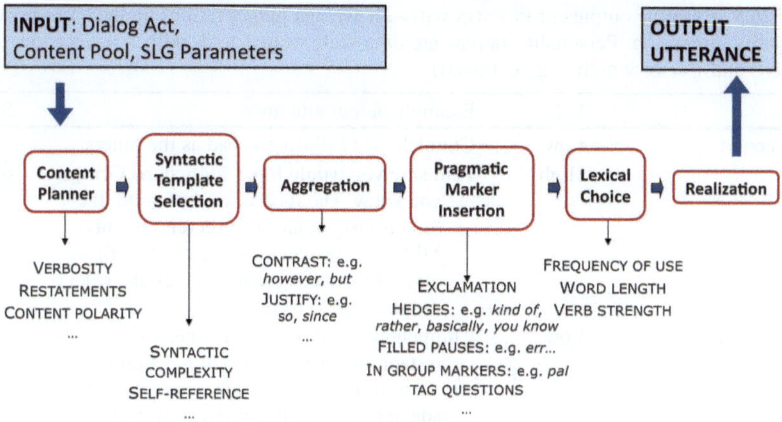

Fig. 26.7 The architecture of the PERSONAGE base generator

Table 26.4 Average personality ratings for the utterances generated with the low and high personality models for each trait on a scale from 1 to 7. The ratings of the two extreme utterance sets differ significantly for all traits ($p < 0.001$, two-tailed)

Personality trait	Low	High
Extraversion	2.96	5.98
Emotional stability	3.29	5.96
Agreeableness	3.41	5.66
Conscientiousness	3.71	5.53
Openness to experience	2.89	4.21

Table 26.4 provides the quantitative results for the rule-based method for each Big Five trait [29]. Perceptual differences between the high and low portrayal of traits were always statistically significant. Table 26.5 show example outputs for one method of control, a rule-based method with groups of parameter settings suggested from the psychological findings on the expression of each Big Five trait. Table 26.5 includes examples for the high and low ends of the scale for each Big Five trait. For example, the **high extraversion** utterance in Table 26.5 is more verbose and chatty than the **low extraversion** utterance; it includes more information possibly relevant to making the decision of which restaurant to choose, and it includes a bit of wordy paraphrasing (*food is kind of good, the food is tasty*). The utterances for the traits of **high agreeableness** and **high conscientiousness** in Table 26.5 illustrate the use of tag questions, a new parameter from the pragmatic marker insertion module shown in Fig. 26.7, while the **low openness to experience** utterance illustrates the use of filled pauses and hedges from the pragmatic marker insertion module.

Table 26.5 Example outputs of PERSONAGE with average judges ratings on the corresponding personality dimension. Personality ratings are on a scale from 1 to 7, with 1 = very low (e.g., introvert) and 7 = very high (e.g., extravert)

Trait	Set	Example output utterance	Score
Extraversion	Low	Chimichurri Grill isn't as bad as the others.	1.00
	High	I am sure you would like Chimichurri Grill, you know. The food is kind of good, the food is tasty, it has nice servers, it's in Midtown West and it's a Latin American place. Its price is around 41 dollars, even if the atmosphere is poor.	6.33
Emotional stability	Low	Chimichurri Grill is a Latin American restaurant, also it's located in Midtown West. It has quite friendly waiters. It offers adequate food. I imagine you would appreciate it.	2.92
	High	Let's see what we can find on Chimichurri Grill. Basically, it's the best.	6.00
Agreeableness	Low	I mean, Chimichurri Grill isn't as bad as the others. Basically, the staff isn't nasty. Actually, its price is 41 dollars. It's damn costly.	2.00
	High	You want to know more about Chimichurri Grill? I guess you would like it buddy because this restaurant, which is in Midtown West, is a Latin American place with rather nice food and quite nice waiters, you know, okay?	5.75
Conscientiousness	Low	I am not kind of sure pal. Err... Chimichurri Grill is the only place I would advise. It doesn't provide unfriendly service! This restaurant is damn expensive, its price is 41 dollars.	3.00
	high	Let's see what we can find on Chimichurri Grill. I guess you would like it since this eating house, which offers sort of satisfying food and quite satisfactory waiters, is a Latin American eating place.	6.00
Openness to experience	Low	Err... I am not sure. Mmhm... I mean, Chimichurri Grill offers like, nice food, so I would advise it, also the atmosphere is bad and its price is 41 dollars.	3.50
	high	You want to know more about Chimichurri Grill? I believe you would love it, you know. I guess it's in Midtown West. Although this eating house's price is around 41 dollars, the food is rather satisfactory. This eating place, which provides kind of second-rate atmosphere, is a Latin American restaurant, alright?	5.00

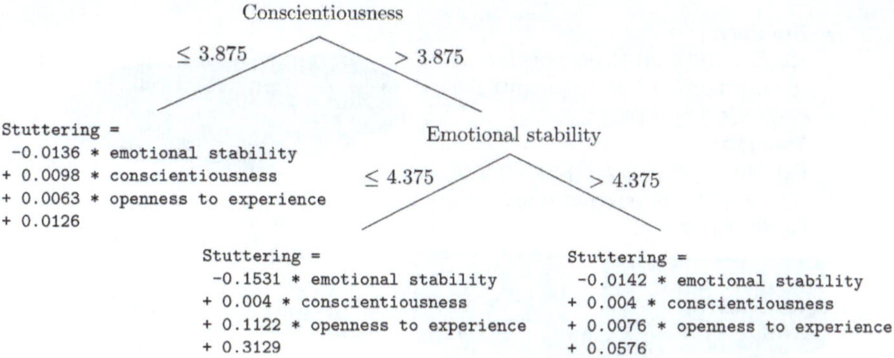

Fig. 26.8 Parameter estimation model learned using tree regression for controlling the stuttering parameter

Fig. 26.9 Alfred agent used for experiments on personality, eye gaze, and blinking

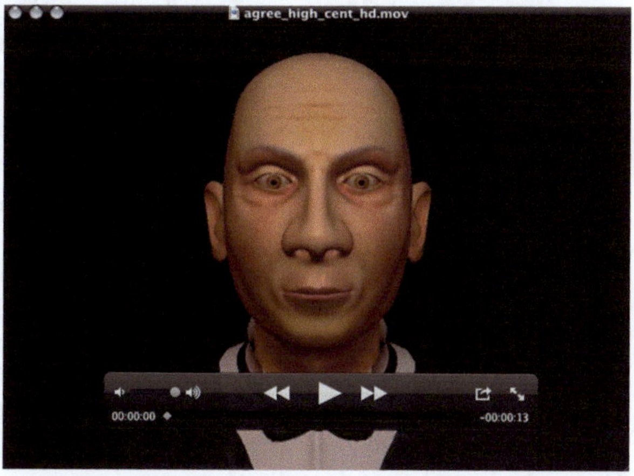

We also developed a new training method called parameter estimation (PE), which allowed us to learn models for expressing combinations of traits [31]. This method involves training from user feedback, as in our previous work on SPoT and SPaRKy, but in this case we invert the training method to learn models for the setting of an individual parameter as a function of a Big Five personality model with a scalar value for each of the five traits. For example, a tree regression model that was learned for the control of the stuttering parameter is shown in Fig. 26.8.

In this context, we also carried out two sets of experiments using the same methods for controlling nonverbal parameters (head, face, body) that were intended to express personality as well and showed that findings from psychology on the nonverbal expression of personality could be used to control eye gaze, blinking, head position, and a set of body parameters, such as gesture extent shape and speed, shoulder position, posture, and hip movement [3,34,35]. An example dialogue agent from our work on head position and eye behavior is shown in Fig. 26.9

Fig. 26.10 SpyFeet story line: interactive narrative game

We then began working on dialogue generation and interaction for interactive stories and games with the aim of testing whether the parameters that we had in PERSONAGE could support a wide range of expressive character types. Our prototype system was an interactive narrative system SpyFeet that involved solving a mystery by talking to various characters, as summarized by the story line in Fig. 26.10. In order to get more nuanced character differences, we developed a new statistical method for learning arbitrary character models based on film characters. Figure 26.11 shows the flow of the new method for learning from dialogue corpora. The method consists of the following steps:

1. Collect movie scripts from The Internet Movie Script Database (IMSDb).
2. Parse each movie script to extract dialogic utterances, producing an output file containing utterances of exactly one character of each movie (e.g., *pulp-fiction-vincent.txt* has all of the lines of the character Vincent).
3. Select characters from those with more than 60 turns of dialogue.
4. Extract features representing the linguistic behaviors of each character.
5. Learn models of character linguistic styles based on the features.
6. Use character models to control parameters of the PERSONAGE generator.
7. Evaluate human perceptions of dialogic utterances generated using the character models.

The results of this new method are provided in Table 26.6 for some sample utterances for the SpyFeet Tortoise character. To date, this method has not required extending the set of parameters that PERSONAGE can control; it just provides a different way of controlling the existing parameters. An evaluation experiment shows that the current method produces perceivable differences in character expression

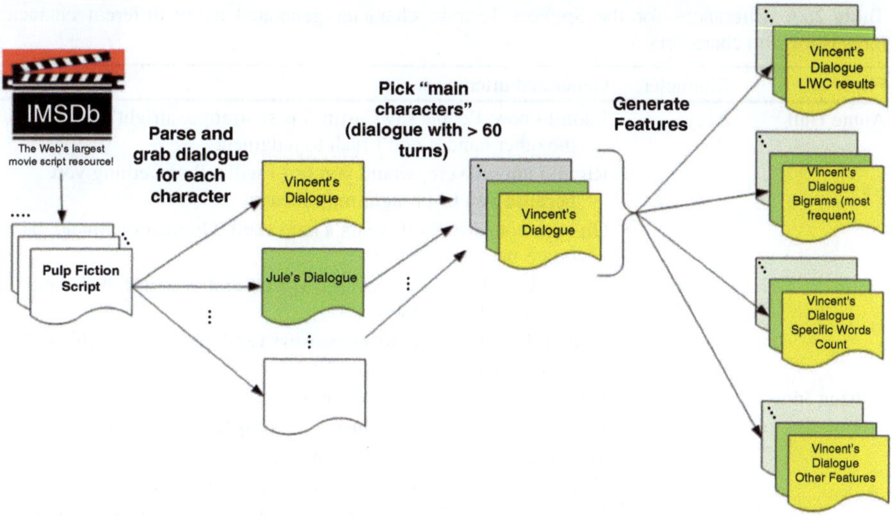

Fig. 26.11 Method

and that in the main, people perceive the generated utterances in a new domain, as being more similar to the character they were modeled on, than to another random character. We are currently running an experiment for a new game called Heart of Shadows that uses these models. The goal of this experiment is to show that we can increase story author creativity and productivity and make more interesting varied interactive stories through integrating expressive generation with an interactive story authoring tool.

26.5 Discussion and Conclusions

In sum, we have developed four methods for generating dialogue, one rule-based method and three statistical methods: (1) a novel version of the overgenerate and rank method [23, 30], (2) parameter estimation [31], and (3) a novel hybrid corpus-based method for learning character models from dialogue in film corpora [27, 57]. We have shown that we can parameterize a range of behaviors, both verbal and nonverbal [3, 34, 35]. We have produce a dynamic, real-time controllable generation engine, and through a number of experiments we have shown that characters and personalities are perceived as the dialogue agent intended.

There is still considerable research to be done. An outstanding question is whether we can show that expressive generation has clear task-related effects as suggested by previous work. We feel that we have not yet fully explored the range of applications that expressive generation could be useful for. Our previous work suggests that the ability to have agents vary their personality might be useful simply

Table 26.6 Utterances for the SpyFeet Tortoise character generated using different character models for film characters

Film	Character	Generated utterances
Annie Hall	Alvy	I don't know. People say Cartmill is st-strange, alright? Err... on the other hand, I don't rush to judgment .
		Right, I am not sure, would you be? I will tell something you because you br-brought me cabbage .
		Oh I am not sure. Wolf wears a hard shell. On the other hand, he is ge-ge-gentle, isn't he?
		I see, I don't know. I respect Wolf, wouldn't you? He, however, isn't my close friend.
		Yeah, I don't know. Sparrow conveys excitement to my life, so I am fr-fr-friends with her.
Indiana Jones	Indy	I don't rush to judgment, but people say Cartmill is strange.
		I will tell something you since you brought me cabbage.
		Wolf is gentle but he wears a hard shell.
		Wolf isn't my close friend. But I respect him.
		I am friends with Sparrow since she brings excitement to my life.
Annie Hall	Annie	Come on, I don't know, do you? People say Cartmill is strange while I don't rush to um.. judgment .
		I don't know. I think that you brought me cabbage, so I will tell something to you, alright?
		Yeah, I am not sure, would you be? Wolf wears a hard shell but he is really gentle
		I see, I am not sure. Obviously, I respect Wolf. However, he isn't my close friend, is he?
		Come on, I am not sure. Because Sparrow brings excitement to my life, I am friends with her, you see?
Pulp Fiction	Vincent	Basically, I don't rush to judgment. On the other hand, people say Cartmill is strange; he is strange.
		Yeah, I can answer since you brought me cabbage that.
		Everybody knows that Wolf wears a hard shell. He, however, is gentle.
		I respect Wolf. However, he isn't my damn close friend.
		Oh God I am friends with Sparrow because she brings excitement to my life.

in the context of having different kinds of agents for different kinds of tasks [29]. However, we believe that the most powerful use of this technology might be in adaptation over time, when dialogue agents function as companions or "relational agents" over longer periods [6,7,12]. Our current experiments involve (1) attempting to show that we can increase story author creativity and productivity and make more interesting varied interactive stories through integrating expressive generation with an interactive story authoring tool and (2) that we can adapt over time to a user in a task-oriented dialogue application.

Acknowledgements Thanks to the organizers of IWSDS 2012 for organizing such a wonderful occasion for discussing work on dialogue systems and for inviting me to give a keynote at the workshop. This paper has benefited from their feedback.

References

1. Allport, G.W., Odbert, H.S.: Trait names: a psycho-lexical study. Psychol. Monogr. **47**(1), (Whole No. 211) 171–220 (1936)
2. André, E., Rist, T., Susanne van Mulken, Klesen, M., Baldes, S.: The automated design of believable dialogues for animated presentation teams. Embodied Conversational Agents, pp. 220–255. MIT Press, Cambridge (2000)
3. Bee, N., Pollock, C., André, E., Walker, M.: Bossy or wimpy: expressing social dominance by combining gaze and linguistic behaviors. In: Intelligent Virtual Agents, pp. 265–271. Springer, New York (2010)
4. Beskow, J., Cerrato, L., Granström, B., House, D., Nordenberg, M., Nordstrand, M., Svanfeldt, G.: Expressive animated agents for affective dialogue systems. In Proceedings of the Tutorial and Research Workshop on Affective Dialogue Systems (ADS'04), pp. 301–304 (2004)
5. Beutnagel, M., Conkie, A., Schroeter, J., Stylianou, Y., Syrdal, A.: The AT&T Next-Generation Text-to-Speech System. In: Meeting of ASA/EAA/DAGA in Berlin, Germany (1999)
6. Bickmore, T.W.: Relational agents: Effecting change through human-computer relationships. PhD thesis, MIT Media Lab (2003)
7. Bickmore, T., Schulman, D.: The comforting presence of relational agents. In: CHI'06 extended abstracts on Human factors in computing systems, pp. 550–555. ACM, Montreal (2006)
8. Bouchard, T.J., McGue, M.: Genetic and environmental influences on human psychological differences. J. Neurobiol. **54**, 4–45 (2003)
9. Brennan, S.E.: Conversations with and through computers. User Modeling and User-Adapted Interaction **1** 67–86 (1991)
10. Brennan, S.E.: Lexical entrainment in spontaneous dialog. In: 1996 International Symposium on Spoken Dialogue, pp. 41–44 (1996)
11. Brennan, S.E., Clark, H.H.: Lexical choice and conceptual pacts in conversation. J. Exp. Psychol.: Learn., Mem. Cognit. **22**(6), 1482–1493 (1996)
12. Cassell, J., Bickmore, T., Negotiated collusion: Modeling social language and its relationship effects in intelligent agents. User Modeling and User-Adapted Interaction **13**, 89–132 (2003)
13. Danlos, L: G-TAG: A lexicalized formalism for text generation inspired by tree adjoining grammar. In: Abeillé, A., Owen Rambow, O., (eds.) Tree Adjoining Grammars: Formalisms, Linguistic Analysis, and Processing. CSLI Publications (2000)
14. Eysenck, S.B.G., Eysenck, H.J., Barrett, P.: A revised version of the psychoticism scale. Personality and Individual Differences, **6**(1), 21–29, (1985)
15. Feng, J., Bangalore, S., Rahim, M.: Webtalk: Mining websites for automatically building dialog systems. In: Proceedings of the IEEE ASR Workshop (2003)
16. Forbes-Riley, K., Litman, D.: Investigating Human Tutor Responses to Student Uncertainty for Adaptive System Development. Lect. Notes Comput. Sci. **4738**, 678–689 (2007)
17. Forbes-Riley, K., Litman, D., Rotaru, M.: Responding to Student Uncertainty During Computer Tutoring: An Experimental Evaluation. Lect. Notes Comput. Sci. **5091** 60–69 (2008)
18. Giles, H., Coupland, N., Coupland, J.: 1. Accommodation theory: Communication, context, and consequence. Contexts of accommodation: Developments in applied sociolinguistics, p. 1, (1991)
19. Goldberg, L.R.: An alternative "description of personality": The Big-Five factor structure. J. Per. Soc. Psychol. **59** 1216–1229 (1990)

20. Hirschberg, J.: Speaking more like you: Lexical, acoustic/prosodic, and discourse entrainment in spoken dialogue systems. In: Proceedings of the 8th SIGdial Workshop on Discourse and Dialogue, p. 128 (2008)
21. Johnston, M., Bangalore, S., Vasireddy, G., Stent, A., Ehlen, P., Walker, M., Whittaker, S., Maloor, P.: MATCH: An architecture for multimodal dialogue systems. In: Annual Meeting of the Association for Computational Linguistics, ACL (2002)
22. Kittredge, R., Korelsky, T., Rambow, O.: On the need for domain communication knowledge. Computat. Intell. **7**(4), 305–314 (1991)
23. Langkilde, I., Knight, K.: Generation that exploits corpus-based statistical knowledge. In: Proceedings of COLING-ACL (1998)
24. Lester, J.C., Converse, S.A., Kahler, S.E., Barlow, S.T., Stone, B.A. Bhogal, R.S.: The persona effect: affective impact of animated pedagogical agents. Proceedings of the SIGCHI conference on Human factors in computing systems, pp. 359–366 (1997)
25. Levelt, W.J.M., Kelter, S.: Surface form and memory in question answering. Cognit. Psychol. **14**, 78–106 (1982)
26. Levin, E., Pieraccini, R.: A stochastic model of computer-human interaction for learning dialogue strategies. In: EUROSPEECH 97 (1997)
27. Lin, G.I., Walker, M.A.: All the world's a stage: Learning character models from film. In: Proceedings of the Seventh AI and Interactive Digital Entertainment Conference, AIIDE '11. AAAI (2011)
28. Litman, D.J., Pan, S.: Predicting and adapting to poor speech recognition in a spoken dialogue system. In: Proc. of the Seventeenth National Conference on Artificial Intelligence, AAAI-2000 pp. 15–21, Austin (2000)
29. Mairesse, F., Walker, M.A.: Towards personality-based user adaptation: psychologically informed stylistic language generation. User Modeling and User-Adapted Interaction **20**(3), 1–52 (2010)
30. Mairesse, F., Walker, M.A.: PERSONAGE: Personality generation for dialogue. In: Proceedings of the 45th Annual Meeting of the Association for Computational Linguistics (ACL), pp. 496–503 (2007)
31. Mairesse, F., Walker, M.A.: Trainable generation of Big-Five personality styles through data-driven parameter estimation. In: Proceedings of the 46th Annual Meeting of the Association for Computational Linguistics (ACL) (2008)
32. Moore, J.D., Paris, C.L.: Planning text for advisory dialogues. In: Proceedings of 27th Annual Meeting of the Association for Computational Linguistics, Vancouver (1989)
33. Mott, B., Lester, J.: Narrative-centered tutorial planning for inquiry-based learning environments. In: Intelligent Tutoring Systems, pp. 675–684. Springer, New York (2006)
34. Neff, M., Toothman, N., Bowmani, R., Tree, J.E.F., and Walker, M. A.: Dont scratch! self-adaptors reflect emotional stability. In: Intelligent Virtual Agents, vol. 6895, Springer, New York (2011)
35. Neff, M., Wang, Y., Abbott, R., Walker, M.: Evaluating the effect of gesture and language on personality perception in conversational agents. In: Intelligent Virtual Agents, pp. 222–235. Springer, New York (2010)
36. Nenkova, A., Gravano, A., Hirschberg, J.: High frequency word entrainment in spoken dialogue. In: Proceedings of ACL-08: HLT. Association for Computational Linguistics (2008)
37. Niederhoffer, K.G., Pennebaker, J.W.: Linguistic style matching in social interaction. J. Lang. Soc. Psychol. **21**, 337–360 (2002)
38. Norman, W.T.: Toward an adequate taxonomy of personality attributes: Replicated factor structure in peer nomination personality rating. J. Abnorm. Soc. Psychol. **66**, 574–583 (1963)
39. Peabody, D., Goldberg, L.R.: Some determinants of factor structures from personality-trait descriptor. J. Pers. Soc. Psychol. **57**(3), 552–567 (1989)
40. Pickering, M., Garrod, S.: Towards a mechanistic theory of dialogue. Behav. Brain Sci. **7**, 77–83 (2003)

41. Pieraccini, R., Levin, E.: A learning approach to natural language understanding. In: Speech Recognition and Coding, New Advances and Trends,NATO ASI Series, pp. 139–155. Springer, New York (1995)
42. Pieraccini, R., Levin, E., Eckert W.: AMICA: The AT&T mixed initiative conversational architecture. In: Eurospeech (1997)
43. Porayska-Pomsta, K., Mellish, C.: Modelling politeness in natural language generation. In: Proceedings of the 3rd Conference on INLG, pp. 141–150, Springer, New York (2004)
44. Rambow, O., Rogati, M., Walker, M.: Evaluating a trainable sentence planner for a spoken dialogue travel system. In: Proceedings of the Meeting of the Association for Computational Lingustics, ACL 2001 (2001)
45. Rambow, O., Korelsky, T.: Applied text generation. In: Proceedings of the Third Conference on Applied Natural Language Processing, ANLP92, pp. 40–47 (1992)
46. Reeves, B., Nass, C.: The Media Equation. University of Chicago Press, Princeton (1996)
47. Reiter, E., Dale, R.: Building Natural Language Generation Systems. Cambridge University Press, Cambridge (2000)
48. Reitter, D., Keller, F., Moore, J.D.: Computational modelling of structural priming in dialogue. Proceedings of Human Language Technology conference-North American chapter of the Association for Computational Linguistics annual mtg, New York (2006)
49. Revelle, W.: Personality processes. Annu. Rev. Psychol. **46**, 295–328 (1991)
50. Scheffler, K., Young, S.: Automatic learning of dialogue strategy using dialogue simulation and reinforcement learning. In: Human Language Technology Conference (2002)
51. Scott, D.S., de Souza, C.S.: Getting the message across in RST-based text generation. In: Dale, R., Mellish, C., Zock, M. (eds.) Current Research in Natural Language Generation. Academic Press, London (1990)
52. Stenchikova, S., Stent, A.: Measuring adaptation between dialogs. Proceedings of the 8th SIGdial Workshop on Discourse and Dialogue (2007)
53. Stent, A., Prasad, R., Walker, M.A.: Trainable sentence planning for complex information presentation in spoken dialog systems. In: Proceedings of the 42nd Annual Meeting of the Association for Computational Linguistics (ACL) (2004)
54. Tapus, A., Tapus, C., Mataric, M.J.: User robot personality matching and assistive robot behavior adaptation for post-stroke rehabilitation therapy. Intell. Serv. Robot. **1**(2), 169–183 (2008) this is really not a very good paper.
55. Tapus, A., Mataric, M.: Socially assistive robots: The link between personality, empathy, physiological signals, and task performance. In: AAAI Spring Symposium, Palo Alto (2008)
56. Van Santen, J., Black, L., Cohen, G., Kain, A., Klabbers, E., Mishra, T., de Villiers, J., Niu, X.: Applications of computer generated expressive speech for communication disorders. In: Proceedings of Interspeech–Eurospeech, pp. 1657–1660, (2003)
57. Walker, M.A., Grant, R., Sawyer, J., Lin, G.I., Wardrip-Fruin, N., Buell, M.: Perceived or not perceived: Film character models for expressive nlg. In: International Conference on Interactive Digital Storytelling, ICIDS'11, Vancouver (2011)
58. Walker, M.A., Rambow, O., Rogati, M.: Training a sentence planner for spoken dialogue using boosting. Comput. Speech Lang. : Special Issue on Spoken Language Generation, **16**(3-4), 409–433 (2002)
59. Walker, M.A., Rambow, O.: Spoken language generation. Comput. Speech Lang., Special Issue on Spoken Language Generation **16**(3-4), 273–281 (2002)
60. Walker, M.A., Stent, A., Mairesse, F., Prasad, R.: Individual and domain adaptation in sentence planning for dialogue. J. Artif. Intell. Res. (JAIR) **30**, 413–456 (2007)
61. Wang, N., Johnson, W.L., Mayer, R.E., Rizzo, P., Shaw, E., Collins, H.: The politeness effect: Pedagogical agents and learning gains. Front. Artif. Intell. Appl. **125**, 686–693 (2005)

Chapter 27
Application and Evaluation of a Conditioned Hidden Markov Model for Estimating Interaction Quality of Spoken Dialogue Systems

Stefan Ultes, Robert ElChab, and Wolfgang Minker

Abstract The interaction quality (IQ) metric has recently been introduced for measuring the quality of spoken dialogue systems (SDSs) on the exchange level. While previous work relied on support vector machines (SVMs), we evaluate a conditioned hidden Markov model (CHMM) which accounts for the sequential character of the data and, in contrast to a regular hidden Markov model (HMM), provides class probabilities. While the CHMM achieves an unweighted average recall (UAR) of 0.39, it is outperformed by regular HMM with an UAR of 0.44 and a SVM with an UAR of 0.49, both trained and evaluated under the same conditions.

27.1 Introduction

Different measures exist for evaluating the quality of spoken dialogue systems (SDSs). *Task completion*, *dialogue duration*, and other objective metrics are unfortunately not human-centered. Subjective measures compensate for this by modeling the user's subjective experience. This subjective information may be used to increase the dialogue system's performance. For example, during each dialogue management cycle, quality information may be used as additional dependency for selecting the next system action (cf. Ultes et al. [15]).

In human-machine dialogues, however, there is no easy way of deriving the user's satisfaction level. Additionally, asking real users for answering questions about the performance of the system requires them to spend more time talking to the machine than necessary for completing the desired task. It can be assumed that a regular user does not want to do this as human-machine dialogues usually have no conversational character but are task oriented.

S. Ultes (✉) • R. ElChab • W. Minker
Institute of Communications Technology, Albert-Einstein-Allee 43, 89081 Ulm, Germany
e-mail: stefan.ultes@uni-ulm.de; robert.el-chab@uni-ulm.de; wolfgang.minker@uni-ulm.de

J. Mariani et al. (eds.), *Natural Interaction with Robots, Knowbots and Smartphones: Putting Spoken Dialog Systems into Practice*, DOI 10.1007/978-1-4614-8280-2_27,

Famous work on determining the satisfaction level automatically is the PARADISE framework by Walker et al. [16]. Assuming a linear dependency between objective measures and user satisfaction (US), a linear regression model is applied to determine US on the *dialogue level*. This is not only very costly, as dialogues must be performed with real users, but also inadequate if quality on a finer level is of interest, e.g., on the *exchange level*. To overcome this issue, work by Schmitt et al. [11] introduced a new metric for measuring the performance of an SDS on the exchange level called interaction quality (IQ).

Human-machine dialogues may be regarded as a process evolving over time. A well-known statistical method for modeling such processes is the hidden Markov model (HMM). Since HMMs do not provide class probabilities, we present an approach for determining IQ using conditioned hidden Markov models (CHMMs). They were originally introduced by Glodek et al. [5] who applied the model to laughter detection on audiovisual data.

For estimating interaction quality, it is important to take into account the IQ value of the previous time step. This is shown by calculating the information gain ratio where the IQ value of the previous time step achieves an IGR of 1.0. While approaches by Schmitt et al. [11] lack of accounting for the temporal dependency on the previous exchange of the new IQ value, Markovian approaches like the CHMM have the ability to deal with this inherently.

In Sect. 27.2, we discuss other work on determining qualitative performance of SDSs and in Sect. 27.3 we present details about the definition of IQ and the data we use. Further, Sect. 27.4 presents a formal description of the CHMM. Evaluation is described in Sect. 27.5 and, finally, Sect. 27.6 concludes this work.

27.2 Related Work

Work on determining user satisfaction using HMMs was performed by Engelbrecht et al. [3]. They predicted US at any point within the dialogue on a five-point scale. Evaluation was performed based on labels the users applied themselves during a Wizard-of-Oz experiment. The dialogue course paused during labeling. They achieved a mean squared error of 0.086.

Further work which incorporates HMMs was presented by Higashinaka et al. [7] using user satisfaction ratings for overall dialogues. HMMs trained for each dialogue rating separately using dialogue act sequences were combined to predict user satisfaction ratings for dialogue acts. By using the Viterbi algorithm for determining the most probable state sequence, the rating assigned to each dialogue act was the overall user satisfaction rating of the HMM this state belongs to. Evaluating in three user satisfaction categories (i.e., "smoothness," "willingness," "closeness") with ratings ranging from 1 to 7, best performance was achieved of 0.192 match rate per rating (MR/r), which is equal to unweighted average recall (UAR, see 27.5.1). The authors compare their approach with HMMs trained on manually annotated exchanges achieving a better performance for the latter.

Higashinaka et al. [6] also presented work on the prediction of turn-wise ratings for human-human (transcribed conversation) and human-machine (text dialogue from chat system) dialogues using both HMM and conditional random fields (CRFs). Ratings ranging from 1 to 7 were applied by two expert raters labeling "smoothness," "closeness," and "willingness" not achieving a MR/r of more than 0.2–0.24 for HMMs with CRFs even performing worse. This only slightly outperforms the random baseline of 0.14.

Dealing with true user satisfaction, Schmitt et al. presented their work about statistical classification methods for automatic recognition of US [12]. The data was collected in a lab study where the users themselves had to rate the conversation during the ongoing dialogue. Labels were applied on a scale from 1 to 5. By applying a support vector machine (SVM), they achieved an unweighted average recall (UAR) of 0.49.

27.3 Interaction Quality

The LEGO corpus, which was published by Schmitt et al. [13] and which contains calls to the "Let's Go Bus Information System" of the Carnegie Mellon University in Pittsburgh [10], was used. It contains 347 calls recorded in 2006, out of which 200 calls have been labeled with interaction quality by three expert raters. By that, an IQ label on a scale from 1 (extremely unsatisfied) to 5 (satisfied) was assigned to a total of 4,885 system-user exchanges. As the users are expected to be satisfied at the beginning, each dialogue's initial rating is 5.

Parameters used as input variables for the IQ model have been derived from the dialogue system modules automatically for each exchange. Further, parameters on three levels have been created: the *exchange level*, the *dialogue level*, and the *window level*. As parameters like ASRCONFIDENCE or UTTERANCE can directly be acquired from the dialogue modules, they constitute the *exchange level*. Based on this, counts, sums, means, and frequencies of *exchange level* parameters from multiple exchanges are computed to constitute the *dialogue level* (all exchanges up to the current one) and the *window level* (the three previous exchanges).

Schmitt et al. [11] performed IQ recognition on this data using SVMs. They achieved an unweighted average recall (UAR) of 0.58.

27.4 CHMM

Conditioned hidden Markov models [5] are an extension of regular HMMs. They provide probabilities for multiple classes. A sequence diagram illustrating the principle operation method of the CHMM in the time domain is shown in Fig. 27.1.

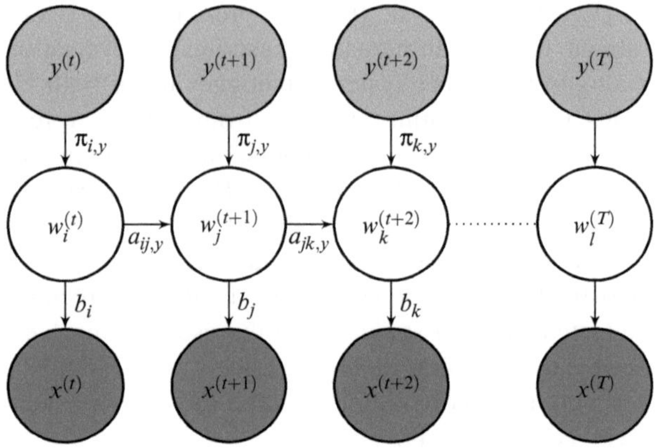

Fig. 27.1 General graphical representation of the CHMM model in the discrete time domain. For each time step t, $y^{(t)}$ represents the most likely label and $w_i^{(t)}$ the most likely hidden state given observation $x^{(t)}$. b_i represents the probability for the observation and $\pi_{i,y}$ the label probability. $a_{ij,y}$ defines the probability of transitioning from state $w_i^{(t)}$ to state $w_j^{(t+1)}$

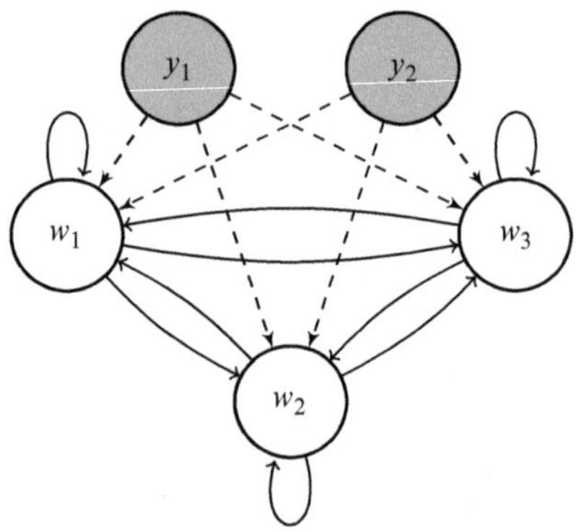

Fig. 27.2 This is an example of a CHMM with 2 labels and 3 hidden states. The dashed lines represent the label dependence of the hidden states, while the full lines illustrate state transitions. Please note that state transitions also depend on the labels which is not shown here

27.4.1 Model Description

Like the continuous HMM, the CHMM also consists of a discrete set of hidden states $w_i \in W$ and a vector space of observations $\mathbb{X} \subseteq \mathbb{R}^n$. A separate emission probability $b_i(\mathtt{x}^{(t)})$ is linked to each state defining the likelihood of observation $\mathtt{x}^{(t)} \in \mathbb{X}$ at time t while being in state w_i. Further, $a_{ij,y} = p(\mathtt{w}^{(t)} = w_j | \mathtt{w}^{(t-1)} = w_i, \mathtt{y}^{(t)} = y)$ defines the transition probability of transitioning from state w_i to w_j. In contrast to the regular HMM, the transition probability distribution also depends on the class label $y \in Y$. This results in the transition matrix $\mathbf{A} \in \mathbb{R}^{|W| \times |W| \times |Y|}$.

Furthermore, the meaning of the initial probability $\pi_{i,y} = p(\mathtt{w}^{(1)} = w_i | \mathtt{y}^{(1)} = y)$ for state w_i is altered. It additionally represents the label probability for label y at any time with the corresponding matrix $\pi \in \mathbb{R}^{|W| \times |y|}$. An schematic example of a CHMM with two labels and three hidden states is illustrated in Fig. 27.2.

According to Glodek et al. [5], the likelihood of an observation sequence $\mathtt{x}^{(n)}$ with corresponding label sequence $\mathtt{y}^{(n)}$ is given by

$$\begin{aligned}
& p(\mathtt{x}^{(n)}, \mathtt{w}^{(n)} | \mathtt{y}^{(n)}, \lambda) \\
& \quad = \sum_{w \in W} p(\mathtt{w}^{(1)} = w | \mathtt{y}^{(1)}, \pi) \\
& \quad \cdot \prod_{t=2}^{T} p(\mathtt{w}^{(t)} = w_j | \mathtt{w}^{(t-1)} = w_i, \mathtt{y}^{(t)}, \mathbf{A}) \\
& \quad \cdot \prod_{t=1}^{T} p(\mathtt{x}^{(t)} | \mathtt{w}^{(t)} = w_j, \theta),
\end{aligned} \tag{27.1}$$

where $\mathtt{w}^{(n)}$ denotes the sequence of the hidden states. Further, in this work, the emission probability $b_j(\mathtt{x}^{(t)}) = p(\mathtt{x}^{(t)} | \mathtt{w}^{(t)} = w_j, \theta)$ is modeled as a Gaussian mixture model (GMM) with the parameter set $\theta = \{\{\phi_{j,k}\}_k^K, \{\mu_{j,k}\}_k^K, \{\Sigma_{j,k}\}_k^K\}$. The parameter set λ describing the complete CHMM is defined by $\lambda = \{\pi, \mathbf{A}, \theta\}$.

27.4.2 *Learning*

The learning phase consists of two parts: *initialization* and *training*.

For *initialization*, the k-means algorithm [4] is used and the number of clusters k corresponds to the number of hidden states. After clustering initial observation sequences with their corresponding label sequences, the transition probabilities are updated according to the transitions between the clusters, given the labels. The initial probabilities are updated according to the cluster and the corresponding label that each element belongs to.

Training is performed using the Baum-Welch algorithm, which is heavily dependent on the initialization. When comparing the HMM explained by Rabiner et al. [9] to the CHMM, several changes[1] must be applied to the Baum-Welch algorithm.

The αs and βs of the forward-backward algorithm according to Glodek et al. [5] are

$$\alpha_{t,y}(j) = b_j(\mathtt{x}^{(t)}) \cdot \sum_{i \in W} a_{ij,y} \cdot \alpha_{t-1,y}(i) \tag{27.2a}$$

$$\alpha_{1,y}(j) = b_j(\mathtt{x}^{(1)}) \cdot \pi_{j,y} \tag{27.2b}$$

$$\beta_{t,y}(i) = \sum_{j \in W} a_{ij,y} \cdot b_j(\mathtt{x}^{(t+1)}) \cdot \beta_{t+1,y}(j) \tag{27.3a}$$

$$\beta_{T,y}(i) = 1 \tag{27.3b}$$

[1]Changes in Eq.: 19, 20, 24, 25, 27, 37, 40a, 40b, and 40c from [9]

$$\beta_{0,y}(i) = \sum_{j \in W} \pi_{j,y} \cdot b_j(\mathrm{x}^{(1)}) \cdot \beta_{1,y}(j) \tag{27.3c}$$

The state beliefs $\gamma_{t,y}(j)$ and the transition beliefs $\xi_{t-1,t,y}(i,j)$ are then computed by using

$$\gamma_{t,y}(j) = \frac{\alpha_{t,y}(j) \cdot \beta_{t,y}(j)}{p(X)} \tag{27.4}$$

$$\xi_{t-1,t,y}(i,j) = \frac{\alpha_{t-1,y}(i) \cdot b_j(\mathrm{x}^{(t)}) \cdot a_{ij,y} \cdot \beta_{t,y}(j)}{p(X)} \tag{27.5}$$

where $\sum_{t=1}^{T-1} \gamma_{t,y}(i)$ is the expected number of transitions from w_i given y and $\sum_{t=1}^{T-1} \xi_{t-1,t,y}(i,j)$ is the expected number of transitions from w_i to w_j given y.

Parameter learning is performed after evaluation of N sequences, updating the initial probabilities using the following formula:

$$\begin{aligned}\pi_{i,y} &= \frac{\text{expected number of times of being in } w_i \text{ at time } t = 1 \text{ given } y}{\text{expected number of times of being in all } w \text{ at } t = 1 \text{ given } y} \\ &= \frac{\sum_{l=1}^{N} \delta_{\mathbf{y}^{(1)(l)}=y} \gamma_{1,\mathbf{y}^{(n)(l)}}(i)}{\sum_{l=1}^{N} \sum_{j \in w_1} \delta_{\mathbf{y}^{(1)(l)}=y} \gamma_{1,\mathbf{y}^{(n)(l)}}(j)}\end{aligned} \tag{27.6}$$

where $\sum_{i=1}^{n} \pi_{i,y} = 1$ and δ is the Kronecker delta.

The update for the transition probabilities after evaluating N sequences is

$$\begin{aligned}a_{ij,y} &= \frac{\text{expected number of transitions from } w_i \text{ to } w_j \text{ given } y}{\text{expected number of transitions from } w_i \text{ given } y} \\ &= \frac{\sum_{l=1}^{N} \sum_{t=0}^{T-1} \xi_{t-1,t,\mathbf{y}^{(n)(l)}}(i,j) \delta_{\mathbf{y}^{(t)(l)}=y}}{\sum_{l=1}^{N} \sum_{t=0}^{T-1} \gamma_{t,\mathbf{y}^{(n)(l)}}(j) \delta_{\mathbf{y}^{(t)(l)}=y}}\end{aligned} \tag{27.7}$$

where

$$\forall_{y \in Y} \sum_{j=1}^{n} a_{ij,y} = 1.$$

The emission probabilities can be computed in accordance with the methods presented by Rabiner et al. [9]. As the state beliefs depend on y, a sum over all labels has to be applied in order to create label-independent emission probabilities.

27.5 Evaluation

The Viterbi algorithm generates a sequence of expected labels which are evaluated with the metrics described in the following.

27.5.1 Metrics

The **unweighted average recall** (UAR) is defined as the sum of all class-wise recalls r_c divided by the number of classes $|C|$:

$$UAR = \frac{1}{|C|} \sum_{c \in C} r_c, \tag{27.8}$$

where recall r_c for class c is defined as

$$r_c = \frac{1}{|R_c|} \sum_{i=1}^{|R_c|} \delta_{h_i r_i} \tag{27.9}$$

with δ as the Kronecker delta, h_i and r_i as the corresponding hypothesis-reference pairs of ratings, and $|R_c|$ as the total number of all ratings of class c. In other words, UAR for multi-class classification problems is the accuracy corrected by the effects of unbalanced data.

To measure the relative agreement between two corresponding sets of ratings we apply **Cohen's kappa** [1]. It is defined by the number of label agreements corrected by the chance level of agreement divided by the maximum proportion of times the labelers could agree is computed. In order to take account for ordinal scores, a weighting factor w is introduced reducing the discount of disagreements the closer the ratings are together [2]:

$$w = \frac{|r_1 - r_2|}{range} \tag{27.10}$$

Here, r_1 and r_2 denote the rating pair and *range* the maximum distance which may occur between two ratings. This results in $w = 0$ for agreement and $w = 1$ if the ratings differ the most.

For measuring the correlation between two variables, *Spearman's rank correlation coefficient* is used [14]. It is a nonparametric method assuming a monotonic function between the two variables, defined by

$$\rho = \frac{\sum_i (x_i - \bar{x})(y_i - \bar{y})}{\sqrt{\sum_i (x_i - \bar{x})^2 \sum_i (y_i - \bar{y})^2}}, \tag{27.11}$$

Table 27.1 Results for CHMM experiments according to the number of hidden states along with results for regular HMM and SVM classification. The '∗' indicates the best result

Class	# states	UAR	Kappa	Rho
SVM	–	0.49	0.61	0.77
HMM	5	0.44	0.56	0.72
CHMM	5	0.38	0.40	0.56
	6	0.38	0.39	0.57
	7	0.35	0.40	0.59
	8	0.37	0.41	0.59
	9*	0.39	0.43	0.60
	10	0.37	0.39	0.55
	11	0.36	0.41	0.58

where x_i and y_i are corresponding ranked ratings and $\bar{x}$ and $\bar{y}$ the mean ranks. Therefore, two sets of ratings may have total correlation even if they never agree. This would happen if all ratings are shifted by the same value, for example.

27.5.2 Setup and Results

For the experiments, we used the LEGO corpus presented in Sect. 27.3. Since the values of multiple parameters are constant for most exchanges, they are excluded. Otherwise, this would have resulted in rows of zeros during computation of the covariance matrices of the feature vectors. A row of zeros in the covariance matrix will make it irreversible, which will cause errors during the computation of the emission probabilities.

The model operated with a vector of 29 dimensions. The results of the experiments are presented in Table 27.1. The data is ranked according to the number of hidden states used for the model. The accuracy decreased remarkably after passing the threshold of nine states, where the highest values for UAR, κ, and ρ could be achieved.

The results are computed using sixfold cross validation. Best performance was achieved for nine states with an UAR of 0.39, Cohen's κ of 0.43, and Spearman's ρ of 0.60.

To define a baseline, we rely on the approach by Schmitt et al. [11]. Using the same features, we trained a SVM with a linear kernel. The results are shown in Table 27.1. Unfortunately, the CHMM approach was not able to outperform the baseline. This is most likely caused by the fact that only little training data was available. For 9 hidden states, a total of 8,280 parameters (initial probabilities, transition probabilities, and mean and covariance matrices of the emission probabilities) have to be learned. Calculating these with a total of 3,908 training vectors (per fold) results in less than one vector per parameter on average. This has shown to be insufficient training data to create good estimates for the parameters. In order to further evaluate the CHMM on this task, additional data has to be acquired.

Furthermore, we conducted an experiment using regular HMMs. Using 5 hidden states, we assigned a label to each hidden state. As depicted in Table 27.1, the CHMM performed worse than the HMM approach. Though performance of both models is dramatically influenced by the lack of data, the CHMM is rather prone to this, as additional probability models have to be estimated to take account for the additional dependency on the class labels.

For future work, additional attribute selection based on IGR, for example, may be applied in order to reduce the dimension of the input vector and therefore reduce the number of parameters which have to be trained. Moreover, other estimation approaches like conditional random fields [8], which have shown to work well on sequential data in other applications, may be more suitable for the given task.

27.6 Conclusions

As dialogues have a sequential structure, an approach for estimating interaction quality on the exchange level has been evaluated using conditioned hidden Markov models. Experiments were conducted for measuring its performance. The best result with 9 hidden states is outperformed vastly by previously presented methods based on SVM classification. While attribute selection may have an effect on the performance, we identified the lack of training data as the major cause for these results, as each parameter had to be estimated by an average of less than one training vector. Further studies investigating the effect of attribute selection as well as using more data should be undertaken in order to get more significant results.

References

1. Cohen, J.: A coefficient of agreement for nominal scales. In: Educational and Psychological Measurement, vol. 20, pp. 37–46 (1960)
2. Cohen, J.: Weighted kappa: Nominal scale agreement provision for scaled disagreement or partial credit. Psychol. bull. **70**(4), 213 (1968)
3. Engelbrecht, K.P., Gödde, F., Hartard, F., Ketabdar, H., Möller, S.: Modeling user satisfaction with hidden markov model. In: SIGDIAL '09: Proceedings of the SIGDIAL 2009 Conference, pp. 170–177. Association for Computational Linguistics, Morristown, (2009)
4. Faber, V.: Clustering and the continuous k-means algorithm. Los Alamos Science (22), 138–144 (1994)
5. Glodek, M., Scherer, S., Schwenker, F.: Conditioned hidden markov model fusion for multimodal classification. In: Proceedings of the 12th Annual Conference of the International Speech Communication Association (INTERSPEECH 2011), pp. 2269–2272. International Speech Communication Association (2011)
6. Higashinaka, R., Minami, Y., Dohsaka, K., Meguro, T.: Issues in predicting user satisfaction transitions in dialogues: Individual differences, evaluation criteria, and prediction models. In: Lee, G., Mariani, J., Minker, W., Nakamura, S. (eds.) Spoken Dialogue Systems for Ambient Environments, Lecture Notes in Computer Science, vol. 6392, pp. 48–60. Springer, Berlin (2010). 10.1007/978-3-642-16202-2_5

7. Higashinaka, R., Minami, Y., Dohsaka, K., Meguro, T.: Modeling user satisfaction transitions in dialogues from overall ratings. In: Proceedings of the SIGDIAL 2010 Conference, pp. 18–27. Association for Computational Linguistics, Tokyo (2010)
8. Klinger, R., Tomanek, K.: Classical probabilistic models and conditional random fields. Tech. rep., Algorithm Engineering, Faculty of Computer Science, Dortmund (2007). TR07-2-013
9. Rabiner, L.R.: A tutorial on hidden Markov models and selected applications in speech recognition. Morgan Kaufmann Publishers Inc., San Francisco (1989)
10. Raux, A., Bohus, D., Langner, B., Black, A.W., Eskenazi, M.: Doing research on a deployed spoken dialogue system: One year of lets go! experience. In: Proceedings of the International Conference on Speech and Language Processing (ICSLP) (2006)
11. Schmitt, A., Schatz, B., Minker, W.: Modeling and predicting quality in spoken human-computer interaction. In: Proceedings of the SIGDIAL 2011 Conference. Association for Computational Linguistics, Portland (2011)
12. Schmitt, A., Schatz, B., Minker, W.: A statistical approach for estimating user satisfaction in spoken human-machine interaction. In: Proceedings of the IEEE Jordan Conference on Applied Electrical Engineering and Computing Technologies (AEECT). IEEE, Amman (2011)
13. Schmitt, A., Ultes, S., Minker, W.: A parameterized and annotated corpus of the cmu let's go bus information system. In: International Conference on Language Resources and Evaluation (LREC) (2012)
14. Spearman, C.: The proof and measurement of association between two things. Am. J. Psychol. **15**, 88–103 (1904)
15. Ultes, S., Schmitt, A., Minker, W.: Towards quality-adaptive spoken dialogue management. In: NAACL-HLT Workshop on Future directions and needs in the Spoken Dialog Community: Tools and Data (SDCTD 2012), pp. 49–52. Association for Computational Linguistics, Montréal, (2012). URL http://www.aclweb.org/anthology/W12-1819
16. Walker, M., Litman, D., Kamm, C.A., Abella, A.: Paradise: a framework for evaluating spoken dialogue agents. In: Proceedings of the eighth conference on European chapter of the Association for Computational Linguistics, pp. 271–280. Association for Computational Linguistics, Morristown (1997). DOI 10.3115/979617.979652

Chapter 28
FLoReS: A Forward Looking, Reward Seeking, Dialogue Manager

Fabrizio Morbini, David DeVault, Kenji Sagae, Jillian Gerten, Angela Nazarian, and David Traum

Abstract We present FLoReS, a new information-state-based dialogue manager, making use of forward inference, local dialogue structure, and plan operators representing subdialogue structure. The aim is to support both advanced, flexible, mixed initiative interaction and efficient policy creation by domain experts. The dialogue manager has been used for two characters in the SimCoach project and is currently being used in several related projects. We present the design of the dialogue manager and preliminary comparative evaluation with a previous system that uses a more conventional state chart dialogue manager.

28.1 Introduction

In this paper we present a new information-state-based dialogue manager called FLoReS (Forward Looking, Reward Seeking). FLoReS has been designed to provide flexible, mixed initiative interaction with users, while at the same time supporting the efficient creation of dialogue policies by domain experts.

The development of new frameworks and techniques that can streamline the creation of effective dialogue policies is an important issue for spoken dialogue systems. One common approach in practical applications is to adopt a strong system initiative design, in which the system steers the conversation and generally prompts the user for what to say at each point. System initiative policies can be relatively easy for authors to create and can also simplify the system's language understanding problem at each point. For authoring, a system that will behave as desired can be specified with simple structures, such as a call flow graph [12] and branching

F. Morbini (✉) • D. DeVault • K. Sagae • J. Gerten • A. Nazarian • D. Traum
Institute for Creative Technologies, University of Southern California,
Los Angeles, CA 90094, USA
e-mail: morbini@ict.usc.edu; devault@ict.usc.edu; sagae@ict.usc.edu; gerten@ict.usc.edu; nazarian@ict.usc.edu; traum@ict.usc.edu

J. Mariani et al. (eds.), *Natural Interaction with Robots, Knowbots and Smartphones: Putting Spoken Dialog Systems into Practice*, DOI 10.1007/978-1-4614-8280-2_28,

narrative for interactive games [15]. A strong system initiative system also reduces the action state space and generally reduces the perplexity of the understanding process, since user actions are only allowed at certain points in the dialogue and usually are limited to a reduced set of options.

Strong system initiative systems can work well if the limited options available to the user are what the user wants to do, but can be problematic otherwise, especially if the user has a choice of whether or not to use the system. In particular, this approach may not be well suited to a virtual human application like SimCoach [13]. In the SimCoach system, which we describe in Sect. 28.2, a virtual human is designed to be freely available to converse with a user population on the web, but its users may choose not to use the system at all if it does not respond to their needs and alleviate their concerns in the dialogue in a direct and efficient way.

At the other extreme, pure user initiative systems allow the user to say anything at any time, but have fairly simple dialogue policies, essentially just reacting to each user utterance individually, in a way that is not very sensitive to context, e.g., [8]. These systems can be easy for authors to design and can work well when the user is naturally in charge, such as in a command and control system, or interviewing a character [9], but may not be suitable for situations like SimCoach, in which the character should sometimes (but not always) take the initiative and ask the user questions, or generally, when mixed initiative is desired. The Tactical Questioning system architecture and authoring environment [4] provides an alternative for systems in which the user is generally in charge and asking the system questions, but policy rules can be easily authored to provide some context sensitivity to the system's responses.

True mixed initiative is notoriously difficult for a manually constructed call flow graph, in which the same sorts of options may appear in many places, but the system might want to take different actions, depending on local utilities. An alternative approach is for authors to develop complex hand-authored rules to achieve a dialogue policy with the desired mixed initiative behavior. These rules will govern system initiative decisions and information state updates [7]; however, in practice, this development often involves a number of formal modeling tasks that require substantial dialogue modeling expertise as well as programming skills, which many potential dialogue system authors and domain experts do not possess [1]. Reinforcement learning approaches [2, 18] can be very useful at learning local policy optimizations from data rather than hand-authored rules, but they require large amounts of training data, possibly using simulated users [5]. While this approach can take some of the burden off the author, it also removes some of the control, which can be undesirable [11]. Moreover, it is very difficult to apply for large state-spaces.

Our approach, embodied in the FLoReS dialogue manager, is to create a forward looking, reward seeking agent with support for complex, mixed initiative dialogue interaction and rich dialogue policy authoring. FLoReS combines several methods of dialogue reasoning, to promote the twin goals of flexible, mixed initiative interaction and tractable authoring by domain experts and creative authors. Authoring involves design of local subdialogue networks (called *operators*) for specific conversation topics. Within a subdialogue network, authors can craft the specific structure of interaction. The higher-level structure of the dialogue, which determines the flow of initiative and topics as the dialogue progresses, is determined at run-time by the dialogue manager. This is realized in a reward-sensitive plan-based paradigm:[1] the subdialogues are given preconditions and effects, to decide where they may be applicable, and also qualitative reward categories (goals), which can be assigned to quantitative reward values. The dialogue manager locally optimizes its policy decisions by calculating the highest overall expected reward for the best sequence of subdialogues from a given point.

The rest of this paper is structured as follows. Section 28.2 describes two of the systems currently using FLoReS. Section 28.3 describes the dialogue manager components, while Sect. 28.4 describes the dialogue manager execution. Section 28.5 describes the evaluation (currently still in progress), comparing the FLoReS version of Bill Ford with a previous version of the character, using a more traditional state chart and directed dialogue, and current limitations of FLoReS. We conclude in Sect. 28.6, describing current work.

28.2 Virtual Human Applications Using FLoReS

SimCoach [13] is the first system to use FLoReS for dialogue management. One example of a SimCoach character, named Bill Ford, is shown in Fig. 28.1. SimCoach is motivated by the challenge of empowering troops and their significant

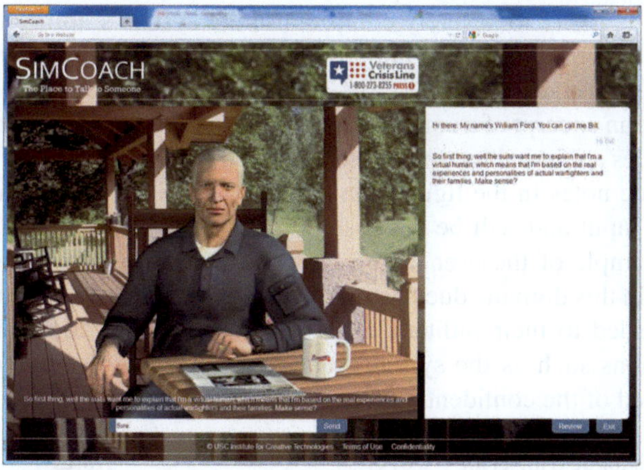

Fig. 28.1 Bill Ford, a SimCoach character. SimCoach virtual humans are accessible through a web browser. The user enters natural language input in the text field on the bottom of the screen. The SimCoach responds with text, speech, and character animation. The text area to the *right* shows a transcript of the dialogue

[1]This approach is similar at a high level to [10], though they do not address its use in dialogue planning specifically.

others in regard to their healthcare, especially with respect to issues related to the psychological toll of military deployment. The SimCoach project is developing virtual human dialogue systems to engage soldiers and their families who might otherwise not seek help for possible mental and behavioral health issues. The reluctance to seek help may be due to stigma or lack of awareness. SimCoach virtual humans are not designed to act as therapists or dispense medical advice, but rather to encourage users to explore available options and seek treatment when needed. To achieve this goal, we aim to provide a safe and anonymous environment, where users can express their concerns to an artificial conversational partner without fear of judgment or possible repercussions.

SimCoach virtual humans present a rich test case for the FLoReS dialogue manager. To be convincing to a military user population, aspects of the system's content need to be authored by someone familiar with the military domain. At the same time, psychological expertise is needed in order to offer appropriate suggestions and helpful information. To do this, the system assesses the user's needs and potential issues using variants of standard psychometric questionnaires such as PCL, the PTSD checklist [17]; the use of such standard instruments requires psychological expertise for authoring of both policies and utterances. Authoring therefore needs to be feasible by military and psychological experts, who cannot be expected to have detailed knowledge of dialogue modeling and natural language processing.

Finally, the dialogue system must be able to take initiative when trying to collect the information it needs to help the user, such as responses to questionnaire questions. But it also must respond appropriately when the user takes initiative. Given the sensitive nature of the topics covered in the system, responding poorly to user initiative is likely to severely compromise the system's ability to accomplish its goal to help the user. Mixed initiative is therefore key to the design goals of SimCoach.

We provide an excerpt of a conversation with Bill Ford in the following table. The user's utterances are in italics, and the subdialogue networks are highlighted in gray boxes. The side notes in the figure summarize FLoReS' decision making for each received user input and will be discussed further as we present FLoReS's design, below. An example of the user taking the initiative is the user's utterance *Is this confidential?* In this domain, due to stigma, users often worry that their conversation could be revealed to their military chain of command, and they might refuse to answer questions such as the system's question, *Have you been deployed?*, until they are assured of the confidentiality of their conversation.

Dialogue transcript	Notes
Greetings and Introduction ...What do you think?	This network is activated by the initial event sent when a new conversation is started. No other networks are available to handle that event.
User: I'm always tired.	This is a possible symptom and is stored in the information state

Continued on next page

continued from previous page

Dialogue transcript	Notes
Feedback This must be hard for you.	Here the system could pick the prequestionnaire network you see later in the conversation or give empathetic feedback to the symptom just mentioned. The DM selects to give feedback because the reward is available only if the network is used immediately after a symptom and that path in the search graph achieves a higher expected reward.
Deployed QA Have you been deployed?	In the SimCoach domain, being tired could be an indicator of two conditions; to disambiguate them, the system needs to know whether the user has been deployed. Even though asking that kind of personal question to the user has a negative reward, the system selects this action because it enables future networks with high reward.
User: Is this confidential?	
Confidentiality QA All your information will be kept extremely confidential. I'm not interested in your name, just in how I can help. Did that answer your questions? *User: Yes it did.* Great.	Here the user doesn't respond to the system question, but asks his own question. The system computes the expected rewards of answering the user and that of ignoring the question. In this case, answering the user comes out on top.
Deployed QA Have you been deployed? *User: Yes.*	After the user completes the confidentiality network, the system recomputes the rewards and decides to ask one more time the deployment question to the user.
Prequestionnaire Can you tell me a bit more about that? *User: Not really.* Okay, you may not be real excited about it, but I've got some questions to ask you so I can help you better. First off, has your sleeping been troubled lately? *User: Sure I have nightmares* I know you mentioned it before, but I want to make sure I got it right. You're having nightmares? *User: Yes* • • •	Answering the deployment question disambiguates the initial symptom and that enables the prequestionnaire network. Here you can see the event listeners in action to update the information state when the user replies to the question about sleeping problems with having nightmares. That reply modifies the way in which the system asks the next question about nightmares to a confirmation.

Fig. 28.2 Ellie is a virtual character similar in purpose to Bill Ford but equipped with multimodal sensing to improve both the interaction with the user and the detection of possible health problems

Another virtual human using the FLoReS DM is Ellie (see Fig. 28.2). Ellie is in an early stage of development, and is designed to address similar problems as those targeted by the SimCoach project. However, Ellie will use speech input and multimodal sensing using a Kinect sensor and cameras in order to have a more compelling conversation with the user, and also to more effectively detect signs of possible health problems (e.g., signs of depression and post-traumatic stress disorder). Information extracted both from the audio (e.g., recognized speech) and visual signals (e.g., movements, facial expression, gaze) will be integrated and used by FLoReS to select what the virtual human will do.

28.3 FLoReS Dialogue Manager Components

In this section, we describe the main components in the FLoReS dialogue manager (DM), which are:

- An **information state** [16], including information about what has happened in the dialogue so far, and information relevant for continuing the dialogue.
- A set of **inference rules** that allows the system to add new knowledge to its information state, based on logical reasoning.
- An **event handling** system, that allows the information state to be updated based on user input, system action, or other classes of author-defined events (such as system timeouts).
- A set of **operators**, that allow the system to model local and global dialogue structure and plan the best continuation.

We will discuss dialogue policy execution in Sect. 28.4.

28.3.1 Inference Rules

FLoReS allows the dialogue system author to specify a set of implications that can be used to infer new knowledge given the current information state. For example, toward the end of the above dialogue excerpt, the user says *Sure i have nightmares*. In the information state, there is a variable that captures whether the user has nightmares, and another variable that captures whether the user has sleeping problems more generally. An author can define an implication that if the user has nightmares, then they also have sleeping problems. These implications are generally expressed in the form of if-then-else conditionals that can test and set the values of information state variables. The implications are evaluated, and the information state is updated, each time the information state is changed.

28.3.2 Event Handling and Dialogue Acts

Most updates to the information state are ultimately triggered by *events*. There are several different kinds of events, including those corresponding to user input (produced by the system's NLU module), system generated events (corresponding to decisions and actions made by the system), and external events (such as timeouts or other perceptual information).

Events received by FLoReS have a name and can have a content. For example, for user events, the name is a dialogue act, and the content is a set of key-value pairs representing the semantic content of the utterance.

Event listeners have a matching expression and a resulting action. The matching expression defines the events being listened for, which can include full event names or a regular expression, focusing on one or more fields of the name. The resulting action is an assertion of a variable value or an if-then-else conditional. For example, when the user says *I'm always tired.* in the example above, an event handler matches the recognized NLU dialogue act (which is represented as `answer.observable.tired`), and increments a counter representing the number of times the user has mentioned being tired.

28.3.3 Operators

The main part of a FLoReS dialogue policy is a set of subdialogue networks, which are formalized as *operators*. The gray boxes in the above dialogue excerpt indicate the different operators that are active during the dialogue. Our definition of operators is motivated by the desire to encourage re-usability of subdialogues across multiple policies, allow authors to craft short portions of conversations, and maintain local coherence. Operators represent local dialogue structure, and can also be thought of as reusable subdialogues. Each operator contains a subdialogue structure, composed of a tree of system and/or user actions and resulting states.[2] A system action can be (1) an update to the information state, (2) a generated utterance to perform back to the user (using speech and animation), or (3) a command to send a particular event to the DM itself (which is later handled by an event handler).

Each state within the subdialogue can be associated with a *goal* and thus a *reward* for reaching that state.[3] The system's goals, such as providing the user with relevant information, are the main factor used to decide what to do when there is more than one applicable operator. Each goal is associated in the information state with a specific numeric reward value. Goals are attached to states within the subdialogue, and their corresponding rewards are thus associated with those states.

Like AI planning operators,[4] operators can have preconditions and effects. Effects specify changes to the information state and can occur at any state within the operator. The preconditions define when an operator can be activated. Preconditions are divided into three types (an operator can have any number of each type):

1. **System initiative** preconditions specify that the system may elect to activate this operator on system initiative, regardless of what the user has recently said.
2. **User initiative** preconditions specify when an operator can be used as part of the processing of a user utterance.[5]
3. **Re-entrance** preconditions enable the DM to resume an operator that was previously interrupted (e.g., by user initiative).

Because an operator can contain a complex subdialogue, each precondition, in addition to specifying when the operator can be initiated (or continued), also defines the state at which it enters the operator's subdialogue tree.

Operators can also be associated with topics, and preconditions can test the current topic (for example) in order to decide whether an operator can be used.

28.4 Dialogue Policy Execution

When each event is received, the event listeners are checked and matching listeners execute their resulting actions, leading to updates to the information state. For each change to the information state, the inference rules are evaluated repeatedly until the information state is stable.

Then the dialogue manager decides which operator to use to best deal with the received event. At any point, the set of available operators is divided in three:

---

[2]The tree structure excludes cycles within subdialogue networks.

[3]Currently each operator must include at least one reward, somewhere in the subdialogue.

[4]Our operators are nonparametric like propositional STRIPS operators [3].

[5]If an operator has a satisfied user initiative precondition for a particular event *e*, the operator is said to *handle e*.

(1) the currently active operator (if any), (2) a set of paused operators; these operators were once active but have been interrupted before completion and put in a paused state, and (3) the set of inactive operators: the remaining operators that are neither active nor paused.

If the current active operator can handle the received event, the DM just continues its execution, traversing the subdialogue of the operator to the next state. An example of this is in the **Deployed QA** operator in the above dialogue excerpt, where the user's answer *Yes.* is handled by the active operator.

Otherwise, the dialogue manager computes the expected rewards for the following cases:

- *Ignore the received event:* here the dialogue manager searches for the most promising system initiative operators. Two sub-searches are executed: one considers keeping the current active operator as it is, and the other considers switching to any of the other system initiative operators.
- *Handle the received event:* here the dialogue manager searches for the most promising operator among those that are paused or inactive and that handle the received event.

An example in the above dialogue excerpt arises with the user's utterance of *Is this confidential?*, which cannot be handled by the active **Deployed QA** operator. The dialogue manager then switches to the inactive **Confidentiality QA** operator, which can handle it.

The expected reward of a given operator, O, starting from the current information state I, is computed by simulating future dialogues that can happen if O is activated. The simulation is executed breadth first and builds a graph where the nodes are possible information states and the edges are operators.

Because an operator represents a subdialogue, it can produce multiple resulting information states depending on the subdialogue branch that is traversed. The multiple possible information states are weighted by the probability with which each state can be reached.[6]

To compute the possible information states that an operator can produce when executed, the DM also considers the updates executed by event listeners. The simulation constructs a graph instead of a tree because the nodes corresponding to information states with the same content are merged into a single node. Arcs that would cause loops are not added. The simulation that builds this graph of possible information states continues until a termination criterion is satisfied. Currently this criterion is based on a maximum depth of the graph and a timeout. The expected reward is computed for all operators in this graph whose head is the current information state (i.e., the root node) and the operator with the highest expected reward is made active and executed. The currently active operator is made paused (if possible, otherwise is made inactive).

[6]Because multiple branches can produce the same final information state, this weight is not simply the uniform weight obtained by $1/|leaves|$).

The formula to compute the expected utility of operator O_i in information state I is

$$E[O_i,I] = \sum_{I_i \in I_r} (\alpha \cdot P(I_i) \cdot R(O_i,I_i) + \underset{O}{\text{argmax}}(E[O,I_i]))$$

where α is a discount factor for future rewards; I_r is the set of possible information states that can be reached from I by executing the subdialogue contained in O_i. $P(I_i)$ is the probability of reaching the information state I_i from I when executing O_i. Currently this probability is just based on uniformly distributing the probability across all paths of the subdialogue of a given operator and merging the paths that produce the same final information state. $R(O,I_i)$ is the reward realized by traversing the path in the subdialogue associated with the operator O that produces the final information state I_i. The final term in the formula calculates the maximum expected reward from operators that could be activated after operator O_i. This allows the immediate selection of operators that have low immediate reward, but whose effects enable higher reward operators in the future. As noted in the above dialogue excerpt, an example of this is when the system asks *Have you been deployed?* in the excerpt.

28.4.1 Mixed Initiative

This architecture allows for mixed initiative and opportunistic action selection based on reward. When the user takes the initiative, as in asking *Is this confidential?* in the above dialogue excerpt, the system can choose whether to follow the user's initiative or not according to the expected rewards associated with each of these options. The system can also take the initiative, as when the system asks the user *Have you been deployed?* in the above dialogue excerpt. Which speaker will have the initiative at each point does not need to be hard coded, as in a call flow graph, but rather is determined by the system's expected reward calculation at each point.

28.5 Evaluation and Discussion

We have explored the various features and design goals for this dialogue manager during the creation of the Bill Ford policy and testing of the system.

We explored the feasibility of nonexperts to author nontrivial dialogue policies by working with a creative writer to define Bill Ford's dialogue policy. The resulting Bill Ford policy is composed of 285 operators, of which about 150 were automatically generated (they are operators that answer simple factual questions). This policy was built in about 3 months by one creative writer, with help from the first author.

To evaluate the FLoReS dialogue manager, we are currently conducting a comparative study between the current version of the SimCoach character Bill Ford, using FLoReS, and an older version of Bill Ford based on a dialogue policy encoded as a finite-state machine. The policy for the older version was encoded in SCXML[7], and focused strongly on system initiative.

To find appropriate users for our study, we selected ROTC (Reserve Officer Training Corps) students in four college campuses in Southern California. Although ROTC members are not the target user population for the SimCoach dialogue system, they serve as a suitable approximation due to their military background and familiarity with the issues faced by our target users. We have so far collected data from over 30 users in these ROTC programs. Each user interacted with one version of Bill Ford (either the one using FLoReS or the one using a simpler finite-state machine), and rated several aspects of the interaction with Bill Ford. Following the methodology in [14], users rated Bill on these items using values from 1 (strongly disagree) to 7 (strongly agree).

Although we have not yet collected data from enough users to determine statistically significant differences between the two systems, we note trends that for the FLoReS version, users had higher ratings for "Bill understood what I said" and "Bill let me talk about what I wanted to talk about" (addressing a user sense that they can take the initiative), but also, unexpectedly "Bill asked too many questions for no reason". Further analysis will examine the role of aspects of the FLoReS dialogue manager versus the specific subdialogues and reward structures used.

In terms of limitations, while we were successful in enabling the new Bill Ford dialogue policy to be almost completely designed by non-programmers, it remains a hard task that requires multiple iterations, and occasional intervention from dialogue system experts, to get a dialogue policy to work as desired. One of the main problems we noticed is that even though FLoReS allows authors to relax preconditions and allow ordering to be determined at run-time based on rewards, they still often preferred to use logical preconditions. This seems to be caused by the fact that predicting the behavior based on rewards is challenging, whereas logical conditions, although hard to write clearly, make it easier to foresee when an operator will be executed. We also observed that deciding on the specific reward value to associate to certain goals was a nontrivial task for our author.

Another limitation is that the simulation currently can make some inaccurate predictions, due to time and resource limits in the search process.

28.6 Conclusion

We have described FLoReS, a mixed initiative, information state, and plan-based DM with opportunistic action selection based on expected rewards that supports non-expert authoring. We presented examples to demonstrate its features,a summary

[7] http://www.w3.org/TR/scxml/

of our experience enabling a creative writer to develop a FLoReS dialogue policy, and encouraging preliminary results from a user evaluation.

Next we plan to work on ways to simplify or eliminate the need to assign numeric rewards by possibly using partial ordering constraints on operators, or inverse reinforcement learning to set the reward values. We also plan to further develop and study the efficacy of a dialogue policy visual editor, including developing a library of templates to quickly build common operators.

Finally, the FLoReS dialogue manager will be made available for other researchers through the ICT Virtual Human Toolkit [6][8].

References

1. DeVault, D., Leuski, A., Sagae, K.: Toward learning and evaluation of dialogue policies with text examples. In: 12th annual SIGdial Meeting on Discourse and Dialogue (2011)
2. English, M., Heeman, P.: Learning mixed initiative dialogue strategies by using reinforcement learning on both conversants. In: HLT-EMNLP (2005)
3. Fikes, R.E., Nilsson, N.J.: Strips: A new approach to the application of theorem proving to problemwilliams-young:2007 solving. Artif. Intell. **2**(3–4), 189–208 (1971). DOI 10.1016/0004-3702(71)90010-5. URL http://www.sciencedirect.com/science/article/pii/0004370271900105
4. Gandhe, S., Whitman, N., Traum, D.R., Artstein, R.: An integrated authoring tool for tactical questioning dialogue systems. In: 6th Workshop on Knowledge and Reasoning in Practical Dialogue Systems. Pasadena (2009). URL http://people.ict.usc.edu/%7Etraum/Papers/krpd09authoring.pdf
5. Georgila, K., Henderson, J., Lemon, O.: User simulation for spoken dialogue systems: Learning and evaluation. In: Interspeech, Pittsburgh (2006)
6. Kenny, P.G., Hartholt, A., Gratch, J., Swartout, W., Traum, D., Marsella, S.C., Piepol, D.: Building interactive virtual humans for training environments. In: Proceedings of Interservice/Industry Training, Simulation, and Education Conference (I/ITSEC). Orlando (2007)
7. Larsson, S., Traum, D.: Information state and dialogue management in the TRINDI dialogue move engine toolkit. Nat. Lang. Eng. **6**, 323–340 (2000)
8. Leuski, A., Patel, R., Traum, D., Kennedy, B.: Building effective question answering characters. In: Proceedings of the 7th SIGdial Workshop on Discourse and Dialogue, pp. 18–27 (2006)
9. Leuski, A., Traum, D.R.: Practical language processing for virtual humans. In: Twenty-Second Annual Conference on Innovative Applications of Artificial Intelligence (IAAI-10) (2010). URL http://www.ict.usc.edu/%7Eleuski/publications/papers/iaai-10-npceditor.pdf
10. Liu, D., Schubert, L.K.: Combining self-motivation with logical planning and inference in a reward-seeking agent. In: Filipe, J., Fred, A.L.N., Sharp, B. (eds.) ICAART (2), pp. 257–263. INSTICC Press (2010)
11. Paek, T., Pieraccini, R.: Automating spoken dialogue management design using machine learning: An industry perspective. Speech Comm. **50**(8–9), 716–729 (2008). DOI 10.1016/j.specom.2008.03.010. Evaluating new methods and models for advanced speech-based interactive systems
12. Pieraccini, R., Huerta, J.: Where do we go from here? Research and commercial spoken dialog systems. In: Proceedings of the 6th SIGdial Workshop on Discourse and Dialogue. Lisbon (2005). URL http://www.sigdial.org/workshops/workshop6/proceedings/pdf/65-SigDial2005_8.pdf

[8] http://vhtoolkit.ict.usc.edu

13. Rizzo, A.A., Lange, B., Buckwalter, J.G., Forbell, E., Kim, J., Sagae, K., Williams, J., Rothbaum, B.O., Difede, J., Reger, G., Parsons, T., Kenny, P.: An intelligent virtual human system for providing healthcare information and support. In: Studies in Health Technology and Informatics, Victoria (2011)
14. Silvervarg, A., Jönsson, A.: Subjective and objective evaluation of conversational agents in learning environments for young teenagers. In: 7th IJCAI Workshop on Knowledge and Reasoning in Practical Dialogue Systems, Barcelona (2011)
15. Tavinor, G.: The art of videogames. New Directions in Aesthetics. Wiley-Blackwell, Oxford (2009)
16. Traum, D., Larsson, S.: The information state approach to dialogue management. In: van Kuppevelt, J., Smith, R. (eds.) Current and New Directions in Discourse and Dialogue, pp. 325–353. Kluwer (2003)
17. Weathers, F., Litz, B., Herman, D., Huska, J., Keane, T.: The PTSD checklist (PCL): Reliability, validity, and diagnostic utility. In: the Annual Convention of the International Society for Traumatic Stress Studies (1993)
18. Williams, J., Young, S.: Scaling POMDPs for spoken dialog management. IEEE Trans. Audio, Speech, Lang. Process. **15**(7), 2116–2129 (2007)

Chapter 29
A Clustering Approach to Assess Real User Profiles in Spoken Dialogue Systems

Zoraida Callejas, David Griol, Klaus-Peter Engelbrecht, and Ramón López-Cózar

Abstract Evaluation methodologies for spoken dialogue systems try to provide an efficient means of assessing the quality of the system and/or predicting the user satisfaction. In order to do so, they must be carried out over a corpus of dialogues which contains as many possible prospective or real user types as possible. In this paper we present a clustering approach to provide insight on whether user profiles can be automatically detected from the interaction parameters and overall quality predictions, providing a way of corroborating the most representative features for defining user profiles. We have carried out different experiments over a corpus of 62 dialogues with the INSPIRE dialogue system, from which the clustering approach provided an efficient way of easily obtaining information about the suitability of distinguishing between different user groups to complete a more significative evaluation of the system.

29.1 Introduction

In order to provide a positive user experience, spoken dialogue systems should ideally adapt to the behaviour of individual users. Some systems automatically adapt to the users by identifying them and changing their dialogue strategies according

Z. Callejas (✉) • R. López-Cózar
Department Languages and Computer Systems, CITIC-UGR,
University of Granada, Granada, Spain
e-mail: zoraida@ugr.es; rlopezc@ugr.es

D. Griol
Department Computer Science, University Carlos III of Madrid, Leganés, Spain
e-mail: dgriol@inf.uc3m.es

K.-P. Engelbrecht
Quality and Usability Lab, Deutsche Telekom Laboratories, TU Berlin, Berlin, Germany
e-mail: klaus-peter.engelbrecht@telekom.de

J. Mariani et al. (eds.), *Natural Interaction with Robots, Knowbots and Smartphones: Putting Spoken Dialog Systems into Practice*, DOI 10.1007/978-1-4614-8280-2_29,

to what they had previously learned in a corpus of interactions of such users. For example, [4, 6] presented spoken dialogue systems for ambience intelligence which adapt to their users. As in this case, such tailored adaptation is efficient mainly in domains in which the system is used very frequently by a reduced number of users. In more open domains in which the system may be used by thousands of not frequent callers, some other strategies must be implemented. For example, the Let's Go system included special strategies for users with low proficiency in the interaction language [9].

Despite of the systems that are specially designed for specific population groups such as children or handicapped people (e.g. [5]), the decision of which groups must be taken into account to adapt the system strategies is not trivial in most application domains and there are few previous studies on the evaluation of the appropriateness of such decision.

Previous works have obtained promising results in applying clustered-based techniques to build simulated users which show consistent behaviours with respect to real users [1, 10]. In this paper, we present an approach based on clustering to assess whether real user groups considered to implement a system establish meaningful differences in their interaction behaviour, which provides designers with a valuable feedback about the appropriateness of considering such user grouping. We propose to carry out a clustering based on interaction parameters and subjective judgements of real users of an interaction corpus, in order to study whether different real user groups are balanced between the clusters or not. We have used this approach with the INSPIRE corpus to obtain feedback about grouping users according to age and technical affinity in its domain.

The rest of the chapter is organized as follows. Section 29.2 presents the experimental setup describing briefly the corpus employed, the parameters computed, and the experiments carried out. Section 29.3 presents a discussion of the results obtained. Finally, Sect. 29.4 presents the conclusions and some guidelines for future work.

29.2 Experimental Setup

We have used a corpus of 62 dialogues of real users interacting with the INSPIRE system. INSPIRE (INfotainment management with SPeech Interaction via REmote microphones and telephone interfaces; IST 2001-32746) was designed to control the following domestic devices via speech: three lamps, an electronic program guide (EPG), a TV, a video recorder, a fan, and an answering machine [7].

As shown in Table 29.1, for our experiments we employed three types of parameters: interaction parameters, user judgements, and user profile parameters. Interaction parameters were computed for each dialogue so that the total, user and system turn duration (in milliseconds) and the number of words per utterance were averaged along all user or system utterances in the dialogue. User judgements corre-

Table 29.1 Summary of parameters

Type of parameters	Parameters used
Interaction parameters	Turn duration, user turn duration, system turn duration, number of turns, number of words per user's utterance, number of words per system's utterance, number of help requests in the dialogue, task success, concept error rate, number of no matches per dialogue, number of repetitions per dialogue, number of barge-in per dialogue
User judgements	Task rate, overall impression with the interaction, overall impression of the presented system
User profile	Technical affinity, age

Table 29.2 Summary of experiments

Parameters for computing distance	Parameter studied	Experiment
Interaction parameters	Task rate	1
Interaction parameters	Overall impression with the interaction	1
Interaction parameters	Overall impression of the presented system	1
Interaction parameters	Technical affinity group (low or high)	2
Interaction parameters	Age group (young or senior)	2
Subjective parameters	Technical affinity group (low or high)	3
Subjective parameters	Age group (young or senior)	3
Interaction and subjective parameters	Technical affinity group (low or high)	4
Interaction and subjective parameters	Age group (young or senior)	4

spond to a number between 1 and 5 (1=bad, 2=poor, 3=fair, 4=good, 5=excellent) with which the users rated several aspects of the system after interacting with it [3].

Finally, regarding the user profile parameters, we were interested in studying the appropriateness of distinguishing age groups (young or senior) and their self-perceived technical affinity (low or high). In the system there were 32 young and 30 senior users and 26 users with low and 36 with high technical affinity.

For the experiments we employed the X-means clustering algorithm, an extended version of k-means which efficiently estimates the number of clusters to be used [8]. Using the Weka software, we established a minimum of 2 and maximum of 5 clusters in up to 1,000 interactions and computed the Euclidean distance between centroids employing different features as summarized in Table 29.2.

In the first group of experiments, we used interaction parameters to compute the distance measures and studied whether different user judgements were balanced between the clusters or were classified in different clusters and thus are easily distinguishable. The user judgements studied were task rate, overall impression with the interaction, and overall impression of the system. The second group of experiments considered the same features for clustering but studied the balance of user profile features, concretely technical affinity group and age group. The third group of experiments considered only user judgements for clustering and studied the same user profile features. With the fourth group of experiments, we studied the user profile measures by employing both interaction parameters and subjective measures for clustering.

29.3 Discussion of the Experimental Results

For all the experiments carried out, there were 2 clusters generated. The first group of experiments showed that the overall subjective impressions about the system and the interaction are not distinguished by the clustering algorithm, as shown by the fact that the clusters are composed of a balanced number of dialogues belonging to the different categories (bad, poor, fair, good, and excellent).

However, for the task rate the clustering algorithm clearly differentiated the dialogues in the extremes. As can be observed in Fig. 29.1, the worst rated dialogues were classified in cluster 1 and the best in cluster 2.[1] This might indicate that the judgement of task rate can be somehow derived from the interaction parameters, whereas overall impression might not be so much affected by the real performance of the system during the interaction.

In the second group of experiments we studied whether the user profile features were classified in different clusters when using only interaction parameters to compute the distances. As shown in Fig. 29.2, the clusters did not clearly distinguish between users with different technical affinity or age groups, although for age groups a slightly better separation was found.

In the third group of experiments we studied whether the user profile features can be determined by user judgements. The experiments revealed that such subjective

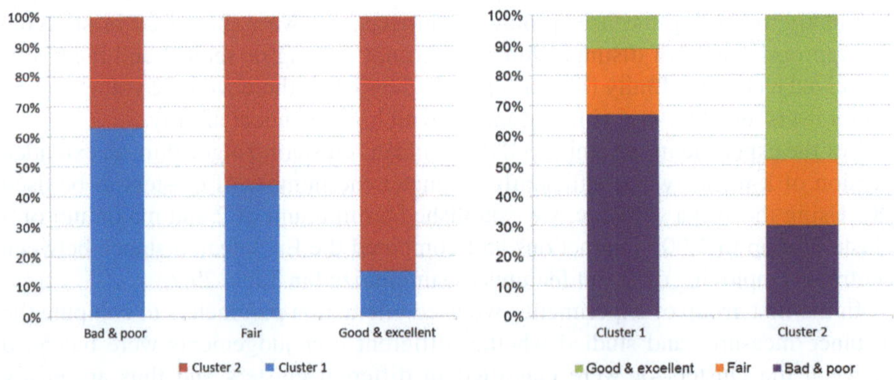

Fig. 29.1 Results for the task rate when clustering interaction parameters

[1]For illustration purposes, we have grouped the five categories into three: *bad&poor*, *fair, and good&excellent.*

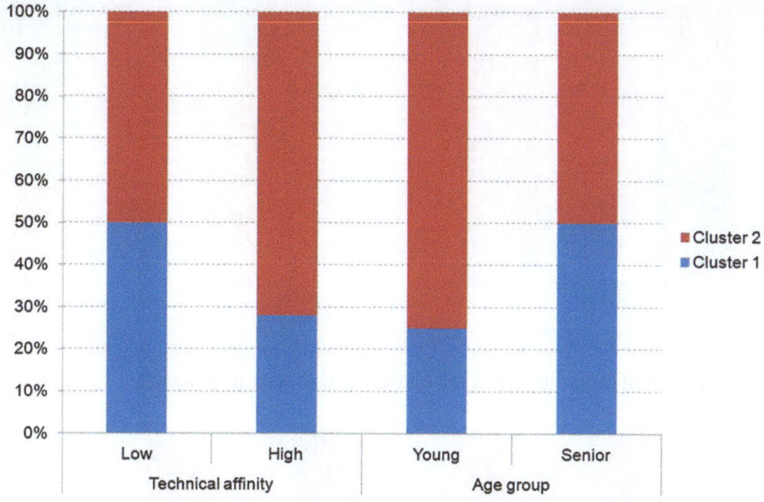

Fig. 29.2 Distribution of technology affinity and age groups when clustering with interaction parameters

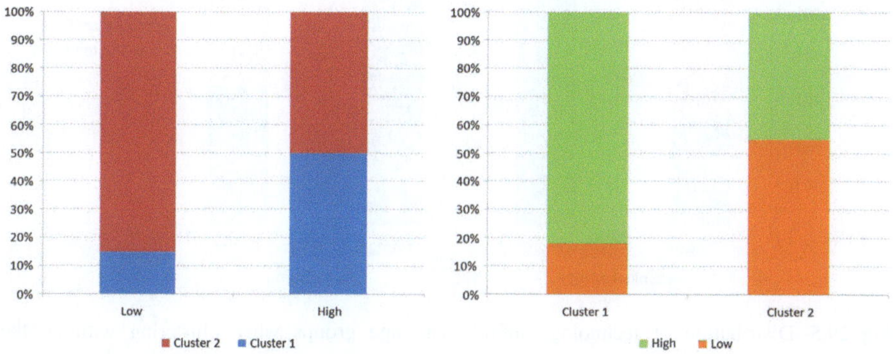

Fig. 29.3 Distribution of technical affinity when clustering with user judgements

features distinguish low technical affinity dialogues. As shown in Fig. 29.3, the 85% of the dialogues with low affinity users were classified in Cluster 2, although the 50% of dialogues is classified in each cluster for high-affinity users. The reason for this result may be that users with low affinity systematically evaluate the system with worse rates whereas high-affinity users provide more varied judgements. A close study of the data revealed that the standard deviation for the overall impression with the interaction rating for the low affinity users was 1.1 and the most frequent judgement 2, whereas for the high-affinity users the deviation was 0.6 and the mode value was 3.

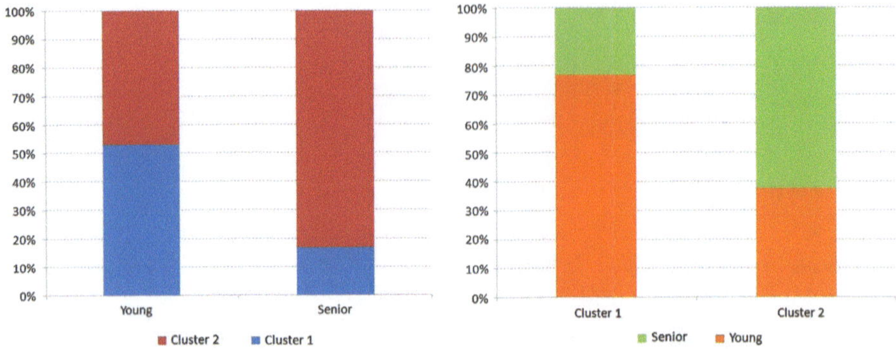

Fig. 29.4 Distribution of age groups when clustering with user judgements

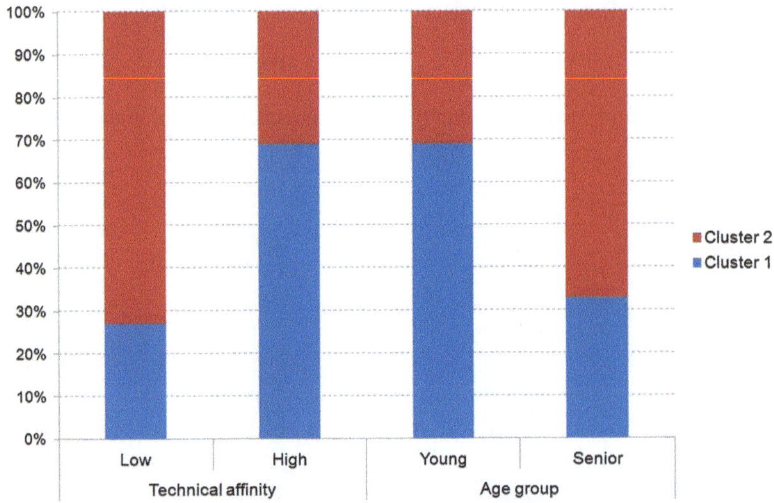

Fig. 29.5 Distribution of technology affinity and age groups when clustering with all the parameters

A similar result was obtained for the age group of the users. As shown in Fig. 29.4, the 83% of the senior users were classified in cluster 2, whereas the dialogues corresponding to young users were balanced between both clusters. A close study of the data revealed that while only the 19% of the young users had a low technical affinity, the 67% of the senior users had low technical affinity.

For the fourth group of experiments we used both interaction parameters and user judgements and obtained that, although low technical affinity and senior age group are separated slightly worse than in the previous experiments, high technical affinity and young users are separated better as shown in Fig. 29.5.

Finally, we focused on the relationship between the age and the technical affinity parameters and their presence in the clusters and classified the results of the previous

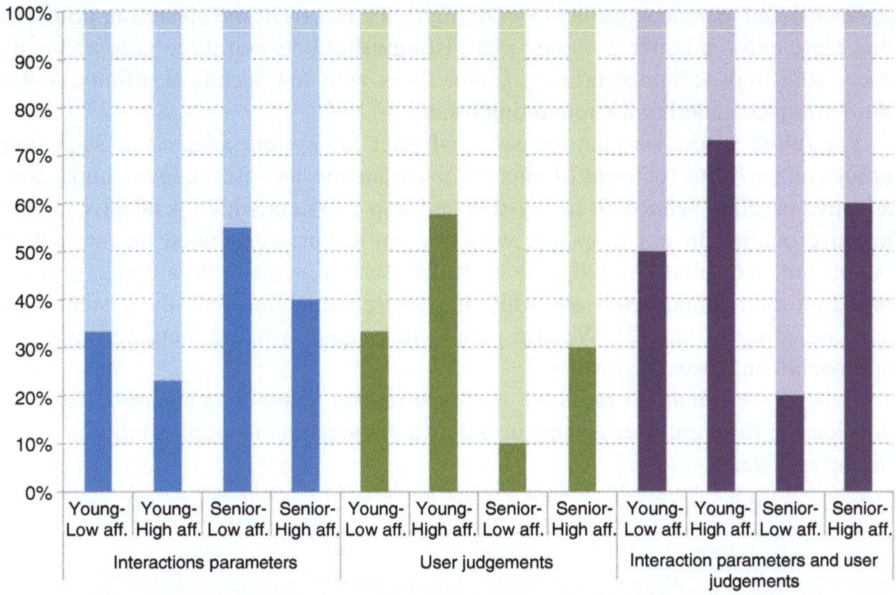

Fig. 29.6 Combinations of the age technical affinity groups with the different clusterings

experiments according to the possible combinations of both features. As shown in Fig. 29.6, the difference really strives between young users with high technical affinity and senior users with low technical affinity. Youngsters with low affinity and elderly with high affinity were not differentiated by the clustering mechanism.

This might be explained by the fact that senior users with low affinity provide a lower assessment of the system despite of how the interaction developed in terms of the interaction parameters. Moreover, interaction parameters along with the judgements can help to differentiate them from young, high-affinity users.

29.4 Conclusions and Future Work

We have used a corpus of real user conversations with the INSPIRE system and assessed the appropriateness of grouping the users by a combination of age (senior or young) and self-perceived technical affinity (low or high) by using a clustering approach. From the experiments we can conclude that the profiles of the users elicited different subjective judgements. Senior users and users with low technical affinity rated the system more consistently and with lower rates, whereas the rates of the young and high-affinity users depended more on how the interaction had developed. Thus, the clustering approach provided an efficient way of easily obtaining information about the suitability of distinguishing between these user groups. In our case, instead of considering the four combinations of age (young or senior) and technical affinity (low or high), the results of the clustering point out that there exists a better grouping that distinguishes between three groups: young users with high technical affinity, senior users with low technical affinity, and a third group considering the remaining users.

For future work, we plan to assess whether the system adapted to such user groups outperforms (in terms of interaction parameters and user judgements) a non-adaptive baseline version of the system and also a system which is adaptive to the initial grouping. In our proposal, we carry out a clustering based on interaction parameters and subjective judgements of the users of an interaction corpus and not based on their actual behaviour while they were interacting with the system such as in other works like [2]. We also plan to investigate whether merging these two approaches enhances the results.

Finally, we intend to replicate the experiments in other application domains considering different user groups in order to demonstrate its applicability to other dialogue systems.

Acknowledgements Research funded by the Spanish project ASIES TIN2010-17344.

References

1. Callejas, Z., Griol, D., Engelbrecht, K.P.: Assessment of user simulators for spoken dialogue systems by means of subspace multidimensional clustering. In: Interspeech (2012)
2. Chandramohan, S., Geist, M., Lefevre, F., Pietquin, O.: Clustering Behaviors Of Spoken Dialogue Systems Users. In: ICASSP (2012)
3. Engelbrecht, K.P., Quade, M., Moeller, S.: Analysis of a new simulation approach to dialog system evaluation. Speech Comm. **51**, 1234–1252 (2009)
4. Espejo, G., Aztiria, A., Augusto, J.C., López-Cózar, R.: Creating adaptive intelligent environments by means of multimodal dialogue and learning systems. In: HCIAmI'11, pp. 362–373. Nottingham (2011)
5. Helal, A.: The engineering handbook of smart technology for aging, disability and independence. Wiley (2008)
6. Lucas-Cuesta, J., Ferreiros, J., Aztiria, A., Augusto, J., McTear, M.: Dialogue-based management of user feedback in an autonomous preference learning system. In: ICAART , pp. 330–336 (2010)
7. Moeller, S., Engelbrecht, K., Schleicher, R.: Predicting the quality and usability of spoken dialogue services. Speech Comm. **8–9** (2009)
8. Pelleg, D., Moore, A.W.: X-means: Extending K-means with Efficient Estimation of the Number of Clusters. In: ICML , pp. 727–734 (2000)
9. Raux, A., Langner, B., Black, A., Eskenazi, M.: Let's go: Improving spoken dialog systems for the elderly and non-natives. In: Eurospeech (2003)
10. Rieser, V., Lemon, O.: Cluster-based User Simulations for Learning Dialogue Strategies. In: Interspeech, pp. 1766–1769 (2006)

Chapter 30
What Are They Achieving Through the Conversation? Modeling Guide–Tourist Dialogues by Extended Grounding Networks

Etsuo Mizukami and Hideki Kashioka

Abstract In goal-oriented or task-oriented conversations, the participants have to share many things to achieve collaboration. The ideas of *common ground*, *shared knowledge*, and similar concepts are important to understand the process of achievement. In this study, to model guide–tourist dialogues considering such grounding process, we proposed the idea of *extended grounding networks* by introducing the concept of *contribution topics* and applied it to different data collected from dialogues between a human guide and tourists.

30.1 Introduction

To engage conversations, especially those which have certain goals or tasks, the participants have to share many things to achieve collaboration.

For example, in case of telephone-based guidance service for tourists, the operator (guide **G**) has to understand what the tourist (user **U**) wants to know. To comprehend and answer it, **G** has to know where **U** is or what kind of transportation **U** is taking.

To share a proposition p, e.g., "**U** is at point **X**," between **G** and **U** via spoken dialogue, it is necessary that **U** presents an utterance (or a sequence of utterances) u which means p and that **G** gives an evidence e to show that **G** understands what **U** meant by u, so that **U** believes that **G** understood p and that **G** believes that **U** believes that **G** understood p.

If all of these are shared between **G** and **U**, we can say they have achieved a *common ground* with respect to the knowledge p to move their conversation forward. They would have to repeat a similar process to accomplish their goal, that is, **U** gets

E. Mizukami (✉) • H. Kashioka
National Institute of Information and Communications Technology, 3-5, Hikaridai, Keihanna, Kyoto, 619-0289, Japan
e-mail: etsuo.mizukami@nict.go.jp; hideki.kashioka@nict.go.jp

J. Mariani et al. (eds.), *Natural Interaction with Robots, Knowbots and Smartphones: Putting Spoken Dialog Systems into Practice*, DOI 10.1007/978-1-4614-8280-2_30,

the answer he/she wanted, e.g., "the way to the nearest Japanese restaurant." To develop a collaborative dialogue system, it is important that the system is designed to consider the grounding process.

Clark and his collaborators [1, 2] discussed this and formulated a *contribution model*. In their model a *contribution* consists of two phases, *presentation* and *acceptance*. However, their model was too simple to describe the dynamical structure of real conversations, since the model could not manage the completeness of each contribution. Developing their idea, [3, 4] proposed a computational model of *grounding* as a transition network by introducing a notion of *state*. Traum's model consists of *discourse unit* and seven *grounding acts*. In the model, a *discourse unit* **DU** is initiated by the participant's *initiate*, one of the grounding acts, from the start state **S** to the middle state **1**, and is grounded (completed) by the responder's *ack* (acknowledgement "yes" or "uh-huh") to the final state **F**. Four other states (**2**, **3**, **4**, **D**) can be transitioned depending on some subsequence of interaction such as *repair* for a misunderstanding or a misspoken word.

Traum's model is strong and sophisticated enough to deal with the grounding states in the context of computation. However, the model does not aim at representing what was achieved by each contribution and in what order the contributions were arranged through the conversation. Moreover, it cannot deal with the degree of grounding correspondence to the understanding level of the responder. In this paper, we propose *extended grounding networks* by introducing the concept of *contribution topic* and some additional states concerning the grounding levels. We tried to apply this idea to the guide–tourists natural dialogue data in order to model them and evaluated them from the viewpoint of redundancy and humanness (or naturalness).

30.2 Method

To understand the structure throughout the dialogue and what is achieved in each dialogue phase, especially in a goal-oriented or task-oriented conversation, when one person tries to describe something in order to explain or instruct another person, the theory of discourse structure based on intention [5] is very useful. According to the theory, a conversation consists of hierarchical and multiple discourse segments. In the same manner that the conversation has a discourse purpose to reach the goal between the speakers, each discourse segment has a discourse segment purpose. That is, to achieve the highest level of discourse purpose, each lower level of discourse segment purposes needs to be properly achieved by the collaborators. Each discourse segment purpose has relationships such as dominance or satisfaction-precedence. For example, imagine a situation where one person illustrates how to turn on the power button of a PC to someone else. To turn the power on, one needs to know its place, if it is at a difficult-to-find position, and then he has to push it. In this case, the discourse segment purpose to turn the power on (**DSP1**) would be the dominance while knowing the place of the button (**DSP2**) would be the satisfaction-precedence to actually push it (**DSP3**) (see also [6] about the discourse segmentation). By merging this idea with Clark's contribution model and Traum's grounding model, we would like to propose the idea of *contribution topics* and *extended grounding networks*.

A *contribution topic* is a unit of achievement corresponding to a discourse segment which has a certain preposition to be shared with collaborators for a certain dialogue period. It also has a hierarchical structure where a contribution topic of higher levels contains the topics from middle and lower levels. For instance, in a guide–tourist conversation through telephone, the largest part is usually occupied by the topic of the tourist's request such as "I'm looking for a good Japanese restaurant nearby." In this case, the top-level contribution topic can be labeled as "request & answer" (we abbreviate this as <request> for descriptive purpose), and once this topic is accomplished, it indicates that a proposition (or fact) "the tourist may be able to achieve necessary information about the restaurant including the way to get there" is mutually shared between the guide and the tourist. However, preceding the achievement at the top level of the contribution topic, every contribution topic of the lower level needs to be achieved such as understanding the tourist's request, grasping his present location, leading him there, and so on. These are also labeled as contribution topics at the lower level. In the expression of the finite state transition, an upper level of contribution topic is initiated from the start state **S** to the final state **F** through middle states n ($n = \mathbf{1}, \mathbf{2}, \ldots$) by a lower level of contribution topic's achievement.

In the bottom level of contribution topics, Traum's seven grounding acts, *initiate, continue, ack, reqrepair, reqack, repair*, and *cancel* (see [3] for details), are labeled to each utterance. In Traum's model, when one's presentation is divided into several utterances by each responder's *ack*, each pair of *initiate* and *ack* is regarded as individual contributions because *ack* would be the evidence of understanding for a common ground. In our method, however, if the sequential utterances have the same purposes, they are regarded as constructing one contribution topic since we would like to regard a contribution as an achievement of the topic's discourse purpose.

Concerning strength of grounding, [7] proposed an annotation method for the degrees of grounding based on evidence of understanding for army dialogue corpus. Based on this idea, we considered the degrees of grounding as different grounding states. Each lowest level of contribution topic is initiated from start state **S** to final state **F** (= mutual grounded states ***m***) through each collaborator's mid-grounded states ***g*** (guide's state) or ***u*** (user's state), which means that up to a certain point their conversation is grounded but has not fully achieved the contribution yet. Therefore, while an explanation or instruction is incomplete, they remain in these middle states. We also label the half-grounded states ***g!*** or ***u!***, which means that they comprehend the utterance meaning but do not have any ideas on how to respond to it, and the ungrounded states ***g-*** or ***u-***, which means that they could not understand or missed the utterance, at their state of understandings.

30.3 Results

We applied this model to the data recorded from cell phones; conversations held between a guide and ten tourists in Japanese. We had the tourists traveling around Kyoto City for a full day and had them calling the guide several times during their trip. Thirteen of them were labeled and analyzed as objective data in this paper. Figure 30.1 shows an example of labeled data throughout one conversation. The top-level contributions consist of three contribution topics, <opening>, <request>, and <closing>, and the dialogue transitions from start state **S** to the final state **F** by the successive achievements of these three topics. The <opening> topic itself also consists of three lower topics: <channel> for establishing the channel for communication, <identify> for identifying each other, and <social> for interactions of social obligations. In the <channel> topic, the user initiates the dialogue by calling the guide, then the grounding state is transitioned from **S** to state **u**. The guide (g or G) receives the call, then the state is transitioned to state **g**, but the guide doesn't hear the user's voice at this point, so the contribution is not grounded yet. The user makes his/her first utterance towards the guide's response. It indicates that the user hears the guide's voice, then the state is transitioned to **m** which expresses user's evidence of understanding or recognizing and mutually grounded with respect to this topic at the same time.

Finally, we identified five types of contribution topics from the objective data (<opening>, <request>, <proactive> as topics concerning unprompted proposal by the guide, <appointment> for scheduling the continuous calls, and <closing>) at the top level (depth d), two types (<und:req> for understanding request and <answer>) at the middle level, six types (<channel>, <identify>, <pre:req> for any preface of request, <sub:req> for supplementary request by the user, <probe> for any follow-up question by the guide, and <social>) at the bottom level, and three types (<situation> for confirming the situation of the user, <goal> for a reference to the goal of the trip, and <addition> for additional chat out of the relevant request) at multi-level.

From the communality of all networks, the simplest (or indispensable) network can be conducted as the basic model of the guide–tourist dialogues as shown in Fig. 30.2. It consists of three levels and nine components of main and sub-networks. The networks constructed from the actual dialogue data would be expressed with states and arcs added to this basic network. Figure 30.3 shows an example of network (some sub-networks are abbreviated). There are several additional contribution topics shown as recursive arcs and backward arcs across all level of contributions. In the network of <und:req> at the third level ($d = 3$), two states **g** and **g-** are added and a deeper level ($d = 4$) of contribution topic <situation> is inserted to the process of grounding for the topic <und:req>.

R	Utterance	U-ID	$d = 1$	State	$d = 2$ (1)	State	$d = 3$ (1)	State	$d = 2$ (2)	State	$d = 3$ (3)	State	$d = 2$ (3)	State	$d = 3$ (3)	State
				S		S		S								
U	[Ring]	0	<opening>		<channel>		initiate	u				(S)				
G	Hello, this is Hiromi speaking.	10					ack	g	<identify>		initiate	g				
U	This is Tanaka speaking.	20			</channel>	1	ack	m			ack/initiate	u				
G	How are you?	30							</identify>	2	ack	m/S	<social>		initiate	g
U	Excuse me for calling many times.	40													ack/initiate	u
G	No problem.	50	</opening>	1									</social>	F/S	ack	m/S
U	So, I'm in the grounds of Kiyomizu temple.	60	<request>		<situation>		initiate	u								
G	Yes.	70			</situation>	S	ack	m/S								
U	I have just drawn a paper fortune,	80			<und:req>		initiate	u								
G	uh-huh	90					ack	g								
U	and I forgot whether I should bind a bad one or a good one with a tree branch.	100					continue	u								
G	Please bind bad one.	110			</und:erq>	1	ack	m/S	<answer>		initiate	g				
U	Should I bind bad one, and take good one along?	120									reqack	u				
G	Yes, you should.	130									ack	g				
U	I got it.	140	</request>	2					</answer>	F/S	ack	m/S				
U	Thank you very much.	150	<closing>		<social>		initiate	u								
G	Was it good result?	160			</social>	1	ack	m/S	<addition>		initiate	g				
U	Yes, it was good.	170									ack/initiate	u				
G	Please keep it.	180									ack/initiate	g				
U	Yes, thanks.	190							</addition>	2	ack	m/S	<social>		initiate	u
G	Sure, anytime.	200											</social>	3	ack/	m/S
U	Goodbye.	210											<social>		initiate	u
G	Goodbye.	220	</closing>	F									</social>	F	ack/	m

Fig. 30.1 An example of labeled data

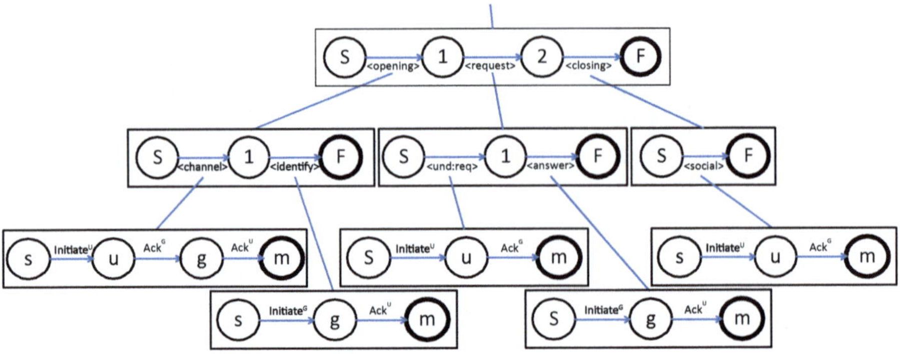

Fig. 30.2 Basic network model of guide–tourist dialogue

Fig. 30.3 A sample of networks of guide–tourist dialogue

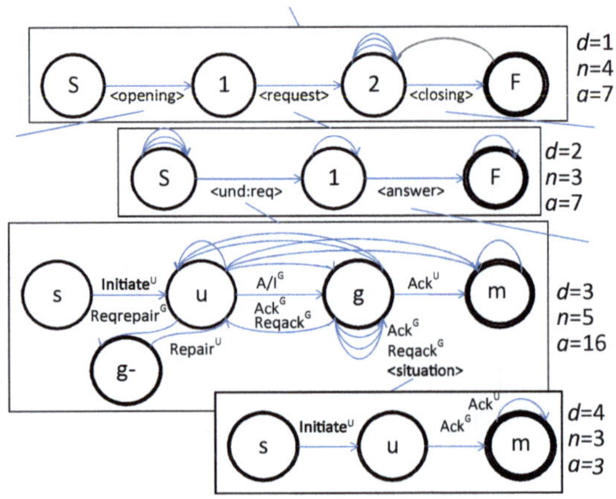

30.4 Discussion and Conclusion

The actual human-to-human dialogues include many redundant exchanges. In the extended grounding networks, those are expressed as additional transitions to the basic network, although most of them are necessary and inevitable transitions for the sequential process of grounding and achieving guide–user collaborations. In a sense, such redundancy can be regarded as naturalness of human-to-human dialogue, and it may be estimated with the network complexity. The network complexity (*NC*) can be approximated as the following equality:

$$NC = \sum_{i=1}^{k} \frac{1}{d_i} \log(n_i - nbase_i + 1)(a_i - abase_i + 1), \tag{30.1}$$

where i is the depth of the network (so d_i=i), n_i is the maximum number of node, a_i is an arc at depth i, and $nbase_i$ and $abase_i$ are those of the basic network. Therefore,

NC would be 0 for the basic network, 1.015 for the network in Fig. 30.3 and the average of all objective dialogues is 0.956. *NC* depends on the types of requests, i.e., it tends to be smaller with Q&A types of requests and larger with navigational types of requests.

It is perfectly possible that the networks be different from one another depending on users and contents of the requests. However, if the fundamental guidance strategy of the guide is not so unstable, then it could be extracted as a particular network. We would have to apply our grounding labels to more conversational data in order to confirm that. Furthermore, the model still has difficulty to deal with parallel topics proceeding simultaneously at the same level. We must continue to develop our idea and extend the framework.

Acknowledgements A part of this work was supported by JSPS KAKENHI Grant Number 24520447.

References

1. Clark, H.H., Brennan, S.A.: Grounding in communication. In: Resnick, L.B., Levine, J.M., Teasley, S.D. (eds.) Perspectives on Socially Shared Cognition, pp. 127–149. APA Books, Washington (1991)
2. Clark, H.H., Schaefer, E.F.: Contributing to discourse. Cogn. Sci. **13**, 259–294 (1989)
3. Traum, D.: A computational theory of grounding in natural language conversation. Ph.D. Thesis (1994)
4. Traum, D.: Computational models of grounding in collaborative systems. In: Working Notes of AAAI Fall Symposium on Psychological Models of Communication, pp. 124–131, November 1999
5. Grosz, B.J., Sidner, C.L.: Attention, intention, and the structure of discourse. Comput. Linguist. **12**, 175–204 (1986)
6. Nakatani, C.H., Traum, D.R.: Coding Discourse Structure in Dialogue (Version 1.0). University of Maryland, College Park (1999)
7. Roque, A., Traum, D.: Degrees of grounding based on evidence of understanding. In: Proceedings of the 9th SIGdial Workshop on Discourse and Dialogue, pp. 54–63 (2008)

Chapter 31
Co-adaptation in Spoken Dialogue Systems

Senthilkumar Chandramohan, Matthieu Geist, Fabrice Lefèvre, and Olivier Pietquin

Abstract Spoken dialogue systems are man-machine interfaces which use speech as the medium of interaction. In recent years, dialogue optimization using reinforcement learning has evolved to be a state-of-the-art technique. The primary focus of research in the dialogue domain is to learn some optimal policy with regard to the task description (reward function) and the user simulation being employed. However, in case of human-human interaction, the parties involved in the dialogue conversation mutually evolve over the period of interaction. This very ability of humans to coadapt attributes largely towards increasing the naturalness of the dialogue. This paper outlines a novel framework for coadaptation in spoken dialogue systems, where the dialogue manager and user simulation evolve over a period of time; they incrementally and mutually optimize their respective behaviors.

S. Chandramohan (✉)
Supelec, MaLIS - IMS Research Group, Metz, France

Université d'Avignon et des Pays de Vaucluse, LIA-CERI, Avignon, France
e-mail: Senthilkumar.Chandramohan@supelec.fr; Senthilkumar.Chandramohan@univ-avignon.fr

M. Geist
Supelec, MaLIS - IMS Research Group, Metz, France
e-mail: Matthieu.Geist@supelec.fr

F. Lefèvre
Université d'Avignon et des Pays de Vaucluse, LIA-CERI, Avignon, France
e-mail: Fabrice.Lefevre@univ-avignon.fr

O. Pietquin
Supelec, MaLIS - IMS Research Group, Metz, France

UMI 2958 (CNRS - GeorgiaTech), Metz, France
e-mail: Olivier.Pietquin@supelec.fr

J. Mariani et al. (eds.), *Natural Interaction with Robots, Knowbots and Smartphones: Putting Spoken Dialog Systems into Practice*, DOI 10.1007/978-1-4614-8280-2_31,

31.1 Introduction

Spoken dialogue systems (SDS) are man-machine interfaces which use spoken language (most often speech but can also be multimodal interaction) as the medium of interaction. The dialogue management module is responsible for navigating the system to accomplish a specific task. Proper functioning of the dialogue management module can be attributed to the so-called dialogue policy. Given the state of dialogue progress (context) and the most recent user response the dialogue policy determines the next action to be performed by the dialogue system. Manually choosing a dialogue policy can perhaps be an option if the problem studied is simple enough. However, for many real-world dialogue problems, handcrafting a dialogue policy is often a complex task. Considering the stochastic behavior of users and various uncertainties involved (e.g., speech recognition errors and channel noise), the complexity of dialogue management increases exponentially with increase in the size of the dialogue problem. To address this issue, dialogue management problems are casted as a Markov Decision Process (MDP) [3, 13, 21] or Partially Observable Markov Decision Process (POMDP) [2, 23], following which dialogue policy optimization can be performed using reinforcement learning (RL) [22].

Generally speaking, in case of human-human interaction, the parties involved in the conversation tend to coadapt and thus mutually evolve over a period of time. Most often, when the conversation begins, the parties assess each other's ability to understand and then continue to interact based on their initial assessment. Let us term this aspect of human-human communication as *dialogue-initiation*. However, once the dialogue progresses, humans tend to evolve mutually (based on the history of conversation). Using an acronym such as MDP (during subsequent references) after defining the Markov Decision Process in oral or written communication is an example for such evolution. Let us term the later stage of human-human communication as *dialogue-evolution*. It may be useful to note that the dialogue-evolution occurs at several levels (such as the amount of information exchanged and terminologies used during the conversation).

In case of man-machine interaction, dialogue-initiation has been carried out by policy optimization. Most existing approaches for dialogue management [7] focus on retrieving some optimal (with respect to the reward function) and user-adaptive (with respect to user simulation) dialogue policy [6]. This can be perceived as the dialogue-initiation stage of human-human interaction. This aspect of man-machine interaction has been studied in detail and is now state of the art. However, from the authors' perspective, much less attention has been paid to dialogue-evolution in man-machine interaction. One of the most relevant works done towards dialogue-evolution is to perform online policy optimization (for instance, [5]). There are two primary drawbacks with regard to online policy optimization: (i) when the dialogue manager tries to evolve or optimize its behavior, the human user also tends to adapt to the dialogue manager (instead of speaking normally, the user tries to provide only information asked by the system). These contradicting efforts when applied simultaneously may often result in suboptimal policies and thus blocking the possibility for dialogue-evolution, (ii) even if dialogue-evolution occurs it may bias the dialogue management to act over confidently [5, 8] and thereby resulting in inferior policies. Changes made directly to the policy used for dialogue management cannot always be guaranteed to improve performance. In the worst-case scenario this may induce very bad user experience. Also in case of online optimization the speed of adaptation is relatively slow considering the fact that users (and hence behaviors) encountered are random in nature.

This paper presents a novel framework for coadaptation in spoken dialogue systems. The primary focus of this work is to introduce coadaptation in man-machine interaction and thereby facilitate dialogue-evolution. The dialogue manager and the user simulation (both casted as an MDP and optimized using RL) are made to interact with each other and subsequently coadapt. Optimizing the dialogue manager and user simulation alternatively over a period of time results in generalization of dialogue policy and user behavior (as a result of back propagation of rewards). Thus coadaptation provides an opportunity to learn policies which can cope with situations unobserved in the dialogue corpus. The layout of the paper is as follows: Sect. 31.2 formally defines the MDP and presents an overview on dialogue optimization. Section 31.3 outlines the process of coadaptation and explains how it can be introduced in spoken dialogue systems. Section 31.4 describes the experimental setup and analyzes the results. Eventually Sect. 31.5 concludes and outlines the future directions of work with regard to coadaptation.

31.2 Spoken Dialogue Optimization

Given a specific task to accomplish (such as town information) the dialogue management engine has to perform a sequence of decisions. Thus in itself the task of dialogue management can be perceived as a sequential decision-making problem. Considering this fact, in recent years, dialogue management problems are often casted as an MDP and policy optimization is carried out using RL.

31.2.1 Markov Decision Process

Statistical frameworks such as MDPs provide a well-defined mathematical paradigm for modeling sequential decision-making problems such as dialogue management. Formally, an MDP [3] is defined as a tuple $\{S, A, P, R, \gamma\}$ where S is the state space, A is the action space, $P : S \times A \rightarrow \mathscr{P}(S)$ is a set of Markovian transition probabilities, $R : S \rightarrow \mathbb{R}$ is the reward or the utility function and γ is the discount factor for weighting long-term rewards. A learning agent has to perform a sequence of decisions and move from one state to another to accomplish the task (succinctly defined by the reward function). At any given time step, the agent is in a state $s_i \in S$ and transits to s_{i+1} according to $p(.|s_i, a_i)$ upon (choosing and) performing an action

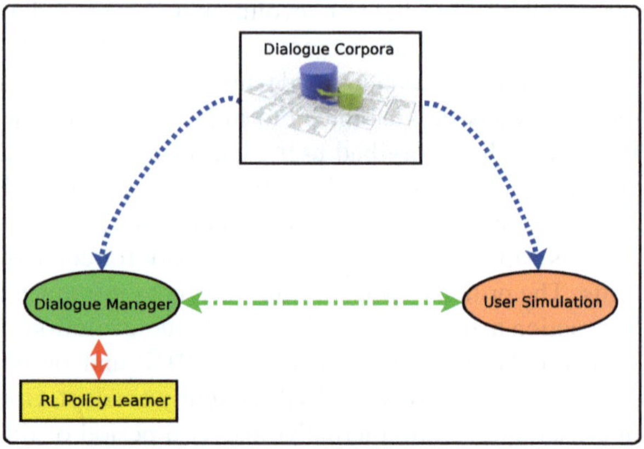

Fig. 31.1 Dialogue optimization using RL and user simulation

$a_i \in A$ according to a policy $\pi : S \rightarrow A$. After each transition the agent receives a reward $r_i = R(s_i)$. The quality of the policy π followed by the agent can be quantified by the state-action value function or Q-function ($Q^\pi : S \times A \rightarrow \mathbb{R}$) defined as

$$Q^\pi(s,a) = E\left[\sum_{i=0}^{\infty} \gamma^i r_i | s_0 = s, a_0 = a\right] \tag{31.1}$$

The optimal policy π^* is the one for which the Q-function is maximum for each state-action pair: $\pi^* \in arg\,max_\pi Q^\pi(s,a)$. The optimal Q-function $Q^*(s,a)$ leads to an optimal policy: $\pi^*(s) = arg\,max_a Q^*(s,a)$. There exist several RL algorithms to compute the optimal policy π^* [22] and the associated $Q^*(s,a)$.

31.2.2 RL-Based Dialogue Optimization

Once the task of dialogue management is modeled as an MDP [13, 21], RL can be used to retrieve the optimal dialogue policy. Since RL-based dialogue optimization is data intensive, user simulations are introduced to cope with the data requirement and also to evaluate the resulting dialogue policies. Dialogue optimization using RL and user simulation is shown in Fig. 31.1. However, there exists a set of sample efficient algorithms for dialogue policy optimization [17]. Despite all these methods, user simulations continue to play a key role in evaluating the quality of the dialogue policy. An overview of dialogue optimization using number of machine learning techniques can be found in [12].

User simulation in dialogue systems [6] aims at generating synthetic dialogue corpus and is often trained using the available dialogue corpus. Most existing methods for building user simulations [9, 16, 18] focus on generating a synthetic dialogue corpus which has the same statistical consistency as observed in the dialogue corpus. Even though there exist goal-directed [15] and agenda-based [20] user simulations, determining the goal or agenda of the user in itself is a complex task. It is important to note that the performance of the user simulation has a direct impact on the optimized dialogue policy as shown in [19]. One possible solution to this problem is to employ inverse reinforcement learning (IRL) [14] to retrieve the utility function of the agent and then use it to perform RL-based optimization. Recently the task of user simulation was casted as an MDP and IRL [1,14] was used for imitating the user behavior [4]. This provides a unique opportunity for the user simulation to generalize (using the reward function learned from the data) and thus evolve over a period of time. Such an evolution would be very similar to that of the dialogue-evolution happening in the human-human interaction. Let us term the task of learning the user behavior as user (behavior) optimization.

31.3 Coadaptation in Dialogue Systems

Naturalness of human-human dialogues can be attributed to several unique abilities of humans (such as mutual evolution during the period of interaction). However, research on man-machine interaction (more specifically on dialogue optimization) primarily focused on retrieving some (initial) optimal policy. The reward function used for dialogue optimization is often handcrafted to retrieve a decent policy which can accomplish the dialogue task in a robust and user-adaptive manner. Thus dialogue managers built using such methods often lag behind in terms of naturalness of the dialogue and thus make it unpleasant for the end users. Coadaptation in dialogue systems can provide an opportunity for the dialogue manager and user simulation to evolve mutually. To begin with the task of dialogue management (mdp-sds) and user simulation (mdp-user) is casted as an MDP. In order to perform policy optimization, reward functions for mdp-sds and mdp-user have to be defined in advance. As a preprocessing step for coadaptation, these reward functions can be learned from the dialogue corpus using IRL. Unlike dialogue optimization shown in Fig. 31.1, coadaptation is an iterative procedure where policy optimization of mdp-user and mdp-sds is performed repeatedly. As shown in Fig. 31.2 in Step 1, policy optimization for mdp-user is performed using a handcrafted dialogue manager and the available dialogue corpus. Using the policy learned in Step 1, dialogue optimization for mdp-sds is performed in the next step. Following which policy optimization for mdp-user is performed using the optimal dialogue policy retrieved in the previous step. Step N and Step N+1 are repeated iteratively until convergence (in other words until the resulting policies cease to evolve).

Even though the resulting policies are deterministic, some amount of stochasticity can be introduced using Gibbs sampling based on state-action values, i.e., Q(s,a). Let the probability of choosing a dialogue act or user act $a_i \in A$ be λ_i such that $\Sigma_{i=1}^{n} \lambda_i = 1$. The values of $\lambda_{1\ldots n}$ can be heuristically determined using a Gibbs distribution: $\lambda_i = e(Q(s,a_i)/\tau)/\Sigma_{j=1}^{n} e(Q(s,a_j)/\tau)$. RL-based optimization is subjective to that of the reward function and the environment the agent is

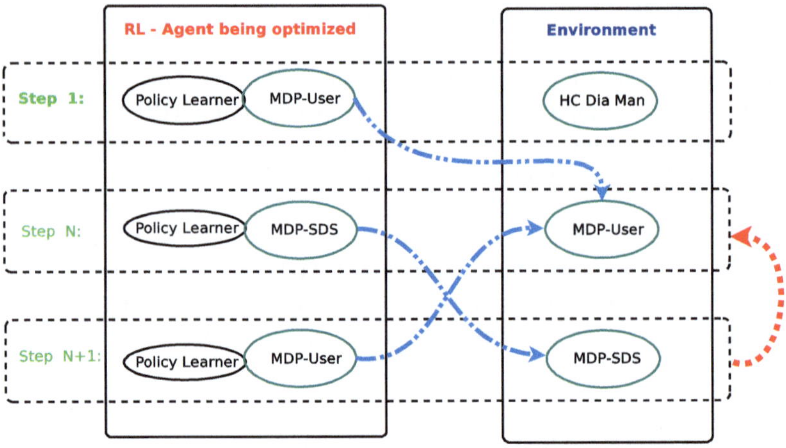

Fig. 31.2 Coadaptation framework for dialogue-evolution

interacting with. In case of coadaptation even though the reward function remains the same, employing Gibbs sampling scheme results in changes in the dynamics of the environment. This very change in the dynamics of the dialogue manager or the user simulation will yield different user or dialogue policies, respectively. It may be useful to note that during the process of coadaptation the rewards obtained by the agents are backpropagated and thus result in some degree of generalization of dialogue manager as well as user simulation modules. The immediate effect of this would be an opportunity for the dialogue manager and user simulation to cope with unseen situations (which are not observed in the dialogue corpus). This in turn will help the dialogue manager to retrieve the real optimal policy from its own capacity (policy which originally may not be present in the corpus but at the same time can cope with noise and changes in user behavior).

31.4 Experiment

This section outlines a simple experiment to exhibit the outcome of coadaptation in spoken dialogue systems. Our primary motivation is to show how coadaptation can be performed in case of man-machine interfaces and analyze the possibility of dialogue-evolution. Generally speaking, more appropriate experiment would have been a two-step process: (i) perform IRL on the available dialogue corpus to retrieve the reward functions and (ii) use the reward function obtained from the data to perform coadaptation. However, for the sake of simplicity, a much smaller dialogue problem and RL-based optimization are performed in the following experiments.

31.4.1 Town-Information Dialogue System (2 Slots)

The dialogue problem studied in this paper is a 2-slot subproblem from the town-information domain [11]. The dialogue manager has to seek and obtain user preferences for price range and location of restaurants in the city. The state of the mdp-sds (dialogue manager casted as an MDP) involves 3 dimensions: (i) 0–2 (corresponding to the price range; 0 (slot is not filled), 1 (slot is filled but yet to be confirmed), 2 (slot is filled and confirmed)), (ii) 0–2 (corresponding to the location), and (iii) 0–1 (indicates whether the user has performed negation: indirect measure of channel noise). The list of dialogue acts includes ask-slot1, ask-slot2, ask-all-slots, confirm1, confirm2, confirm-both, close-dialogue, and two implicit confirmation acts. The state of the mdp-user (user simulation casted as an MDP) involves 4 dimensions: (i) 0–9 (represents the action performed by the dialogue manager), (ii) 0–2 (corresponding to the price range), (iii) 0–2 (corresponding to the location), and (iii) 0–1 (indicates whether speech recognition errors have occurred). The list of user acts includes provide-slot1, provide-slot2, provide-all-slots, confirm1, confirm2, confirm-all-slots, negate, remain-silent, and hangUp. The same reward function was used for both mdp-sds and mdp-user (positive reward of 20 for each correctly filled slot and bonus reward of 60 for successful task completion). The discount factor of both the MDPs is set to 0.95.

31.4.2 Experimental Setup and Results

To begin with, handcrafted (HC) policies for both mdp-sds and mdp-user are defined (these policies attempt to ask/provide and confirm/ascertain one slot after the other) so as to generate data from which a user simulation will be learned (simulating the acquisition of a corpus). During the experiments, dialogue manager and user policy optimization are carried out using Least Squares Policy Iteration (LSPI) [10] considering its sample efficiency and generalization abilities. Coadaptation is performed as explained in Fig. 31.2. First user optimization is performed using the mdp-sds with HC policy. The resulting user policy was then used to perform dialogue optimization. During the initial stages of the experiment, the channel noise is set to zero. At the end of each step, a dialogue/user policy is generated. The following are a set of dialogue episodes generated from these retrieved policies:

```
===============================================
Step 1: HC-DialogueManager vs Train-RL-User
===============================================
  UserState: AskSlot_1 0 0 0
    UserAct: provide_slot_1
  UserState: ExpConfirm_1 1 0 0
    UserAct: confirm_slot_1
  UserState: AskSlot_2 2 0 0
```

```
    UserAct: provide_slot_2
  UserState: ExpConfirm_2 2 1 0
    UserAct: confirm_slot_2
  UserState: CloseDia 2 2 0
    UserAct: hangUp
==================================================
Step 2: Train-RL-DialogueManager vs RL-User
==================================================
  DiaState: 0 0 0   DiaAct: AskSlot_1
    UserResponse: provide_slot_1
  DiaState: 1 0 0   DiaAct: ExpConfirm_1
    UserResponse: confirm_slot_1
  DiaState: 2 0 0   DiaAct: AskSlot_2
    UserResponse: provide_slot_2
  DiaState: 2 1 0   DiaAct: ExpConfirm_2
    UserResponse: confirm_slot_2
  DiaState: 2 2 0   DiaAct: CloseDia
    UserResponse: hangUp
==================================================
Step 3: RL-DialogueManager vs Train-RL-User
==================================================
  UserState: AskSlot_1 0 0 0
    UserAct: provide_slot_1
  UserState: ExpConfirm_1 1 0 0
    UserAct: confirm_slot_1
  UserState: AskSlot_2 2 0 0
    UserAct: provide_slot_2
  UserState: ExpConfirm_2 2 1 0
    UserAct: confirm_slot_2
  UserState: CloseDia 2 2 0
    UserAct: hangUp %
==================================================
Step 4: Train-RL-DialogueManager vs RL-User
==================================================
  DiaState: 0 0 0   DiaAct: AskAllSlots
    UserResponse: provide_all_slots
  DiaState: 1 1 0   DiaAct: ExpConfirmAll
    UserResponse: confirm_all_slots
  DiaState: 2 2 0   DiaAct: CloseDia
    UserResponse: hangUp
==================================================
```

Dialogue episodes generated using policies learned from Step 3 and Step 4 clearly indicate that coadaptation of dialogue management engine and user simulations has indeed happened. A more interesting aspect of this result is the fact

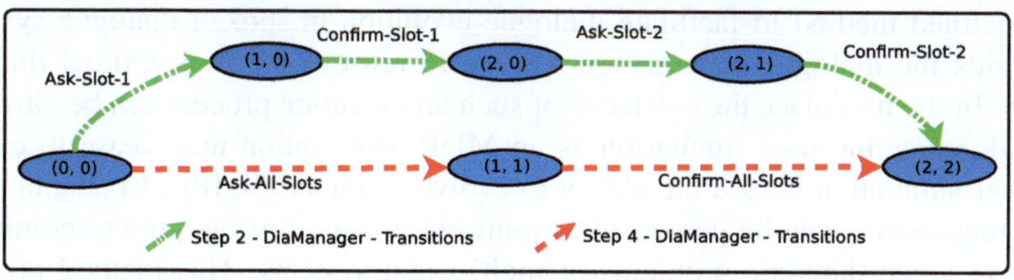

Fig. 31.3 Changes in dialogue transitions due to coadaptation

that such optimal policies can be learned even though they were not observed in the dialogue corpus (recall that the handcrafted policies used in Step 1 only used simple dialogue acts and user acts). This result can be attributed to the generalization ability of dialogue manager and user simulation when casted as interacting MDPs. It may be useful to note that Gibbs sampling is not introduced in Step 2, since the focus was to learn a basic dialogue policy (similar to the handcrafted dialogue manager used in Step 1). Even though Gibbs sampling of dialogue policy (from Step 2) was employed in Step 3, the dialogue episode may look similar to that of Step 1. In this case the evolution of the user simulation is invisible because the episode presented here is pure greedy interaction between the dialogue policy from Step 2 and user policy from Step 3. However, this invisible evolution of user simulation is responsible for the evolution of dialogue policy in Step 4. Changes in transitions caused due to coadaptation of the dialogue management engine are shown in Fig. 31.3.

As a next step artificial noise (error model) is introduced. Ideally speaking if there is some amount of channel noise, performing complex user action (where more information is exchanged) will exponentially increase the possibility for a speech recognition error. Having this in mind, 40 % error is introduced when the user simulation performs a complex user action (i.e. provide-all-slots). The binary field in the user-state is set to 1 if both the user and dialogue manager perform complex acts. When the same set of experiments is repeated after including the error model, the dialogue policy and the user policy remain the same even after coadaptation. These experimental results showcase that the coadaptation framework proposed here is indeed a robust means for facilitating dialogue-evolution. Given the same scenario human users will tend to act in a similar fashion and thus evolve to showcase complex behaviors if the conditions are favorable (as shown in Step 4), if not stick to basic means of communication (as shown in Steps 1–3).

31.5 Conclusion and Future Work

Dialogue-evolution commonly seen in human-human interaction has not been accounted for in the design process of man-machine interfaces. Online dialogue optimization may not always give the opportunity for the dialogue management engine to evolve. Coadaptation framework presented in this paper provides a well-defined method to facilitate dialogue-evolution in spoken dialogue systems. This aids the dialogue management engine to retrieve the real optimal dialogue policy. In the first place the existence of such an evolution process can be attributed towards modeling user simulation as an MDP. This option may cease to exist if the user simulation lagged the ability to evolve (such as [9, 18]). Coadaptation in dialogue systems can be perceived as joint optimization of the policy (commonly used) in game theoretic problems or multi-agent systems. This method provides an opportunity for the dialogue manager and user simulation to interact with each other and thus mutually evolve in an incremental manner over a period of time. Experimental results presented in Sect. 31.4 tend to support these claims.

Even though the work presented here may seem as evolution in machine-machine interaction, it is a viable option to reach equilibrium. The state of equilibrium, however, is defined by the reward function (given suitable reward functions machines will certainly evolve like humans as shown here). Thus in the future it is important to determine how dialogue tasks can be summarized as reward functions similar to the utility function of humans. Future directions of work also include (i) as shown in the experimental setup, it is possible to cope with unseen situations which are not present in the dialogue corpus. This is made possible by the use of batch RL methods for policy optimization (which have good generalization ability) and coadaptation (which provides opportunity to interact and evolve mutually). One interesting direction of future work would be to investigate the effectiveness of such generalizations and quantify the resulting dialogue and user policies; (ii) one other direction of future work is to explore the possibility of performing IRL-based coadaptation as discussed in Sect. 31.4.

Acknowledgements This research was partly funded by the EU INTERREG IVa project ALLEGRO and by the Règion Lorraine (France).

References

1. Abbeel, P., Ng, A.: Apprenticeship learning via inverse reinforcement learning. In: Proceedings of ICML, Banff, Alberta (2004)
2. Astrom, K.J.: Optimal control of markov decision processes with incomplete state estimation. J. Math. Anal. Appl. **10**, 174–205 (1965)
3. Bellman, R.: A markovian decision process. J. Math. Mech. **6**, 679–684 (1957)
4. Chandramohan, S., Geist, M., Lefèvre, F., Pietquin, O.: User simulation in dialogue systems using inverse reinforcement learning. In: Proceedings of Interspeech 2011, Florence (2011)
5. Daubigney, L., Gasic, M., Chandramohan, S., Geist, M., Pietquin, O., Young, S.: Uncertainty management for on-line optimisation of a POMDP-based large-scale spoken dialogue system. In: Proceedings of Interspeech 2011, Florence pp. 1301–1304 (2011)
6. Eckert, W., Levin, E., Pieraccini, R.: User modeling for spoken dialogue system evaluation. In: Proceedings of ASRU, pp. 80–87 (1997)
7. Frampton, M., Lemon, O.: Recent research advances in reinforcement learning in spoken dialogue systems. Knowl. Eng. Rev. **24**(4), 375–408 (2009)

8. Gasic, M., Jurcicek, F., Thomson, B., Yu, K., Young, S.: On-line policy optimisation of spoken dialogue systems via live interaction with human subjects". In: Proceedings of ASRU 2011, Hawaii (2011)
9. Georgila, K., Henderson, J., Lemon, O.: Learning user simulations for information state update dialogue systems. In: Proceedings of Eurospeech, Lisbon (2005)
10. Lagoudakis, M.G., Parr, R.: Least-squares policy iteration. J. Mach. Lear. Res. **4**, 1107–1149 (2003)
11. Lemon, O., Georgila, K., Henderson, J., Stuttle, M.: An ISU dialogue system exhibiting reinforcement learning of dialogue policies: generic slot-filling in the TALK in-car system. In: Proceedings of EACL'06, Morristown (2006)
12. Lemon, O., Pietquin, O.: Machine learning for spoken dialogue systems. In: Proceedings of InterSpeech'07, Belgium (2007)
13. Levin, E., Pieraccini, R.: Using markov decision process for learning dialogue strategies. In: Proceedings ICASSP'98, Seattle (1998)
14. Ng, A.Y., Russell, S.: Algorithms for inverse reinforcement learning. In: Proceedings of ICML, Stanford (2000)
15. Pietquin, O.: Consistent goal-directed user model for realistic man-machine task-oriented spoken dialogue simulation. In: Proceedings of ICME'06, Toronto, pp. 425–428 (2006)
16. Pietquin, O., Dutoit, T.: A probabilistic framework for dialog simulation and optimal strategy learning. IEEE Trans. Audio Speech Lang. Process. **14**(2), 589–599 (2006)
17. Pietquin, O., Geist, M., Chandramohan, S., Frezza-Buet, H.: Sample-efficient batch reinforcement learning for dialogue management optimization. ACM Trans. Speech Lang. Process. **7**(3), 7:1–7:21 (2011)
18. Pietquin, O., Rossignol, S., Ianotto, M.: Training bayesian networks for realistic man-machine spoken dialogue simulation. In: Proceedings of IWSDS 2009, Irsee (2009)
19. Schatzmann, J., Stuttle, M.N., Weilhammer, K., Young, S.: Effects of the user model on simulation-based learning of dialogue strategies. In: Proceedings of ASRU, Puerto Rico (2005)
20. Schatzmann, J., Thomson, B., Weilhammer, K., Ye, H., Young., S.: Agenda-based user simulation for bootstrapping a POMDP dialogue system. In: Proceedings of HLT NAACL, Rochester (2007)
21. Singh, S., Kearns, M., Litman, D., Walker, M.: Reinforcement learning for spoken dialogue systems. In: Proceedings of NIPS, Denver (1999)
22. Sutton, R., Barto, A.: Reinforcement Learning: An Introduction, 3rd edn. MIT, Cambridge (1998)
23. Williams, J.D., Young, S.: Partially observable markov decision processes for spoken dialog systems. Comput. Speech Lang. **21**(2), 393–422 (2007). DOI: http://dx.doi.org/10.1016/j.csl.2006.06.008

Chapter 32
Developing Non-goal Dialog System Based on Examples of Drama Television

Lasguido Nio, Sakriani Sakti, Graham Neubig, Tomoki Toda, Mirna Adriani, and Satoshi Nakamura

Abstract This paper presents a design and experiments of developing a non-goal dialog system by utilizing human-to-human conversation examples from drama television. The aim is to build a conversational agent that can interact with users in as natural a fashion as possible, while reducing the time requirement for database design and collection. A number of the challenging design issues we faced are described, including (1) filtering and constructing a dialog example database from the drama conversations and (2) retrieving a proper system response by finding the best dialog example based on the current user query. Subjective evaluation from a small user study is also discussed.

32.1 Introduction

Natural language dialog systems have so far mostly focused on two main dialog genres: goal-oriented dialog (such as ATIS flight reservation [1], DARPA communicator dialog travel planning [2]) and non-goal-oriented dialog (such as chatterbot systems like Eliza [3] or Alice [4]). Though various techniques have been proposed, data-driven approaches to dialog have become the most common method used in

L. Nio (✉)
Nara Institute of Science and Technology, Ikoma, Japan

Faculty of Computer Science, Universitas Indonesia, Depok, Indonesia
e-mail: lasguido@is.naist.jp

S. Sakti • G. Neubig • T. Toda • S. Nakamura
Nara Institute of Science and Technology, Ikoma, Japan
e-mail: ssakti@is.naist.jp; neubig@is.naist.jp; tomoki@is.naist.jp; s-nakamura@is.naist.jp

M. Adriani
Faculty of Computer Science, Universitas Indonesia, Depok, Indonesia
e-mail: mirna@cs.ui.ac.id

J. Mariani et al. (eds.), *Natural Interaction with Robots, Knowbots and Smartphones: Putting Spoken Dialog Systems into Practice*, DOI 10.1007/978-1-4614-8280-2_32,

dialog agent design. Example-based dialog modeling (EBDM) is one of several data-driven methods for deploying dialog systems. The basic idea of this approach is that a dialog manager (DM) uses dialog examples that are semantically indexed in a database, instead of domain-specific rules or probabilistic models [5]. With various sources of natural conversation examples, the usage of EBDM techniques has great potential to allow more efficient construction of natural language dialog systems.

Many studies have been conducted to develop technologies related to EBDM, such as a back-end workbench for implementing EBDM [6], query relaxation based on correlation for EBDM [7], and confirmation modeling for EBDM [8]. However, tedious and time-consuming design, collection, and labeling of a large set of user-system interactions are often required. Moreover, the scripted design scenarios in a lab typically result in unnatural conversations, with users responding differently from what is found in real situation. Consequently, many studies use EBDM to find the best responses or utilize template from available log databases [9]. To address this problem, some studies have proposed using Twitter data or crowdsourcing over large databases [10]. These techniques are also used by chat bots like Jabberwacky.[1] and Cleverbot[2] However, on the other hand, the issue of how to handle uncontrolled conversation content still remains.

One way to overcome these problems was proposed by [11] IRIS (Informal Response Interactive System), which uses a vector space model to implement a chat-oriented dialog system based on movie scripts [12]. Following their work, we further make improvements on the retrieval system by using a semantic similarity formula [13] with examples from drama television. The aim is to build a conversational agent that could interact with users as naturally as possible, while reducing the time requirement for database design and collection. One of the advantages of using examples from drama television is that the conversation content is more natural than scripted lab dialog design, since it contains some humorous dialog conversation. Yet, it is still within controlled drama scenes. More or less, drama television also affect the way people communicate. To build an example database, we propose a *tri-turn* unit for dialog extraction and semantic similarity analysis techniques to help ensure that the content extracted from raw movie/drama script files forms appropriate dialog examples.

32.2 System Overview

Figure 32.1 shows an overview of our system architecture. The system includes two components: (1) filtering and construction of a dialog example database from the drama conversations and (2) retrieval of a proper system response by finding the

---

[1] Jabberwacky—http://www.jabberwacky.com.

[2] Cleverbot—http://www.cleverbot.com.

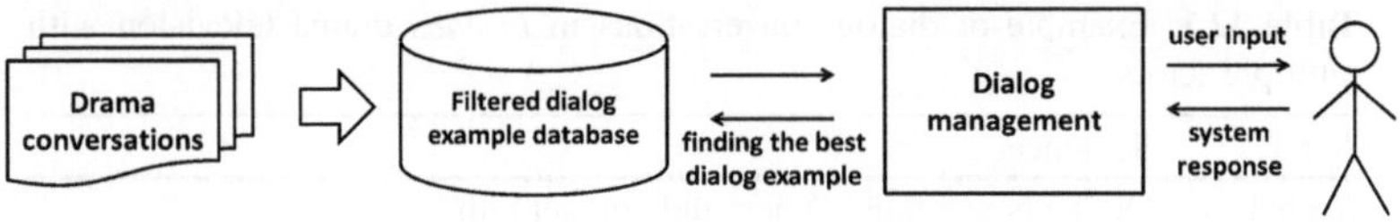

Fig. 32.1 System overview

best dialog example based on the current user query. Each of these components is described in the following sections.

32.3 Filtering Data

In EBDM, one of the important tasks is to filter and construct a dialog example database from the drama conversations. The challenge is that many drama dialog-turn conversations are not two-way "query-and-response" sentences. Even consecutive dialog turns may contain disjoint conversations from more than two persons/actors, which makes identifying the query and response difficult (see Table 32.1). In this study, to make sure the dialog examples are based on two-way "query-and-response" sentences, we select dialog data by proposing a concept called the trigram-turn sequence or *tri-turn*.

An example of a tri-turn dialog is shown in Table 32.2. The first and last utterances of the tri-turn are performed by the same person or actor (i.e., Joey), while the second turn is performed by another actor (i.e., Rachel). When a tri-turn pattern exists, we can generally assume that the two-actor conversation has a two-way "query-and-response" format.

After extracting the tri-turn from a dialog script, all words in all tri-turns were labeled by part of speech (POS) tagger and named entity (NE) recognizer. NE generalization was performed with normalizing all person or place name into general form such as "Joey" to "that man" or "Japan" to "that place."

Semantic similarity matching (similar to the approach introduced in [13]) is performed to ensure a high semantic relationship between each dialog turn in the dialog pair data. The formula requires two sentences (S_1 and S_2) and its synset (S_{syn1} and S_{syn2}) as an input. As shown in Eq. 32.1, the similarity is computed using WordNet[3] synsets in each dialog turn. Finally, the tri-turn dialogs exceeding a similarity threshold are extracted and included into the database

$$sem_{sim}(S_1, S_2) = \frac{2 \times |S_{syn1} \cap S_{syn2}|}{|S_{syn1}| + |S_{syn2}|}. \tag{32.1}$$

[3] Wordnet—http://wordnet.princeton.edu/.

Table 32.1 Example of dialog conversations in *Friends* drama television with multiple actors

Actor	Sentence
Rachel	Oh, he is precious! Where did you get him?
Ross	My friend Bethel rescued him from some lab.
Phoebe	That is so cruel! Why? Why would a parent name their child Bethel?
Chandler	Hey, that monkey's got a Ross on its ass!
Monica	Ross, is he gonna live with you, like, in your apartment?

Table 32.2 Example of a tri-turn with two actors from the *Friends* drama television

Actor	Sentence
Joey	I might know something.
Rachel	I might know something too.
Joey	What's the thing you know?

32.4 Dialog Management

The dialog management consists of two important elements, the dialog template and the response search. Both are described in the following.

32.4.1 Dialog Template

Figure 32.2 shows the overall dialog system template. It mainly consists of three conversations states: the *greeting* state, the *discussion* state, and the *farewell* state. The system responses for *greeting* and *farewell* states will be selected randomly from a handmade template combined with *greeting* and *discussion* examples in the database. For the *discussion* state, every time the system receives a user input, it generates the response with the highest similarity score from the example database. If no example is found, the system will respond "I don't understand what you mean" and send a new topic. To avoid repetitive responses, the system will search responses from dialog turns that have not been selected previously.

32.4.2 Retrieving Proper Response

A proper system response is retrieved by measuring both semantic and syntactic relations. These two measures are combined using linear interpolation as shown below:

$$sim(S_1,S_2) = \alpha \times sem_{sim}(S_1,S_2) + (1-\alpha) \times cos_{sim}(S_1,S_2). \qquad (32.2)$$

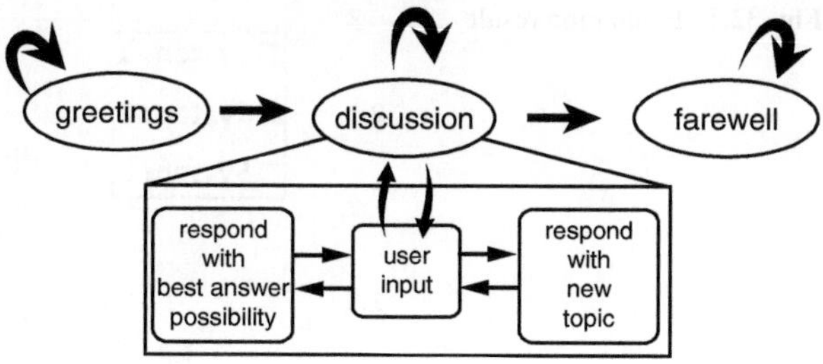

Fig. 32.2 Dialog system template

This value is calculated over the user input sentence (S_1) and every input examples on database (S_2). These values are calculated using semantic similarity in WordNet as a semantic factor and POS tag cosine similarity

$$cos_{sim}(S_1, S_2) = \frac{S_1 \cdot S_2}{\| S_1 \| \| S_2 \|} \tag{32.3}$$

as a syntactic factor. In this study, we assume the semantic factor is more important than the syntactic factor, so we set the interpolation coefficient α to be 0.7. Finally, if there is more than one retrieved example on database, the system will give more priority to the shortest response.

32.5 Experiments and Evaluation

We conduct our experiments using *Friends*[4] TV show scripts. Parsing the HTML data is done with the Perl CPAN HTML-Parser[5] and the system is built in the Python environment using the Python NLTK tools.[6]

The *Friends* TV show's scripts are written in English and contain 5 seasons, with a total of 112 episodes. Each episode contains several scenes and each scene contains several dialog turns. The total number of scenes and dialog turns in the corpus are 1,437 and 26,658, respectively. By applying the filter function, the total number of gathered tri-turns is 18,690. Based on semantic similarity, we finally extract and collect 1,901 tri-turn dialogs with high similarity.

The main objective of the evaluation in this preliminary study is to investigate the usefulness or effectiveness of different approaches in giving responses through the example database. Three different systems are evaluated: (1) only data filtering, (2) data filtering and NE generalization, and (3) data filtering, NE generalization, and priority selection on the shortest related answer.

[4]Friends Scripts—http://ufwebsite.tripod.com/scripts/scripts.htm.

[5]Perl CPAN—http://search.cpan.org/dist/HTML-Parser/Parser.pm.

[6]NLTK—http://nltk.org.

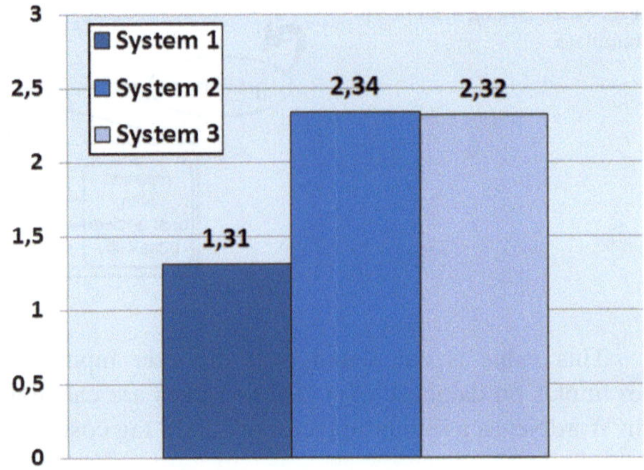

Fig. 32.3 Evaluation result

Table 32.3 Dialog conversations snippet between system and user

Actor	Sentence
System	Hi.
User	How are you?
System	What are you doing here?
User	I just wanted to see how it's going.
System	Well, it's going okay.
User	Where have you been?
System	Oh, just had a baby.
User	This is unbelievable.
System	So, what is this information worth to you, my friend?
User	It's not that bad.
System	Not that bad? Did you hear the hammer story?
User	Which one do you have?

All systems are evaluated by asking a human evaluator to give a score between 1 and 3 for every dialog-turn response. Score 1 means that the system is not giving a natural and relevant response to the user input and score 3 means that the system is giving a natural and relevant response to the user input. Each system is evaluated ten times by ten different evaluators and results are shown in Fig. 32.3.[7] It is expected that system 2 with full capabilities would outperform system 3 and system 1. However, it also reveals that the NE generalization gave a significant improvement in naturalness, while selecting a short controlled response had no significant difference. A conversation snippet between the user and system 2 can be seen in Table 32.3.

[7]The standard deviation from system 1 to 3 is 0.563, 0.768, and 0.764.

32.6 Conclusion

This paper addressed a non-goal EBDM dialog system based on examples from a TV show's dialog scripts. Filtering is performed to capture relevant dialog chat in the example corpus. We compared three different approaches to giving responses using an example database. The results reveal that the NE generalization from conversation in tri-turns gives a significant effect of naturalness, while selecting a short controlled response has no significant difference. However, much of the work shown in this paper is a preliminary work. Many improvements should be done to present a better non-goal dialog system. Future work could be done by adding a learning process to the system, so that it can remember the context of the conversation. Furthermore, compounding other examples from other data sources is also necessary to extend the system response.

References

1. Seneff, E., Hirschman, L., Zue, V.: Interactive problem solving and dialogue in the ATIS domain. In: Proceedings of the Fourth DARPA Speech and Natural Language Workshop, pp. 354–359 (1991)
2. Walker, M., Aberdeen, J., Boland, J., Bratt, E., Garofolo, J., Hirschman, L., Le, A., Lee, S., Narayanan, S., Papineni, K., Pellom, B., Polifroni, J., Potamianos, A., Prabhu, P., Rudnicky, A., Sanders, G., Seneff, S., Stallard, D., Whittaker, S.: DARPA communicator dialog travel planning systems: The June 2000 data collection. In: Proceedings of EUROSPEECH, pp. 1371–1374 (2000)
3. Weizenbaum, J.: Eliza - a computer program for the study of natural language communication between man and machine. Commun. ACM **9**(1), 36–45 (1966)
4. Wallace R.: Be Your Own Botmaster. A.L.I.C.E A.I. Foundation, California (2003)
5. Lee, C., Jung, S., Kim, S., Lee, G.: Example-based dialog modeling for practical multi-domain dialog system. Speech Commun. **51**(5), 466–484 (2009)
6. Jung, S., Lee, C., Lee, G.: Dialog studio: An example based spoken dialog system development workbench. In: Proceedings of the Dialogs on Dialog: Multidisciplinary Evaluation of Advanced Speech-Based Interactive Systems. Interspeech2006-ICSLP Satellite Workshop, Pittsburg (2006)
7. Lee, C., Lee, S., Jung, S., Kim, K., Lee, D., Lee, G.: Correlation-based query relaxation for example-based dialog modeling. In: ASRU, pp. 474–478. Merano (2009)
8. Kim, K., Lee, C., Lee, D., Choi, J., Jung, S., Lee, G.: Modeling confirmations for example-based dialog management. In: SLT, pp. 324–329. Berkeley, California (2010)
9. Murao, H., Kawaguchi., N., Matsubara, S., Yamaguchi, Y., Inagaki, Y.: Example-based spoken dialogue system using WOZ system log. In: SIGDIAL, pp. 140–148. Saporo (2003)
10. Bessho, F., Harada, T., Kuniyoshi, Y.: Dialog system using real-time crowdsourcing and Twitter large-scale corpus. In: SIGDIAL, pp. 227–231. Seoul (2012)
11. Banchs, R.E., Li, H.: IRIS: A chat-oriented dialogue system based on the vector space model. In: ACL (System Demonstrations), pp. 37–42 (2012)
12. Banchs, R.E.: Movie-dic: A movie dialogue corpus for research and development. In: ACL (2), pp. 203–207. The Association for Computer Linguistics (2012). URL http://dblp.uni-trier.de/db/conf/acl/acl2012-2.html
13. Liu, D., Liu, Z., Dong, Q.: A dependency grammar and wordnet based sentence similarity measure. J. Comput. Inform. Syst. **8**(3), 1027–1035 (2012)

Chapter 33
A User Model for Dialog System Evaluation Based on Activation of Subgoals

Klaus-Peter Engelbrecht

Abstract User models have become increasingly popular to conduct simulation-based testing of spoken dialog systems. These models usually describe users' overt behavior, as opposed to the underlying reasons for the observed actions. While such models are useful to generate test data, a causal model might be more generally applicable to different systems and, in addition, allows to derive useful information for data analysis and prediction of user judgments. Thus, a modeling approach trying to explain user behavior is proposed in this paper, which is based on Dörner's PSI theory. The evaluation shows that the utterances generated by this model are similar to those of real users.

33.1 Introduction

Spoken dialog systems (SDS) are supposed to enable an efficient and intuitive interaction between humans and machines. Unfortunately, however, the users of such systems often encounter errors in automatic speech recognition (ASR) or natural language understanding (NLU), which can lead to situations in the dialogs which were not anticipated by the system designer. The development of SDSs is thus complex and requires exhaustive tests of the application.

In research, user models have become increasingly popular to conduct simulation-based testing [3, 8, 10]. Some other researchers, who try learning dialog strategies from data, use user models to generate a large number of training dialogs (e.g., [11, 13]). The user model produces a user action based on the previous system action and the dialog history. System actions are represented as dialog acts, which may be further specified using attributes (e.g., *query(foodtype)* or

K.-P. Engelbrecht (✉)
Quality and Usability Lab, Telekom Innovation Laboratories, TU Berlin, Ernst-Reuter-Platz 7, D-10589 Berlin, Germany
e-mail: Klaus-Peter.Engelbrecht@telekom.de

J. Mariani et al. (eds.), *Natural Interaction with Robots, Knowbots and Smartphones: Putting Spoken Dialog Systems into Practice*, DOI 10.1007/978-1-4614-8280-2_33,

no_match()) or attribute-value-pairs (AVPs) in case of confirmation acts (e.g., *confirm(foodtype=Italian)*). User actions can be represented on the concept level [5, 11, 13], in which case they can be described by AVPs, on the wording level [3], or on the utterance level [3, 8], depending on the system component to optimize.

Currently, most user models describe the overt behavior of users in a probabilistic manner [6, 11, 13] and do not consider cognitive processes inside the user. Previous research has shown that such simple user models are useful to detect design errors in an interface [3, 6] and to optimize different aspects of the system [8] by performing simulation experiments. However, the models used in these studies cannot readily be applied to arbitrary systems, since they can only react properly to dialog contexts defined for the model, and the set of potential dialog situations is very large. Also, the user actions in a dialog context may depend on subtle differences, such as the wording of the system prompt. Thus, a *causal* user model explaining *why* a user action is observed, instead of defining a statistical distribution of actions, is required to simulate dialogs with arbitrary systems.

In addition, while the generation of dialogs is possible with the present user models, the resulting dataset, which may be very large, has to be searched manually for potential problems and design errors. An indication of the users internal state during the dialogs can be useful to find such problematic situations more quickly. For example, a model of the users belief about the current dialog state has been used to detect if the confirmation strategy of the system always ensures correct grounding [7]. Similarly, a causal user model may allow to detect dialog contexts where it is unclear to the user what he/she is required to say or where the system does not proceed through the dialog in a consistent manner. Such information can also be indicative of the perceived quality of a SDS, as it has been shown for grounding-related parameters in [7]. It may thus be useful as input to judgment prediction models, such as PARADISE [14].

The user model proposed in this paper owes much to the PSI theory of Dörner [4]. In a connectionist manner, Dörner models the motivation and affect of an agent based as emergent properties of a network composed of neuron structures. The resulting agents learn autonomously how to survive on an island, which involves finding food and water and avoiding dangers. Their behavior is driven by *needs*, such as hunger, thirst, or physical integrity. For example, the motivation to drink water (or look for some) would be activated by the need *thirst*. Usually, several motifs are active at the same time, but the degree of activation may differ between them. For the selection of a motif, the strength of its activation, and also the expectation of successful accomplishment of the respective action, is evaluated. In addition, a selection threshold moderates how easily the agent switches motifs.

Most of the needs postulated by Dörner are not relevant in the context of human-machine interaction (HCI). Needs for food and water will rarely impact the interaction. However, the PSI theory specifies basic principles and algorithms of problem solving, motivation, and affect and is thus a good starting point for modeling the behavior of users of SDSs.

33.2 Proposed User Model

33.2.1 Needs and Values

Like in the PSI theory, we assume that all user actions are initiated by needs. A need, such as *hunger*, can vary in its strength, depending on the situation the agent is in. For example, the need increases if energy is consumed, e.g., by doing sports. In turn, the need decreases if it is satisfied, e.g., by eating. As this example illustrates, needs can be associated with actions impacting their strength, e.g., *eating* or *doing sports*. Actions which reduce needs (in this case *eating*) get a *value* assigned. The value of an action is proportional to the strength of the associated need. Thus, an action satisfying a strong need is more valuable than an action satisfying a minor need.

Needs governing user behavior in an interaction with a SDS are probably less basic than *hunger* or *thirst*. A reasonable assumption would be that the need motivating the user actions is simply the need to complete a task. There may be larger goals behind this need, for example, the user might look for a train connection in order to visit a friend (the respective need could be *affiliation*). However, since the model aims at explaining the microworld of the interaction, the assumption that the user is motivated by the need to complete the task may be sufficient.

The user's task can be divided into subgoals, which are roughly equivalent with the information units the user wants to enter or obtain from the system. We can then assign a need to each of these subgoals. If a subgoal is completed, the strength of the associated need should be zero, as there should be no motivation to complete it again. On the other hand, before the goal is completed, the need should be stronger (e.g., one), leading to a higher value of actions which will reduce it. The need to complete a goal may be even higher than this if it has to be accomplished very urgently. For example, if a constraint was misunderstood by the system, the need to correct it could be stronger than in the case that it just has not been accomplished.

In order to determine the needs (or, respectively, the values) associated with each subgoal, a model of the belief the user has about the current dialog state, as in [7], is useful. The *dialog state* can be described by a set of system *slots* (i.e., variables which the user can fill by interacting with the system) and the values currently set for each slot. Thus, it captures which constraints were already understood by the system and which were misunderstood. Similar to the dialog state, the *user task* can be described in terms of desired slot-value-pairs or *constraints*. The *believed dialog state* is structured like the actual dialog state; however, it is inferred entirely from the information exchange between the user and the system, that is, only information visible to the user during the interaction is taken into account. In other words, it represents the dialog state from the user's perspective. If the dialog is correctly grounded, the believed system state equals the actual dialog state.

Figure 33.1 illustrates the relation between the above concepts and provides some examples of the strength of the needs associated with each subgoal, depending on the situation. Note that the strength of the needs does depend not on the actual

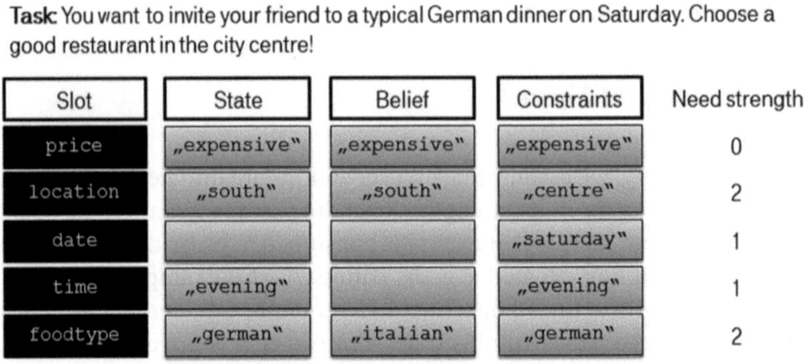

Fig. 33.1 Illustration of the relation between system slots, system state, the user's belief about the system state, and task constraints. The rightmost column shows the strength of the need associated with each constraint in the given situation. The task is described verbally to the users, as displayed at the top of the figure

system state but on the system state believed by the user. Thus, a constraint can be associated with the need strength 1 although the respective system slot is correctly filled.

33.2.2 Probability of Success

Obviously, the value of an action is not the only factor determining if (or when) it is taken by the user. Otherwise, the users would always take all actions leading to a need reduction immediately. Rather, users seem to evaluate a given dialog context with respect to the chances that an utterance will be understood by the system. In other words, they take into account the expected probability that the action will be successful.

Generally, the expected probability of success of an action depends on the dialog context (in particular the system prompt), the previous experience of the user with the system (i.e., the dialog history), and the user's general confidence in interacting with the system (in the terms of the PSI theory, we could speak of the *situation*, the *specific*, and the *general competence* of the user).

The features of the system prompt impacting the expected probability of success are manifold. For example, constraints corresponding to the slot the system asked for will likely be understood more easily than other constraints. Also, the prompt could be formulated in a way indicating the amount of user initiative allowed in this situation. Moreover, on the wording level, vocabulary used by the system will probably also be understandable to it (the user aligns her vocabulary to the system in the sense of *audience design*; cf. [2]).

Regarding the previous experience of the user with the system, recognition errors are probably the main determinant of the expected probability of success. Similarly, the general confidence of the user in interacting with the system may be determined by recognition performance encountered when interacting with other systems or using other speech recognition software. In addition, the general confidence may moderate how events in the dialog history are coped with by the user. For example, a confident user may not be impacted by the experience made in the current dialog as much as a less confident user.

While this brief discussion shows that there are multiple factors which could be incorporated in the calculation of the expected probability of success, at this stage we implemented a simple algorithm only: the constraint the system asked for receives a fixed high probability of success, whereas all other constraints receive a lower probability, whose value varies between users in order to model different degrees of user initiative.

33.2.3 Planning

Some actions which have to be taken during a dialog once in a while are not immediately related to constraints or needs, in particular

- Affirmation or negation of confirmed input
- Acceptance or negation of offers made by the system (e.g., to relax a constraint)
- Dialog control actions (naming slot to fill next, restarting, going back one dialog step)

In order for the users to take such actions, they would have to be associated with a need. However, different to the task-related subgoals, the user should not have a need to pursue these actions in an arbitrary dialog situation. Rather, they should be used in specific situations to support the accomplishment of the actual subgoals, i.e., entering of the task constraints. In fact, in accordance with the PSI theory, we can assume that in situations where no constraint can be entered directly, the user *plans* how to reach a situation where this is possible.

There exist a number of planning algorithms, which are usually based on either logical inference, path search, or learning of action sequences. For example, the PSI agents learn how a given situation is modified by the available actions. In the user model proposed here, planning is currently represented differently and more simply, based on values and success probabilities of actions.

In reference to the concept of information scent, the method used for planning could be termed "need scent." Information scent is used to predict the link on a web page a user would click on, given a verbally described task or concept to look for [1]. Technically, it could be described as the semantic similarity between the link text and the task description. According to the model presented here, it could be said that links may inherit *value* from the actual goal, depending on their similarity to this goal.

Similarly, actions like those mentioned above could be considered to inherit value from the subgoals they are related to. Since actions are selected based on their value and their probability of success, such an action may then be chosen instead of providing the constraint immediately, if its probability of success is higher (e.g., if the system asked for a logical or for a slot name). The value of subgoal-related actions is inherited to other actions as follows:

- Explicit confirmation: This asks for an AVP with the attribute "logical"; the negation action (AVP {*logical="no"*}) receives a value equal to the number of confirmed constraints which are misunderstood, multiplied by 2. The AVP {*logical="yes"*} receives a value equal to the number of correctly confirmed constraints minus the value of {*logical="no"*}. Thus, the values are equal to the need reduction (or increment) achieved when the AVP is successfully communicated to the system.
- Constraint relaxation offer: This occurs when database search failed and asks for a "logical" AVP. In this case, the user model first decides if a constraint will be relaxed, in which case the respective value is changed accordingly. As the new value is then different from the value believed to be stored in the system, this causes the strength of the associated need to provide this constraint to raise. The accept action (AVP {*logical="yes"*}) allows in this situation to reduce this need in the next step and thus receives a value corresponding to this need's strength.
- Naming slots to fill next: In the analyzed system, this action allows to control the dialog flow. It receives a value which is a fraction of the value of the respective constraint, which could be understood as *spreading activation*. The size of this fraction is a user parameter.

33.2.4 Action Selection

Actions are selected based on their *activation*, which can be calculated as the product of their *value* and their *probability of success*. Thus, if either the value or the probability of success is low, the activation will be low, too.

As illustrated in Fig. 33.2, several actions may be activated in a given situation. In most cases, the user will not combine all these actions to a single utterance. Different options for the action selection are thus possible. The two most likely options are that either (1) only the most activated action or (2) all actions whose activation exceeds a (possibly user-specific) threshold are selected. In order to allow for the users to include several actions in an utterance, the second option was followed.

It turned out that two additional modifications to the final distribution of overall activations are useful. Firstly, several actions of the same type, such as {*logical="yes"*} and {*logical="no"*}, often do not make sense in one utterance. Thus, of these actions only the one with the highest activation should be available in the selection, and the remaining ones should be inhibited. Secondly, noise can be imposed on the value of the constraints to simulate unequal importance of the subgoals.

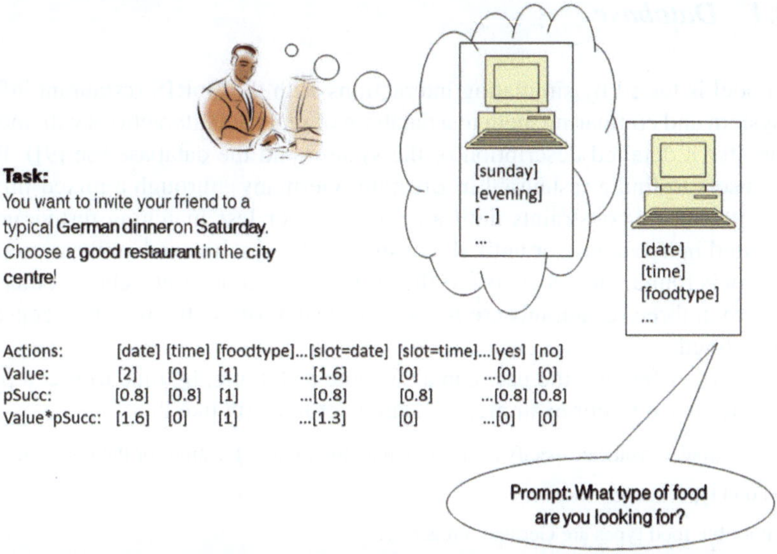

Fig. 33.2 Overview of parameters involved in the action selection. In the example, one of the slots is believed to be set incorrectly (value=2), and one is not set (value=1). These values are propagated to the slot-naming actions. The probability of success (pSucc) is highest for the constraint the system asks for (foodtype). All other actions have an equal probability of success. After multiplying value and pSucc, several actions are activated. The actual choice depends on the action selection procedure

33.3 Evaluation

In the introduction of this paper, it was claimed that a model like the one presented in the previous section would provide useful information for data analysis or prediction of user judgments. Another claim was the improvement in generalizability across different systems. Before the validity of these claims will be evaluated for the proposed model, a proof-of-concept test should be conducted to verify that it does generate behavior similar to that of real users at all.

Showing similarity to real user behavior is tricky, and there are several methods and metrics which all focus on specific aspects of the user behavior or the dialogs simulated. Since the novelty of the model proposed in this paper mainly lies in the way actions are selected to form an utterance, this aspect was focused in the evaluation. Thus, as proposed by Schatzmann et al. [12], utterances produced by the model are compared to those produced by real users. However, while Schatzmann et al. compare the utterances in the context of the current user state, we ignore the user state and simply compare the sets of utterances in both corpora. The reason is simply that currently we are not able to determine the user state in terms of our model for real user data.

33.3.1 Database

The model is tested by simulating interactions with the BoRIS restaurant information system and comparing them to a database of real user interactions with the same system (for a detailed description of the system and the database see [9]). BoRIS allows users to find a restaurant in Bochum (Germany), through a mixed-initiative dialog. It collects constraints until a set of three or less matching restaurants has been found in a database or until all constraints the system can handle are set. If no restaurant is found, the system offers the user to relax constraints. On the other hand, if more than three restaurants are found, the system offers the user that constraints can be refined.

As a further feature, the user can also name a slot and, by this, trigger a prompt targeted to this slot. For example, the dialog could begin like this:

S: Welcome … you can specify a date, daytime, food types, locations or the price range.

U: Food type.

S: Possible food types are German, French,…

The analyzed data stem from an experiment performed to examine the impact of confirmation strategy (explicit or no confirmation), system voice (recorded male, recorded female, or TTS), and ASR performance (target word accuracy of 60, 70, 80, 90, and 100 %) on the user ratings. To achieve the specific ASR performances, user utterances were transcribed by a Wizard-of-Oz, and ASR errors were artificially induced on the transcriptions.

Forty users performed five tasks each. Three dialogs had to be removed due to technical problems, leading to 197 dialogs (2,003 turns). Four tasks were predefined by the experimenter, and one was defined by the user before that trial. Special care was taken for the users to behave naturally and in a variable way: the predefined tasks were partly described nonverbally to avoid priming effects, and some included the specification of a constraint to relax if no restaurant was found. In the latter case, either a new constraint or just the slot to change was specified.

In addition, some tasks were not fully defined, i.e., not for all slots a value was specified in the task description. For example, one task was to find a restaurant that serves duck. For other attributes, like the price range or the location of the restaurant, either the user could provide the value "neutral" or he/she could invent constraints if he/she felt this was necessary. In order to avoid that such invented constraints blur the comparison to the simulated dialogs, the database was restricted to the only task which was fully specified (40 dialogs).

33.3.2 Simulation Settings

Before the user model could be tested, settings for the free model parameters had to be defined. At the current state of the implementation, all parameters are static throughout a dialog, i.e., no learning effects are simulated.

- The *value* of a constraint not specified so far is set to 1 plus a random number sampled from the interval [0; *noiseLevel*]. By adding a random number to each value, a random prioritization of constraints is achieved. This accounts for the observation that the real users occasionally prioritized some subgoals, although the task descriptions given to the test users did not specify the relative importance of the constraints. In case of misunderstanding, the value of the respective constraint is multiplied by 2.
- The *noiseLevel* specifies how strongly the constraints are prioritized (as explained above), which can be considered a user parameter.
- *pSuccAsked* is the expected probability of an action to be successful if the system asked for the respective attribute (e.g., *logical* or *date*) and is always set to 1.
- *pSuccNotAsked* represents a user characteristic which could be termed *degree of initiative* or, in Dörner's terms, *general competence*. Since different types of users should be simulated, different values for this parameter are assigned in each iteration of the simulation.
- *dampEqualAttributes* also impacts the degree of initiative, but was assumed to be equal for all users, as one free parameter to control this should be sufficient. In the simulations presented below, it was set to 0.8.
- *needScent* quantifies the fraction of activation spread from constraints to the respective slot-naming actions. Since plans are encoded in this parameter, it relates to the planning depth of the user. A small value means that the user mostly follows its immediate goals (i.e., submit constraints). Even if the success probability of this immediate action is low and the probability to successfully submit a slot-naming action is high, it would rather say nothing than name the slot.
- *threshold* determines the minimum activation required for an action to be selected. The meaning of this parameter in terms of user characteristics is not well specified at the moment. Thus, no educated guess about an appropriate setting could be made. For our simulation, a different value was used in each iteration.

Since the values of the user and task-specific parameters are not known for the real users, for the evaluation they were sampled randomly from intervals expected to lead to humanlike behavior (all [0; 1], except *threshold*: [0.5; 1]).

33.3.3 Results

Two thousand dialogs were simulated with the model and the parameter settings specified above. Table 33.1 shows that in the simulation, four times as many unique user turns were generated as observed in the real user corpus. Nevertheless, only 53 % of the real user turns were generated during the simulation (*Recall*). The observed *Precision* (i.e., proportion of simulated utterances also observed in the corpus) is even lower; however, over-generation of user turns is somewhat expected,

Table 33.1 Recall and Precision of simulated user turns compared to those found in the empirical corpus

N(emp)	40
N(sim)	161
N(common)	21
Recall	0.53
Precision	0.13
MAD	0.005

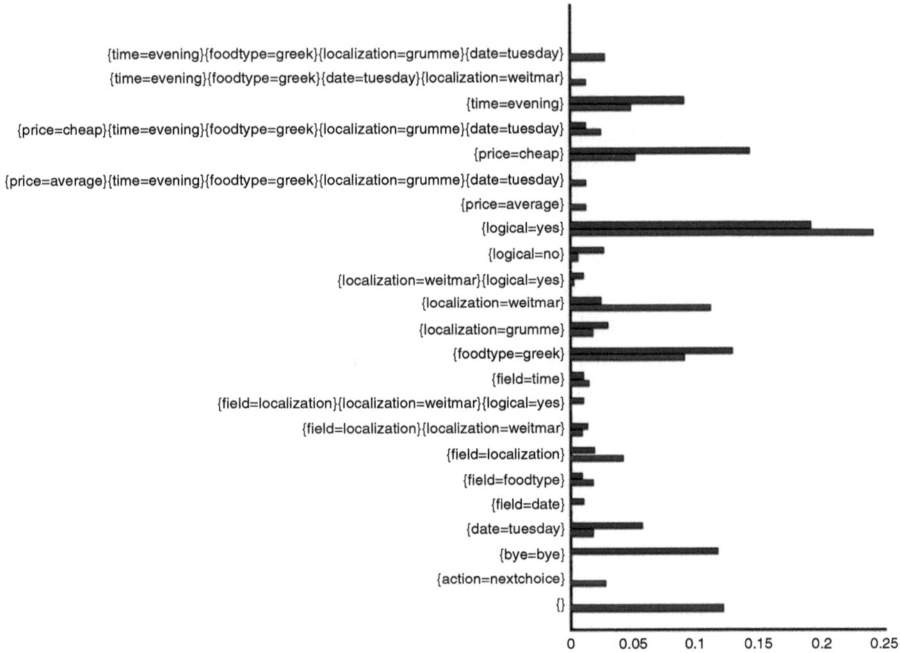

Fig. 33.3 Comparison of probabilities of utterances in empirical (*blue*) and simulated (*red*) corpus

since a small number of real user dialogs was analyzed, which were not expected to represent the entirety of possible user behaviors.

We also calculated the mean absolute difference (*MAD*) of the prior probabilities of each user turn in the empirical and the simulated corpus. This measure is very small, indicating that despite the relatively low numbers in *Recall* and *Precision* the overall distributions of turns are quite similar in both corpora. This indicates that user turns which are present in one corpus only appeared there infrequently. In fact, the majority of utterances were observed to be very infrequent. Figure 33.3 shows a comparison of frequencies of different user turns in both corpora. Only the more frequent turns, with a prior probability above 0.01, are listed. It can be seen that the distributions of these more frequent actions are somewhat similar. Note that in order to simplify the implementation of the simulation the user model ended its dialogs with the {*bye=bye*} action, whereas real users did not reply to the final system prompt ({} in the figure). Thus, the results for these actions are very similar as well.

33.4 Discussion and Conclusion

This paper presented a new approach to user modeling, which is based on connectionist models of human behavior rather than a probabilistic description of user behavior in different dialog contexts. To our knowledge, such connectionist models have not yet been applied to human-machine interaction.

The advantage of the model, as compared to a probabilistic model, is that its parameters are directly related to characteristics of the users and the task. Thus, their modification allows to directly model different types of users. In addition, the model may be suitable to explain *why* a specific behavior of the user model was observed.

The evaluation presented in this paper should be regarded as a proof of concept. A comparison to previous user models should be performed, which however should also take into account the practical advantages claimed for the proposed model, such as improved generalizability to other systems or the possibility to derive useful predictors of user judgments from internal model parameters. Before this can be achieved, considerably more experience with the new model is required.

While at the current state the user behavior depends on a few system features only, extensions to the model are easily thinkable. For example, the concept of *needScent* could be used to model activation of semantically related concepts (*spreading activation*), explaining why some constraints are more likely to be named in the same turn than others. Also, activation could be assumed to gradually decay if a constraint was activated, but not used (or not understood by the system), which in effect would implement a memory. This may then also be related to the user's perception of dialog consistency.

As a more long-term goal, we will try to implement emotional user behavior with this modeling approach. As a first step, the involved entities need to be adapted dynamically during the interaction. We will then analyze which needs are responsible for emotional reactions in dialogs and will try to understand (and model) how these needs manifest in human-machine dialogs.

References

1. Blackmon, M., Kitajima, M., Polson, P.: Tool for accurately predicting website navigation problems, non-problems, problem severity, and effectiveness of repairs. In: Proceedings of CHI '05, pp. 31–40 (2005)
2. Branigan, H., Pickering, M., Pearson, J., McLean, J.: Linguistic alignment between humans and computers. J. Pragmat. **42**, 2355–2368 (2010)
3. Chung, G.: Developing a flexible spoken dialog system using simulation. In: Proceedings of ACL 2004, pp. 93–98 (2004)
4. Dörner, D.: Die Mechanik des Seelenwagens. Eine neuronale Theorie der Handlungsregulation, 1st Edition, Verlag Hans Huber, Bern (2002)
5. Eckert, W., Levin, E., Pieraccini, R.: User modeling for spoken dialogue system evaluation. In: Proceedings of the IEEE Workshop on Automatic Speech Recognition and Understanding, pp. 80–87 (1997)

6. Engelbrecht, K.-P.: Estimating spoken dialog system quality with user models. Ph.D. Thesis, Berlin Institute of Technology, Berlin (2012)
7. Engelbrecht, K.-P., Möller, S.: Correlation between model-based approximations of grounding-related cognition and user judgments. In: Proceedings of Interspeech 2012 (2012)
8. López-Cózar, R., Callejas, Z., McTear, M.: Testing the performance of spoken dialogue systems by means of an artificially simulated user. Artif. Intell. Rev. **26**, 291–323 (2006)
9. Möller, S.: Quality of Telephone-Based Spoken Dialog Systems. Springer, New York (2005)
10. Möller, S., Schleicher, R., Butenkov, D., Engelbrecht, K.-P., Gödde, F., Scheffler, T., Roller, R., Reithinger, N.: Usability engineering for spoken dialogue systems via statistical user models. In: Proceedings of IWSDS '09 (2009)
11. Pietquin, O.: A framework for unsupervised learning of dialogue strategies. Ph.D. Thesis, TCTS Lab, Mons (2004)
12. Schatzmann, J., Georgila, K., Young., S.: Quantitative evaluation of user simulation techniques for spoken dialogue systems. In: Proceedings of SIGDial, pp. 45–54 (2005)
13. Schatzmann, J., Thomson, B., Weilhammer, K., Ye, H., Young, S.: Agenda-based user simulation for bootstrapping a pomdp dialogue system. In: Proceedings of HLT/NAACL, pp. 149–152 (2007)
14. Walker, M., Litman, D., Kamm, C., Abella, A.: Paradise: A framework for evaluating spoken dialogue agents. In: Proceedings of ACL/EACL, pp. 271–280 (1997)

Chapter 34
Real-Time Feedback System for Monitoring and Facilitating Discussions

Sanat Sarda, Martin Constable, Justin Dauwels, Shoko Dauwels (Okutsu), Mohamed Elgendi, Zhou Mengyu, Umer Rasheed, Yasir Tahir, Daniel Thalmann, and Nadia Magnenat-Thalmann

Abstract In this chapter, we present a system that provides real-time feedback about an ongoing discussion. Various speech statistics, such as speaking length, speaker turns and speaking turn duration, are computed and displayed in real time. In social monitoring, such statistics have been used to interpret and deduce talking mannerisms of people and gain insights on human social characteristics and behaviour. However, such analysis is usually conducted in an offline fashion, after the discussion has ended. In contrast, our system analyses the speakers and provides feedback to the speakers in real time during the discussion, which is a novel approach with plenty of potential applications. The proposed system consists of portable, easy to use equipment for recording the conversations. A user-friendly graphical user interface displays statistics about the ongoing discussion. Customized individual feedback to participants during conversation can be provided. Such close-loop design may help individuals to contribute effectively in the group discussion, potentially leading to more productive and perhaps shorter meetings. Here we

S. Sarda • J. Dauwels (✉) • U. Rasheed
School of Electrical and Electronic Engineering, Nanyang Technological University, Singapore, Singapore
e-mail: sanat.sarda@gmail.com; jdauwels@ntu.edu.sg; umer1@e.ntu.edu.sg

M. Constable • Z. Mengyu
School of Art, Design, and Media, Nanyang Technological University, Singapore, Singapore
e-mail: mconstable@ntu.edu.sg; zhou0138@e.ntu.edu.sg

S. Dauwels (Okutsu)
Centre of Innovation Research in Cultural Intelligence and Leadership (CIRCQL), Nanyang Business School, Singapore, Singapore
e-mail: sdauwels@ntu.edu.sg

M. Elgendi • Y. Tahir • D. Thalmann • N. Magnenat-Thalmann
Institute for Media Innovation, Nanyang Technological University, Singapore, Singapore
e-mail: elgendi@ntu.edu.sg; yasir001@e.ntu.edu.sg; danielthalmann@ntu.edu.sg; nadiathalmann@ntu.edu.sg

J. Mariani et al. (eds.), *Natural Interaction with Robots, Knowbots and Smartphones: Putting Spoken Dialog Systems into Practice*, DOI 10.1007/978-1-4614-8280-2_34,

present preliminary results on two-people face to face discussion. In the longer term, our system may prove to be useful, e.g. for coaching purposes and for facilitating business meetings.

34.1 Introduction

People have varying individual characteristics, personality, status, intelligence, maturity and language among others. All these aspects in different combinations result in individual speaking mannerisms, such as how much a person speaks during a conversation or how much he or she interrupts another person while speaking [1]. Talking mannerisms of individuals play an important factor for meetings to be productive and achieve certain objectives. If talking mannerisms become mutually compatible or aligned, the meetings are likely to be more productive and efficient [2]. Our long-term objective is to develop systems that provide real-time feedback about social behaviour in conversations, helping speakers to adjust their talking mannerisms to each other. Such systems may help to boost the effectiveness of job interviews, group discussions, coaching sessions or public speaking.

In the fields of psychology and cognitive science, human behaviour is often studied from the perspective of social interactions [3, 4]. Traditionally, expert observers take notes during conversations. Alternatively, audio and video recordings of conversations are analysed manually by experts [5]. Both approaches are time-consuming and unavoidably subjective. Recent advances in recording equipment and signal processing may ultimately enable automated and real-time analysis of talking mannerisms and social interactions at large, yielding more objective results. Several studies in that direction have been conducted in recent years, to deduce individual characteristics like dominance status [6, 7], emerging leadership [8] and other personality-related traits [9, 10]. In such studies, various statistics of the conversation are extracted, e.g. natural turns, turn duration, speaking percentage, interruptions and failed interruptions. Also, the combination of speech and visual features has been shown to provide increased accuracy for detecting characteristics such as dominance and leadership [11].

However, in none of those studies [6–11], social interactions are analysed in *real time;* instead various corpora of audio and video recordings are analysed offline [12]. Many corpora of audio and video signals are available related to small-group interactions (see [13] for a survey). These corpora are continuously updated with manual annotations, which can be used as gold standard to assess automated analysis methods. The systems presented in [13, 14] have the capability to provide real-time feedback; however, those systems do not use speech or video signals, but rather crude signals. Consequently, subtle talking mannerisms may not be detectable through such approach.

In the present work, we propose a system that provides real-time feedback about talking mannerisms, generated from speech and video signals. The system first extracts numerous speaking statistics from those signals, most of which are similar to features considered in recent studies [6–14]. Machine learning algorithms further process those statistics, to extract higher-level characterization of the speaking mannerisms. That information is eventually exploited to generate real-time feedback for every participant in the meeting. It can inform the speakers about their speaking mannerisms and, if needed, provide guidelines.

In this work, we limit ourselves to automatic analysis of conversations of two persons. Such scenario is relevant for coaching, interviews and business meetings. In the future, we plan to scale our system towards small-group interactions.

This chapter is structured as follows. In Sect. 34.2, we elaborate on the speech statistics extracted by our system. In Sect. 34.3, we explain our implementation, including the recording setup, voice activity detection (from audio and video signals) and our design of graphical user interface. In Sect. 34.4, we outline the proposed framework for offline and real-time feedback. In Sect. 34.5, we present our conclusions and make suggestions for future work.

34.2 Non-verbal Speech Cues

The core of our system consists of simple (and hence fast) signal processing algorithms that detect who is speaking and when and use that information to compute various statistics about non-verbal speaking mannerisms. In this section, we elaborate on the latter statistics. In the next section, we will explain how we determine who is speaking and when, from audio and video signals.

34.2.1 Non-verbal Speaking Statistics

We compute various simple non-verbal speaking statistics (see Fig. 34.1). Each of those statistics can be computed in real time. Specifically, the following non-verbal speaking statistics are considered:

Speaking %: The percentage of time a person speaks in the conversation.
Voicing Rate: The number of syllables spoken per minute.
Pitch: Pitch of speech is calculated using the Voice-box Toolbox [15].
Natural Turn-Taking: The number of times person "A" speaks in the conversation without interrupting person "B".
Silence: The percentage of time when both participants are silent.
Interruption: Person "A" interrupts person "B" while speaking and takes over. Person "B" stops speaking before person "A" does (see Fig. 34.1).
Failed Interruption: Person "A" interrupts person "B" while speaking but stops speaking before person "B" does (see Fig. 34.1).
Interjection: Short utterances such as "no", "ok", "yeah and exactly" (see Fig. 34.1).

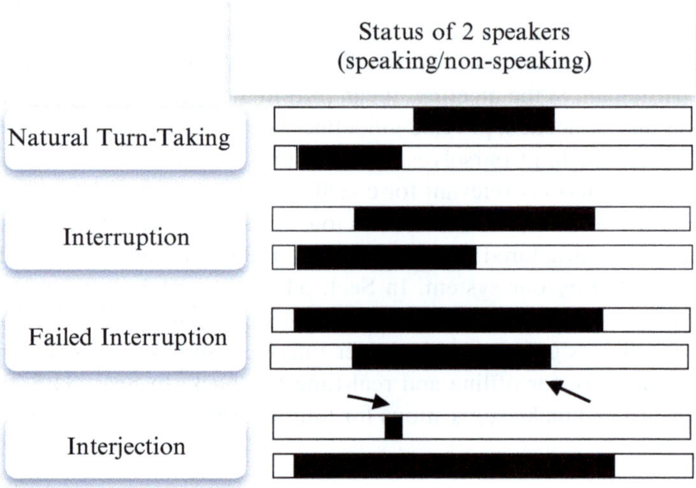

Fig. 34.1 Illustration of turn-taking, interruption, failed interruption and interjection derived from binary speaking status (speaking and non-speaking). Periods of speaking and non-speaking are indicated in *black* and *white* respectively

Speaker Turn Duration: Average duration of each speaker turn.

Overlap Percentage: Percentage of time when both persons speak at the same time during the conversation.

Each of the above statistics, individually and in combination, portrays different characteristics of people's behaviour during the discussion. For instance, high speaking percentage may indicate dominant behaviour of a person, whereas a high interruption count may indicate aggressive behaviour. In other words, we may be able to interpret various personality traits and dynamics of conversations from these basic statistical measures.

34.2.2 Interpretation of Non-verbal Speaking Statistics

It may not always be always straightforward to directly interpret the non-verbal speaking statistics. For instance, the number of interruptions during a debate or friendly conversation may be identical. However, the interpretation of those statistics may be quite different. In other words, the context plays an important role in interpreting the statistical measures. To address this issue, we suggest recording many conversations, in a large variety of settings. Next, from each of those recordings, one may compute the statistical measures. For each setting, histograms of the statistical measures can be generated, and based on those, one can set thresholds. The latter are then used to assess discussions. For instance, if the number of interruptions is above a threshold, the person is considered as highly interrupting; if that number is below another threshold, the speaking behaviour is considered as non-interrupting. The thresholds can be chosen as statistical quantiles. Alternatively, unsupervised learning (e.g. k-nearest neighbours clustering) may be applied to identify clusters in the statistical features across a large number of recordings. Such clusters may correspond to different speaking behaviours. By identifying to which cluster an ongoing discussion belongs, the system may be able to identify speaking behaviours in real time.

34.3 Implementation

In this section, we describe our system for extracting non-verbal speaking statistics from ongoing discussions. First we explain the recording hardware and next we elaborate on speaker segmentation and our graphical user interface (GUI).

34.3.1 Sensing and Recording

In our setup, we use two tabletop microphones, one per person (see Fig. 34.2), and a ZOOM H4n portable voice recorder which allows us to record from multiple speakers simultaneously. That voice recorder has flexible sampling rates and bit resolutions. We selected a sampling rate of 8 kHz; as such low rate suffices for our purposes. The microphones are connected to the recorder via balanced XLR connectors. The recorder acts as an interface and is connected to the laptop via USB (see Fig. 34.2).

For online recording and real-time feedback, recordings are saved directly on the laptop. With the H4n recorder acting as interface, recordings from two separate tabletop microphones are recorded synchronously without any delay. The setup with H4n recorder provides easy, quick, cost-effective, portable and, most importantly, undistorted signal recording. We refer to [16] for an excellent review of computer-based audio recordings.

It is important to use suitable microphones and connectors to acquire undistorted original speech signals. Microphones are required to have a flat frequency response, in order to preserve the original speech energy and spectrum, optimally sensitive to allow talking from comfortable distance. Moreover, the microphones are recommended to bidirectional limit interfering signals and to reduce background noise. The connectors should be balanced to reject line noise interference. Overall, the recording system should not be imposing, in order not to disturb the speakers. For our present recording, we use Sennheiser e845s microphones with XLR connectors, as those components fulfil all the requirements. That solution is limited to two-

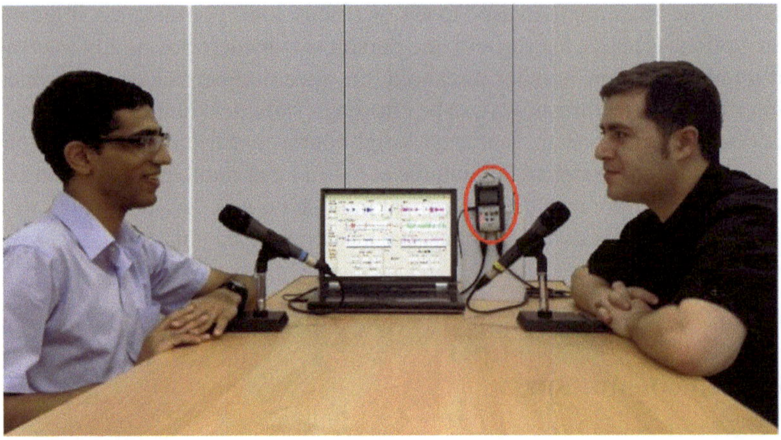

Fig. 34.2 Recording system consisting of the Zoom H4n voice recorder (*circled in red*) and two Sennheiser e845s microphones, one for each participant in the conversation. The laptop can run the GUI (cf. Sect. 34.3.3), which can be used to provide real-time non-verbal speech statistics to an external observer

people recordings. For small-group discussions with more than two people, we will use professional audio interfaces in the future that allow simultaneous multichannel recording.

34.3.2 Voice Activity Detection and Speaking Segmentation

After the audio signals have been acquired and stored on the laptop, pre-processing is conducted, i.e. voice activity detection and segmentation of the speakers. Generally, the objective of voice activity detection is to differentiate between speech and non-speech (including silence and all kinds of noise and signals unrelated to speech). The purpose of speaker segmentation is to extract speaker turns in speech segments. Note that we conduct voice activity detection and segmentation of the speakers in real time, while the discussion is in progress: The recording system continuously writes audio signals on the hard drive, and the pre-processing methods are continuously applied to the audio signals available at any given time.

Voice activity detection algorithms typically use audio features like frequency, energy and spectral entropy to extract speech activity from audio recordings. There are many voice activity detection systems available; we use the algorithm proposed in [17]. For speech segmentation, we use the approach of [18, 19]. For each of the two participants, we extract two binary indicators that show voice status and speaking status at each time instance (see Fig. 34.3). Voice status roughly corresponds to syllables and speaking status corresponds to the speaking time of a person.

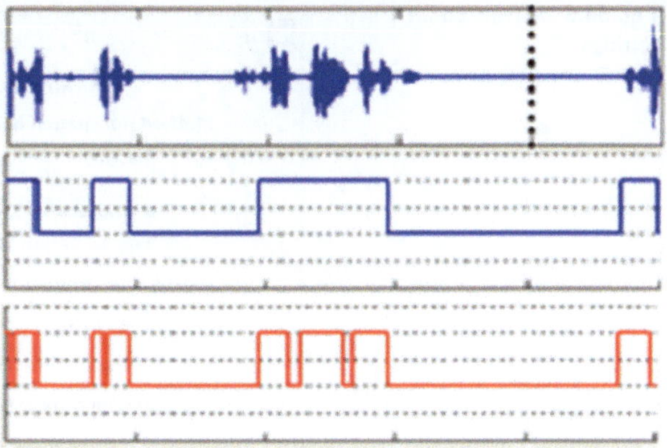

Fig. 34.3 Illustration of voice activity detection and speaker segmentation, (*top*) audio signal, (*middle*) voice activity detection, (*bottom*) speaker segmentation

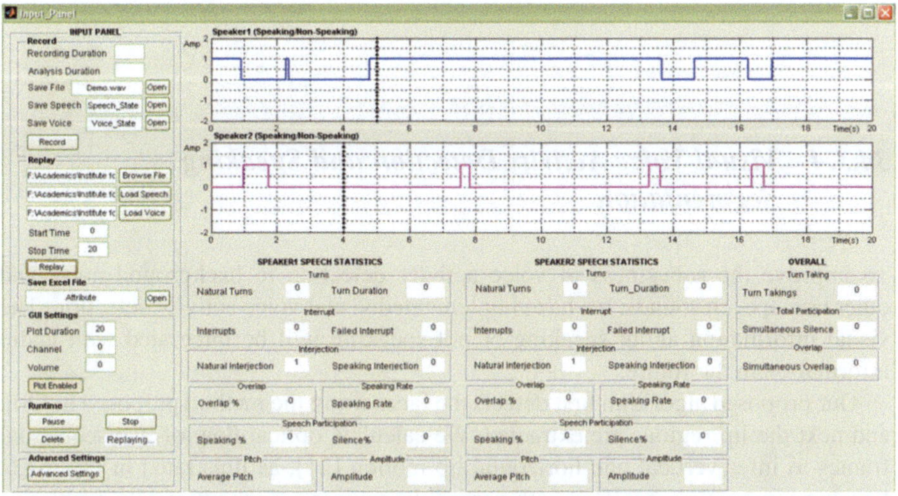

Fig. 34.4 The graphical user interface (GUI) displays the speaker segmentation, and below those plots, various non-verbal speech measures are displayed. On the *left-hand side*, user Input Panel is at the top and the Controls Panel is beneath it. The Advances Settings Panel (see Fig. 34.5) allows the user to specify parameter values for voice activity detection and speaker segmentation

34.3.3 Graphical User Interface

After voice activity detection and speaker segmentation has been carried out, the non-verbal speaking statistics discussed in Sect. 34.2 are computed in real time. We have designed a graphical user interface (see Figs. 34.4 and 34.5) that displays and continuously updates those measures (while those statistics are regularly being stored in data files).

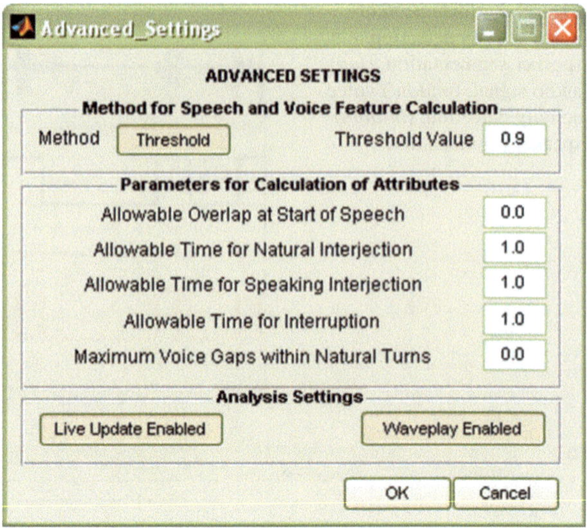

Fig. 34.5 Advanced GUI settings

34.3.4 Visual Voice Activity Detection and Speaking Segmentation

To improve the robustness of voice activity detection to background noise and other non-speech sounds, we have implemented visual voice activity detection. The visual information about speaking or not speaking will be integrated with audio information in the future.

Our proposed algorithm first detects the faces using the method proposed in [20], and next the lip regions are extracted. We calculate optical flow of two sequential frames to infer vertical and horizontal lip motion [21], as illustrated in Fig. 34.6. Speaking is mostly associated with vertical lip motion, which ultimately enables us to detect speaking from video.

34.3.5 Accuracy

To assess the accuracy of speaker segmentation based on audio and video signals, we have recorded 17 two-person conversations. From each of those recordings, we extracted a segment of 4 min. We manually labelled the speakers at each time instant, which will serve as ground truth to assess the speaker segmentation algorithms. The results for audio- and video-based speaker segmentation are summarized in Table. 34.1. Both approaches seem to be reliable, and, not surprisingly, audio-based segmentation is more reliable than video-based. In the future, we will combine both approaches to further improve speaker segmentation.

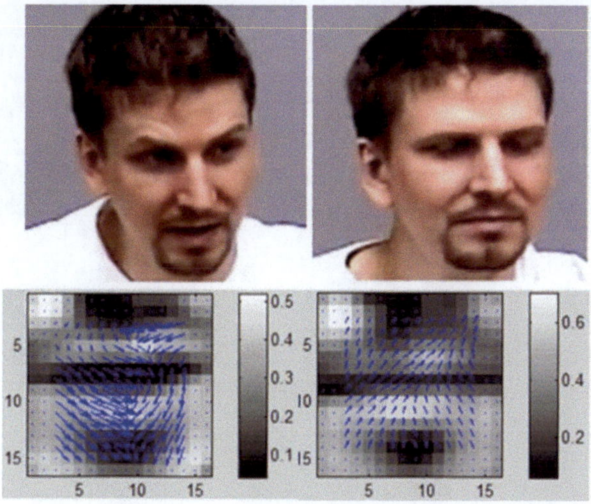

Fig. 34.6 Face detection and optical flow for (*left*) speaking sequence and (*right*) silent sequence

Table 34.1 Accuracy of audio and video-based speaker segmentation

Session	Audio based	Video based
1	96 %	83 %
2	93 %	77 %
3	94 %	85 %
4	94 %	82 %

34.4 Feedback

A crucial component in the proposed system is feedback. The non-verbal speaking statistics, calculated in real time, can be exploited to provide feedback to the speakers, either in real time or offline fashion (after the discussion). The GUI discussed in Sect. 34.3.3 is one potential approach to feedback. It can be used by an external observer to analyse a conversation in real time or after the discussion. However, as the GUI displays numerous statistics, it may not be straightforward to grasp the essential interactions, especially in real time and for non-experts. Also for real-time feedback to the speakers, the GUI is not suitable, as the speakers cannot extract relevant information from the GUI without interrupting the conversation.

So far, we have explored two alternative means of feedback: (a) retrospective (offline) feedback, after the discussion, in the form of animations and (b) real-time feedback to the speakers through emoticons displayed on smartphones. We will now briefly discuss those two forms of feedback.

Fig. 34.7 Retrospective feedback in the form of animations: continuous turn-taking (*top*) and excessive interruptions (*bottom*)

34.4.1 Retrospective Feedback

In this approach, each participant is depicted as a character in an animation. Using a commercial game engine (Unity) data from the conversation is processed automatically to produce the animation. This animation aims to highlight selected significant non-verbal interactions in the discussion, e.g. smooth turn-taking, inappropriate silence, excessive interruptions or unbalanced voicing rates between the speakers (see Fig. 34.7).

This approach can visualize the many complex threads of information in a manner that is relatively intuitive for participants to review and comprehend.

34.4.2 Real-Time Feedback on Smartphones

Real-time feedback to the speakers may help them to adapt their individual behaviour within the group and increase the effectiveness of a conversation. However,

Fig. 34.8 Examples of emoticons. From *left to right*: interrupting; monotonic speech; aggressive behaviour; emerging leader

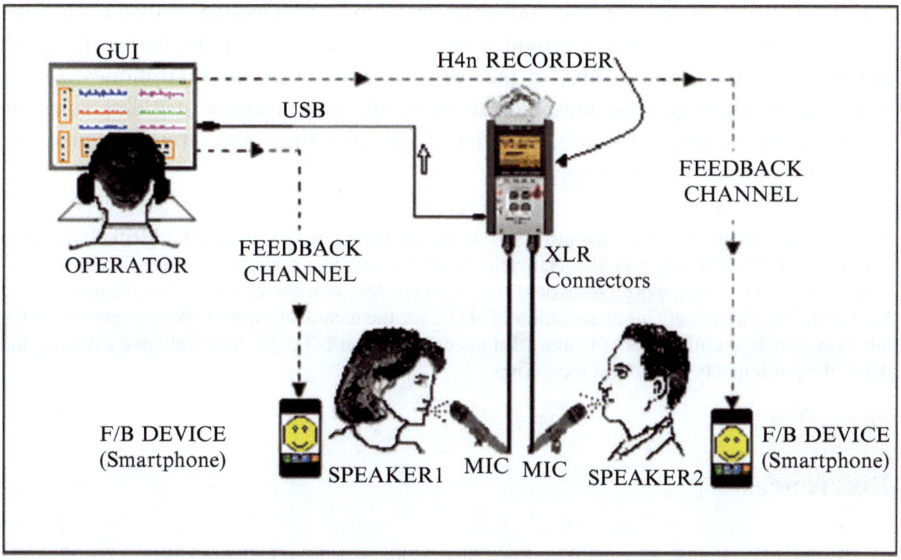

Fig. 34.9 Real-time feedback system. The system provides feedback through the GUI, which may be used by an external observer "operator" to monitor the discussion; it also provides feedback to the speakers in the form of emoticons displayed on smartphones. Note that the system does not rely on any external observer or operator, as it is fully automatic

designing real-time feedback is a challenge, as it can easily interrupt the flow of a discussion. Auditory feedback could easily be disturbing, and therefore, we decided to design graphical feedback instead. Images can easily portray the behaviour of individuals as they can be self-explanatory.

In our approach, every participant is given feedback through emoticons displayed on a smartphone, as illustrated in Figs. 34.8 and 34.9. Whenever a significant event occurs (e.g. excessive interruptions), an emoticon is displayed. As we only trigger such feedback for the most significant interactions, the amount of feedback is limited and does not disturb the flow of a conversation. In future work, we will experiment more with such real-time feedback to assess its effectiveness and effect on conversations.

34.5 Conclusion and Future Work

In this chapter, we have presented preliminary results on our systems for automated real time and offline analysis of conversations from audio and video recordings. We have developed a user-friendly GUI for analysing a discussion, both in real time (for external observer) and in retrospective fashion.

We are in the process of designing animations that summarize salient social interactions during a discussion (e.g. interruptions), for retrospective analysis. We have also introduced a system that provides real-time feedback to individual participants during ongoing conversations, in the form of emoticons on smartphones. In the longer term, such systems may help to boost the effectiveness of a diverse range of social interactions, e.g. job interviews, business meetings, group discussions, coaching sessions or public speaking.

Acknowledgments This research project is supported in part by the Institute for Media Innovation (Seed Grant M4080824) and the Nanyang Business School, both at Nanyang Technological University (NTU), Singapore. We would like to thank Mr. ***Vincent Teo*** and his colleagues at the Wee Kim Wee School of Communication of NTU, for the technical support. We are grateful to the lab managers and colleagues at Control Engineering Lab at NTU for their valuable support, and thank the participants for the test recordings.

References

1. Pentland, A.: Honest Signals: How They Shape Our World. MIT, Cambridge (2008)
2. Pentland, A.: Socially aware computation and communication. IEEE Comput. **38**(3), 33–40 (2005)
3. Poole, M.S., Holligshead, A.B., McGrath, J.E., Moreland, R,L., Rohrbaugh, J.: Interdisciplinary perspectives on small groups, Small Group Res. **35**(1), 3–16 (2004). Sage.
4. Salas, E., Sims, D.E., Burke, C.S.: Is there a big five in teamwork. Small Group Res. **36**(5), 555–599 (2005). Sage
5. Gatica-Perez, D.: Automatic nonverbal analysis of social interaction in small groups: A review. Image Vision Comput. **27**(12), 1775–1787, Dec 2009. Elsevier.
6. Aran, O., Hung, H., Gatica-Perez, D.: A multimodal corpus for studying dominance in small group conversations. In: Proceedings of LREC Workshop on Multimodal Corpora and 7th International Conference for Language Resource and Evaluation, Malta (2010)
7. Rienks, R.J., Heylen, D.: Automatic dominance detection in meetings using easily detectable features. In: Proceedings of the Workshop on Machine Learning for Multimodal Interaction (MLMI), Edinburgh (2005)
8. Sanchez-Cortes, D., Aran, O., Schmid-Mast, M., Gatica-Perez, D.: Identifying emergent leadership in small groups using nonverbal communicative cues. In: 12th International Conference on Multimodal Interfaces and 7th Workshop on Machine Learning for Multimodal Interaction (ICMI-MLMI), pp. 39, Beijing, ACM (2010)
9. Pianesi, F., Mana, N., Cappelletti, A., Lepri, B., Zancanaro, M.: Multimodal recognition of personality traits in social interactions. In: Proceedings of 10th International Conference on Multimodal Interfaces, pp. 53–60, Chania, ACM Oct (2008)
10. Pentland, A.: Social dynamics: signals and behaviour. In: International Conference on Developmental Learning (ICDL), San Diego, vol. 5. IEEE (2004)

11. Aran, O., Gatica-Perez, D.: Fusing audio-visual nonverbal cues to detect dominant people in conversations. In: 20th International Conference on Pattern Recognition (ICPR), IEEE, pp. 3687–3690, Istanbul, Turkey, Aug 23–26 (2010)
12. Carletta, J., et al.: The AMI meeting corpus: A pre-announcement. In: Proceedings of Machine Learning for Multimodal Interaction (MLMI), pp. 28–39, Edinburgh, Jul (2005)
13. Sanchez-Cortes, D., Aran O., Gatica-Perez, D.: An audio visual corpus for emergent leader analysis. In: (ICMI-MLMI), Multimodal Corpora for Machine Learning, Nov 14–18, Alicante. ACM (2011)
14. Kim, T., Chang, A., Holland, L., Pentland, A.: Meeting mediator: enhancing group collaboration with sociometric feedback. In: Proceedings of ACM Conference on Computer Supportive Cooperative Work (CSCW), pp. 457–466, San Diego (2008)
15. Mike Brooks: VOICEBOX: Speech Processing Toolbox for MATLAB, Department of Electrical and Electronic Engineerng, Imperial College, London
16. Chial, M.R.: Suggestions for computer based audio recordings of speech samples for perceptual and acoustic analyses. In: Phonology Project Technical Report No. 13, Department of Communicative Disorders, Phonology Project, University of Wisconsin-Madison, Oct (2003)
17. Basu, S.: Conversation Scene Analysis. PhD Thesis, MIT, Department of Electrical Engineering and Computer Science (2002)
18. Stoltzman, W.T: Towards a Social Signaling Framework: Activity and Emphasis in Speech. Master Thesis. MIT, Sep (2006)
19. Ambady, N., Rosenthal, R.: Thin slices of expressive behaviour as predictors of interpersonal consequences: A meta analysis. Psychological Bulletin, vol. 111(2), pp. 256–274. American Psychological Association (1992)
20. Viola, P., Jones, M.: Rapid object detection using a boosted cascade of simple features. In: Jacobs A., Baldwin, T. (eds.) Proceedings of 2001 IEEE Computer Society Conference on Computer Vision and Pattern Recognition, Kauai, HI, USA vol. 1, pp. 511–518 (2001)
21. Dollar, P.: Piotr's Image and Video Matlab Toolbox (PMT). Available from http://vision.ucsd.edu/~pdollar/toolbox/doc/

Chapter 35
Evaluation of Invalid Input Discrimination Using Bag-of-Words for Speech-Oriented Guidance System

Haruka Majima, Rafael Torres, Hiromichi Kawanami, Sunao Hara, Tomoko Matsui, Hiroshi Saruwatari, and Kiyohiro Shikano

Abstract We investigate a discrimination method for invalid and valid inputs, received by a speech-oriented guidance system operating in a real environment. Invalid inputs include background voices, which are not directly uttered to the system, and nonsense utterances. Such inputs should be rejected beforehand. We have reported methods using not only the likelihood values of Gaussian mixture models (GMM) but also other information in inputs such as bag-of-words, utterance duration, and signal-to-noise ratio to discriminate invalid inputs from valid ones. To deal with these multiple information, we used support vector machine (SVM) with radial basis function kernel and maximum entropy (ME) method and compare the performance. In this paper, we compare the performance changing the amount of training data. In the experiments, we achieve 87.01% of F-measure for SVM and 83.73% for ME using 3,000 training data, while F-measure for GMM-based baseline method is 81.73%.

H. Majima (✉)
Graduate School of Information Science and Technology, Osaka University, Osaka, Japan
e-mail: majima.haruka@ist.osaka-u.ac.jp

R. Torres • H. Kawanami • H. Saruwatari • K. Shikano
Graduate School of Information Science, Nara Institute of Science and Technology, Nara, Japan
e-mail: rafael-t@is.naist.jp; kawanami@is.naist.jp; sawatari@is.naist.jp; shikano@is.naist.jp

S. Hara
Graduate School of Natural Science and Technology, Okayama University, Okayama, Japan
e-mail: hara@cs.okayama-u.ac.jp

T. Matsui
Department of Statistical Modeling, The Institute of Statistical Mathematics, Tokyo, Japan
e-mail: tmatsui@ism.ac.jp

J. Mariani et al. (eds.), *Natural Interaction with Robots, Knowbots and Smartphones: Putting Spoken Dialog Systems into Practice*, DOI 10.1007/978-1-4614-8280-2_35,

35.1 Introduction

Automatic speech recognition (ASR) has been widely applied to dictation, Voice Search, and car navigation, to name a few. In this paper, we describe the speech-oriented guidance system *Takemaru-kun* [1], which aims to realize a natural speech interface using ASR.

Takemaru-kun is a real-environment speech-oriented guidance system adopting an example-based question answering that is flexible to respond to user's questions on demand. An answer to a user's question is selected by referring to the question and answer database, which can be easily maintained without paying particular attention to the scope of the system.

A serious problem in *Takemaru-kun* is that the system receives unintended inputs such as laughter and cough, which decrease the response accuracy. To solve this problem, there had been studies on noise input rejection, using likelihood values of Gaussian mixture models (GMMs) [2]. But there are many more unintended inputs that the system unnecessarily answers, such as background voices between users and nonsense utterances. For voice activity detection, a method using linguistic constraints [3] has also shown an effective result. To discriminate these inputs from valid ones, we utilize the information of bag-of-words (BOW) from ASR results of input utterances. Moreover, we employ it with the likelihood values of GMMs and duration and SNR information of input utterances to improve the total performance of discrimination between invalid and valid inputs [4].

In this paper, we describe effective feature combinations for invalid input discrimination, using the BOW from the 10-best ASR results to train models with either support vector machine (SVM) or ME and consider the amount of training data needed for the best performance. The outline of the paper is as follows. First, we describe the overview of *Takemaru-kun* system. Then, we describe our proposed method with several effective features using SVM or ME and show the experimental results using real users' utterances. Finally, we conclude our proposal.

35.2 Speech-Oriented Guidance System *Takemaru-kun*

The *Takemaru-kun* system [1] (Fig. 35.1) is a real-environment speech-oriented guidance system, placed inside the entrance hall of the Ikoma City North Community Center located in the Prefecture of Nara, Japan. The system has been operated daily since November 2002, providing guidance to visitors regarding the center facilities, services, neighboring sightseeing, weather forecast, news, and about the agent itself, among other information. Users can also activate a Web search feature that allows searching for Web pages over the Internet containing the uttered keywords. This system is also aimed at serving as field test of a speech interface and to collect actual utterance data.

Fig. 35.1 Speech-oriented guidance system *Takemaru-kun*

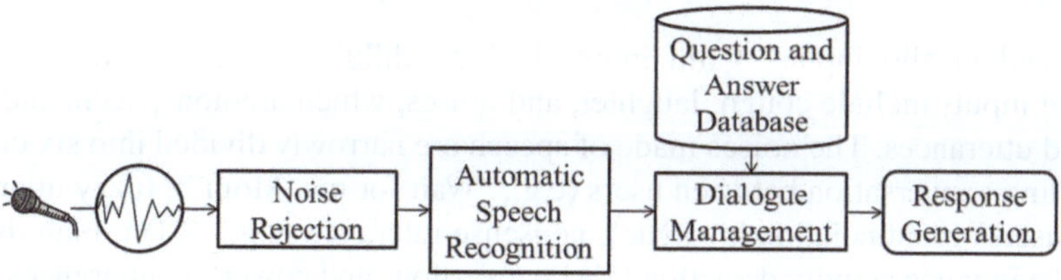

Fig. 35.2 Processing flow of *Takemaru-kun*

The system displays an animated agent at the front monitor, which is the mascot character of Ikoma city, *Takemaru-kun*. The interaction with the system follows a one-question-one-answer strategy, which fits the purpose of responding simple questions from a large number of users. When a user utters an inquiry, the system responds with synthesized voice, an agent animation, and displays information or Web pages at the monitor in the back, if required.

The system structure is illustrated in Fig. 35.2. Speech/noise discrimination using the likelihoods of GMMs is executed in parallel with ASR. In this process, the system can only reject noise inputs, which are part of unintended inputs. However there are many more of them, such as conversation between users and utterances of fillers or nonsense words, as we discussed before. We define these unintended inputs as invalid inputs.

All inputs to the system have been collected from the start of operation. The data of the first two years were manually transcribed with tags concerning the invalid inputs mentioned above and labels about age-group and gender. The tags and labels are given by hearing of four trained labelers. These data were used to construct the GMMs and to adapt acoustic models and language models used in the daily operation.

The real users' inputs to the system contain invalid inputs more than a half of all, as shown in Table 35.1. All inputs to the system are broadly divided into

Table 35.1 Classification result on input data of *Takemaru-kun* (from November 2002 till October 2004)

Category			Utterances	Total
	Adult's utterances		20,436	
Valid inputs	Child's utterances		85,889	106,325
Invalid inputs	Invalid utterances	Background voices	26,319	122,939
		Fussy utterances	13,348	
		Nonsence utterances	11,991	
		VAD mistake	12,937	
		Overflow utterances	1,417	
		Powerless utterances	7,347	
	Cough		727	
	Laughter		6,232	
	Noise		50,756	

valid and invalid inputs. Valid inputs include child's and adult's valid speech. Invalid inputs include cough, laughter, and noises, which are nonspeech, and other invalid utterances. The noises made of speech are narrowly divided into six classes, including conversation between users (e.g., "Wait for me, Mom"), fuzzy utterances (i.e., users' pronunciation is vague), nonsense utterance (e.g., "Blah, blah, blah"), mistake in voice activity detection (VAD), overflow, and powerless utterances. They are difficult to be discriminated as invalid inputs by only using the likelihood values of GMMs, which have no information of the meaning of the utterance. Therefore we employ BOW as a classification feature.

35.3 Discrimination Using Multiple Features

Here we describe the proposed method using multiple features of BOW, GMM likelihood values, duration, and SNR.

35.3.1 Features

A **BOW** vector consists of frequencies of each word in a vocabulary word list which is comprised of the 10-best ASR candidates of training data. The dimension of BOW vector is determined by the number of words in the word list.

GMM likelihood is given as the likelihood values of each utterance to six GMMs. The GMMs are trained using six kinds of data for adult's valid speech, child's valid speech, laughter, cough, noise, and other invalid utterances, respectively. The speech data is recorded through the system operation in a real environment. When the

Table 35.2 Training conditions for GMMs

Amount of training data	Valid input	Adult's utterances	1,053
		Child's utterances	6,554
	Invalid input	Laughter	287
		Cough	29
		Noise	3,318
		Other invalid utterances	3,640
Sampling/quantization	16 kHz/16 bit		
Width/shift length of window function	25 ms/10 ms		
Feature	MFCC (12 dimensions), Δ MFCC		
Number of mixtures	128		

likelihood values of GMMs for laughter, cough, noise, and other invalid utterances are high, the input should be rejected as an invalid input. Table 35.2 lists the training conditions for GMMs.

Duration is the duration of an utterance, determined by voice activity detection of Julius [7] using amplitude and zero crossing count.

SNR is the signal-to-noise ratio of an utterance. We divide an input wave into frames and then consider the top 10% frames having larger power in average as signal and the bottom 10% frames having smaller power in average as noise, conveniently. SNR is calculated by:

$$SNR = 10log_{10}\frac{P_S - P_N}{P_N}$$

where P_S is the average power of signal and P_N is that of noise.

35.3.2 *Discrimination Method*

We used the features described above for the classification, employing SVM and ME as classifiers.

35.3.2.1 SVM-Based Method

SVM is a useful machine for data classification [5]. It is basically a supervised learning binary classifier. When training vectors $x_i \in R^n, i = 1, \cdots, l$ (l: number of examples) and their corresponding classes $y_i \in 1, -1$ are given, the SVM estimates a separating hyperplane with a maximal margin in a higher dimensional space. The soft margin notation, which permits the existence of incorrectly classified data, is also introduced using a slack variable ξ_i and C parameters as:

$$\min_{\mathbf{w},b,\xi} \frac{1}{2} \mathbf{w}^T\mathbf{w} + C\sum_{i=1}^{l} \xi_i$$
$$\text{subject to } y_i(\mathbf{w}^T \phi(\mathbf{x}_i) + b) \geq 1 - \xi_i,$$
$$\xi_i \geq 0, i = 1, \ldots, l.$$

$$\min_{\mathbf{w},b,\xi} \frac{1}{2} \mathbf{w}^T\mathbf{w} + C_+ \sum_{\{i:y_i=+1\}} \xi_i + C_- \sum_{\{i:y_i=-1\}} \xi_i$$
$$\text{subject to } y_i(\mathbf{w}^T \phi(\mathbf{x}_i) + b) \geq 1 - \xi_i,$$
$$\xi_i \geq 0, i = 1, \ldots, l.$$

C_+: cost parameter for positive examples

C_-: cost parameter for negative examples

For this work, LIBSVM [5] is used to apply SVM. Specifically, we are using C support vector classification (C-SVC), which implements soft margin.

35.3.2.2 ME-Based Method

Maximum entropy (ME) method [6] provides a general purpose machine learning technique for classification and prediction, which has been successfully applied to natural language processing, named entity recognition, etc. ME method can integrate features from many heterogeneous information sources for classification. Each feature corresponds to a constraint on the model. Given a training set of (E,D), where E is a set of class labels and D is a set of feature-represented data points, the ME method attempts to maximize the log likelihood

$$\log P(E|D,\Lambda) = \sum_{(e,d)\in(E,D)} \log \frac{\exp \sum_i \lambda_i, f_i(e,d)}{\sum_{e'} \exp \sum_i \lambda_i f_i(e',d)}$$

where $f_i(e,d)$ are feature indicator functions. We use ME method for invalid utterance discrimination. In this context, such features, for instance, could be the utterances' GMM likelihood, duration, or SNR, which will be shown later. λ_i are the parameters that need to be estimated, which reflect the importance of $f_i(e,d)$ in the prediction. For this work, Stanford Classifier [6] is used to apply ME.

35.4 Experiments

To elucidate the effect of BOW and other features, which are GMM likelihood, duration, and SNR, and from the point of view of the relation between amount of training data and classification performance, evaluation experiments were conducted using real users' utterances.

35.4.1 *Speech Database*

We used the speech database of about 3-month data of real users' input received by *Takemaru-kun*. The 15,000 data of from November 2002 to December 2002 are used as training data and August 2003 as test. The amount of data we experimented was 400, 800, 1,600, 2,000, and every 2,000 data from 3,000 to 15,000 as shown in Table 35.3.

35.4.2 *Experimental Conditions*

The experimental conditions on feature extraction are illustrated in Table 35.4. The acoustic model (AM) and the language model (LM) were separately prepared for adults and children. The AMs were trained using The Japanese Newspaper Article Sentence database (JNAS) and adapted by user data spoken to *Takemaru-kun*. The LMs were constructed using the manual transcription of the user's speech to *Takemaru-kun*. Morphological analyzer Chasen is used to split the transcription data into words. In the following experiments, 10-best ASR candidates were used for constructing the word list and BOW vectors.

Table 35.3 Speech database

	Valid inputs	Invalid inputs	Total
Training data	9,509	6,491	15,000
Test data	7,607	7,274	14,881

Table 35.4 Experimental condition

ASR	Engine	Julius 4.2 [7]
	AM, LM	*Takemaru-model* [3]
	Output	10-best candidates
Morphological analyzer		Chasen 2.3.3 [8]
SVM	Tool	LIBSVM [5]
	Kernel function	Radial basis function (RBF)
	Parameter C	$10^{-2}, 10^{-1}, \ldots, 10^{4}$
ME	Tool	Stanford Classifier 2.1.3 [6]

35.4.3 Evaluation Measures

Classification performance of the methods on valid or invalid inputs was evaluated using the F-measure, as defined by:

$$F\text{-}measure = \frac{2 \cdot Precision \cdot Recall}{Precision + Recall}$$

35.5 Experimental Results

The purpose of these experiments was to compare the classification performance of SVM and ME using combination of the features and to consider the amount of training data. In our previous work [4], we conducted an experiment using about 15,000 training data and the combination of features which presented the best performance in classification was SVM using all four features, which are BOW, GMM likelihood, duration, and SNR, with an F-measure of 85.77%, which represents a difference of 4.04% in comparison to the baseline method of SVM using only GMM likelihood with 81.73%. Then we changed amount of training data to observe the effect of amount of data to the classification performance. Relationship of the amount of training data between 400 and 15,000 and calculated F-measures are shown in Fig. 35.3.

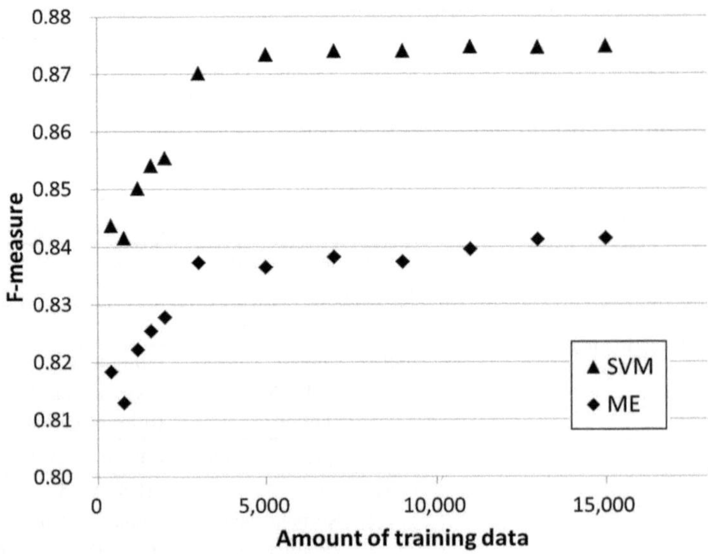

Fig. 35.3 F-measures of differed amount of training data

The result shows that F-measures of SVM are always higher than that of ME. Both F-measures of SVM and ME are saturated using about 3,000 training data; they continue increasing and reached about 0.4% higher F-measures when we used 15,000 training data.

35.6 Conclusions

We investigated discrimination between invalid and valid spoken inquiries using multiple features as the likelihood values of GMMs, BOWs, utterance durations, and SNRs. SVM and ME methods were compared in performance with the F-measure and both methods outperformed the conventional method, which uses the likelihood values of GMMs. The classification performance was better using larger amount of training data; however it saturated using only 3,000 data. Our future work includes further investigation for effective combination methods of different kinds of features and more amount of training data.

Acknowledgements This work was partially supported by CREST (Core Research for Evolutional Science and Technology) and Japan Science and Technology Agency (JST).

References

1. Nisimura, R., Lee, A., Saruwatari, H., Shikano, K.: Public speech-oriented guidance system with adult and child discrimination capability. In: Proceedings ICASSP2004, vol. 1, pp. 433–436 (2004)
2. Lee, A., Nakamura, K., Nishimura, R., Saruwatari, H., Shikano, K.: Noise robust real world spoken dialogue system using GMM based rejection of unintended inputs. In: Proceedings International Conference on Spoken Language Processing, pp. 847–850 (2004)
3. Sakai, H., Cincarek, T., Kawanami, H., Saruwatari, H., Shikano, K., Lee, A.: Voice activity detection applied to hands-free spoken dialogue robot based on decoding using acoustic and language model. In: Proceedings of the 1st International Conference on Robot Communication and Coordination (ROBOCOMM2007), Article No. 16, 8 pp. (2007)
4. Majima, H., Torres, R., Fujita, Y., Kawanami, H., Matsui, T., Saruwatari, H., Shikano, K.: Spoken Inquiry Discrimination Using Bag-of-Words for Speech-Oriented Guidance System. INTERSPEECH2012, Portland, September (2012)
5. Chang, C., Lin, C.: LIBSVM: A Library for Support Vector Machines. http://www.csie.ntu.edu.tw/~cjlin/libsvm (2001)
6. Manning, C., Klein, D.: Optimization, Maxent Models, and Conditional Estimation without Magic, Tutorial at HLT-NAACL 2003 and ACL 2003. http://nlp.stanford.edu/software/classifier.shtml
7. Lee, A., Kawahara, T., Shikano, K.: Julius - an open source realtime large vocabulary recognition engine. In: Proceedings Eurospeech 2001, pp. 1691–1694 (2001)
8. Matsumoto, Y., Kitauchi, A., Yamashita, T., Hirano, Y., Matsuda, H., Takaoka, K., Asahara, M.: Morphological Analysis System Chasen 2.3.3 User's Manual http://chasen-legacy.sourceforge.jp/ (2003)

MIX
Papier aus verantwortungsvollen Quellen
Paper from responsible sources
FSC® C105338

If you have any concerns about our products,
you can contact us on
ProductSafety@springernature.com

In case Publisher is established outside the EU,
the EU authorized representative is:
Springer Nature Customer Service Center GmbH
Europaplatz 3, 69115 Heidelberg, Germany

Printed by Libri Plureos GmbH
in Hamburg, Germany